# THE GLACIATIONS OF WALES
# AND ADJOINING REGIONS

GEOGRAPHIES FOR ADVANCED STUDY

Edited by Professor S. H. Beaver, M.A., F.G.S.

Human Geography
Geography of Population
Urban Geography
Urban Essays—Studies in the Geography of Wales
Statistical Methods and the Geographer
Geomorphology
Central Europe
Eastern Europe
A Regional Geography of Western Europe
An Historical Geography of Western Europe before 1800
The Western Mediterranean World
The British Isles—A Geographic and Economic Survey
The Scandinavian World
The Soviet Union
The Polar World
North America
The Tropical World
West Africa
An Historical Geography of South Africa
Malaya, Indonesia, Borneo and the Philippines
Land, People and Economy in Malaya

# The Glaciations of Wales
# and Adjoining Regions

edited by
Colin A. Lewis

LONGMAN

LONGMAN GROUP LIMITED
London

*Associated companies, branches and representatives throughout the world*

© LONGMAN GROUP LTD 1970

First published 1970

SBN 582 48154 6

*Printed by Spottiswoode, Ballantyne and Co Ltd,
London and Colchester*

# Acknowledgements

We have been unable to trace the copyright holders of a table from 'The problems and contributions of methods of absolute dating within the Pleistocene period' by F. W. Shotton from Quarterly Journal of Geological Society of London, and would appreciate any information that would enable us to do so.

# Contents

# Figures

# Plates

# Folding Charts

# Preface

Appropriately, the idea of writing this book was conceived in a snow storm on the Brecon Beacons on April Fools' Day, 1966. Happily neither Mr Marmion of Longman, Browne and Nolan of Dublin, nor Mr Yglesias of Longman thought our idea too incongruous and, on behalf of all the contributors, I must thank them for their initial interest and encouragement. Since then Professor S. H. Beaver has helped us greatly at all stages of preparation and his advice and aid has been most welcome. Our publisher, Mr Marix Evans, and his editor, Mrs Anne Bassett, have also given us great assistance with the multitude of matters that always seem to worry writers and we greatly appreciate their interest. Similarly, Miss Mary Evans has taken great pains to ensure that the plates and figures have been presented to the best advantage, a most difficult task. We are also very grateful to Mrs Dorothy Stephens for drawing a number of the figures; she was extremely patient with us and we are delighted with the results.

As editor, I should like to thank all the contributors for the promptness with which they produced their scripts and with their forbearance with me and my editorial foibles. I must also express my appreciation to Miss Honóra Cregan for typing a considerable proportion of the book. Additionally, my wife deserves a special word of thanks. She read, and commented on, the initial manuscript and has helped in all the subsequent stages of preparation and with the arduous tasks of proof-reading and indexing.

Many details of the glaciations of Wales are still uncertain, whilst others are probably completely unknown. It is not our intention to present our readers with a complete history of glacial events, for it will be many years hence before such a task may even be attempted. Instead, we have tried to present a synopsis of the knowledge available to us as we enter the 1970's, in the hope that it will encourage, and possibly guide, further research in the years ahead.

Nid byd heb wirionedd.

Colin A. Lewis (Editor)

University College, Dublin.
St. David's Day, 1970.

# CHAPTER ONE

# Introduction

Colin A. Lewis, B.A., Ph.D.

Today we have no difficulty in accepting that much of the surface of the earth has, at various stages in the past, been covered with glacier ice. We accept, without too much hesitation, that thicknesses of 1000 m (3000 ft) or more of ice covered many areas of Wales. Yet two hundred years ago such ideas would have seemed preposterous. Of course, as far back as 1200 Saxo Grammaticus was writing about glacier movement in Iceland (see Thorarinsson, 1960), but, like many earlier authors, he was describing existing ice, and not the spread of ice in former times. At a later date the results of glacial action, the spectacular erosive landforms and the equally impressive glacial deposits, were the subjects of much attention, speculation and, especially in the last century, heated debate.

During the early nineteenth century it was believed, not only in Britain, but throughout the world, that the deposits we now know to be of glacial origin were due to marine submergence. The notion that they might be glacial had occurred to no one of acknowledged scientific renown, or, if it had, it had been veiled in silence. Yet in the Swiss Alps a cleric by the name of Bernard Kuhn had, in 1787, suggested 'that debris lying far beyond the present glacial limits had been deposited by the swollen glaciers of some bygone era' (Davies, 1969, p. 264). Perhaps it was not surprising that the glacial hypothesis was first voiced by a cleric, for glacier advances were not uncommon in the Alps during the seventeenth and eighteenth centuries, and Grove (1966) has drawn attention to the fact that 'processions to exorcise the local glaciers' (p. 132) had been undertaken by various priests. From exorcising an advancing glacier to the recognition that glaciers had previously been even larger than at present was an obvious step, although that does not detract from Kuhn's achievement, or, let it hastily be said, necessarily mean that Kuhn indulged in the ceremonies of exorcism!

1                                    1

Before the eighteenth century ended it became widely recognised, especially by naturalists like de Luc and de Saussure, that existing glaciers could transport debris. The scene had therefore been set for the general evolution of the glacial theory, and in 1821 Ignace Venetz prepared a memoir 'suggesting that the snouts of the Alpine glaciers had once lain several miles downstream of their present positions', although publication was delayed until 1833. In 1829 Venetz read a paper at a meeting of the Société Helvétique des Sciences Naturelles in which he propounded a number of his ideas relating to glaciation. Venetz was not alone in this early glacial work, for he was supported by, and in turn supported, de Charpentier. In 1836 they won a powerful ally, for Louis Agassiz, after accompanying Venetz and de Charpentier on a field excursion, became converted to the theory that glaciers had once been widespread. The following year Agassiz read a famous paper, the 'Discours de Neuchatel', to the Société Helvétique in which he stated that moraines, erratics, polished and striated surfaces, were evidence of former glaciation.

The glacial ideas which were developing on the Continent were weakly echoed by Robert Jameson in Edinburgh, who, Davies considers (1969), lectured on 'Proofs of former glaciers in countries where they are no longer met with', perhaps as early as 1827. In Scandinavia, too, these ideas were nurtured, and an English translation, published in 1826, of a paper by Jens Esmarsk, showed that he believed that Denmark may have formerly been glaciated. In 1838 the British geologist, Buckland, travelled to the Alps, partly, no doubt, to see firsthand evidence of glaciation, and became converted to the new hypothesis. Two years later, at his invitation, Agassiz came to Britain and, in the autumn of 1840, travelled extensively in England, Scotland and Ireland. Davies (1968) has given a full account of his travels and of the papers he read to the British Association, meeting in Glasgow, and to the Geological Society of London. Suffice it to say that Agassiz found plentiful evidence for former glaciation in Britain, and the paper by Buckland which was read to the Geological Society immediately after that of Agassiz, confirmed that finding. Of course, there were many who doubted Agassiz, and many who openly contradicted and ridiculed him, but the glacial theory had come to stay.

In spite of all the interest in glaciation the theory of submergence remained tenaciously in existence. Even Agassiz was not prepared to reject it fully, for whilst he stated that the glacial tills were the product of land-ice, he believed the sands and gravels that commonly divide the tills were the products of marine submergence. He suggested that, when the glaciers melted, their meltwaters caused the level of the sea to rise, thereby engendering submergence. Agassiz, in 1840, had not visited Wales, but in the following year Buckland rectified that omission and was thrilled to find striae, evidence of glaciation, in the Llanberis Pass. Nevertheless it was still necessary for him to invoke a marine submergence to account for the shell-bearing till found in the Vale of Clwyd and on the slopes of Moel Tryfaen (see p. 23). Over twenty years later, in 1863, Sir Charles Lyell inspected the Moel Tryfaen and other sites in Snowdonia, and, finding marine shells of Arctic provenance in the drift, believed that 'these shells show that Snowdon and all the highest hills which are in the neighbourhood of Moel Tryfaen were mere islands in the sea at a comparatively late period' (quoted from Davies, 1969). For many years it was held that erratics found in the drifts had come from icebergs, floating in this icy sea. Old ideas die hard and often, once a theory appears in print and is accepted, it long outlives its usefulness: reader, beware!

By the 1850s enough was known of the glacial deposits in North Wales for Ramsay to annunciate his tripartite division: (1) Lower boulder clay, deposited by local Welsh glaciers; (2) Middle sands and gravels and the shelly till of Moel Tryfaen, deposited as a result of marine submergence; (3) Upper boulder clay, deposited when, the waters abating, there was a readvance of 'small valley glaciers which had lingered in the highlands throughout the submergence' (Davies, 1969, p. 298). Of course, we now know that the sands and gravels so commonly found in glacial deposits were laid down by meltwaters issuing from the ice, and that the shelly drifts of North Wales (and other areas) were dredged off the sea-floor by ice advancing south down the bed of what is now the Irish Sea, later to be deposited in their present positions. It is still well to remind ourselves that, as late as 1910, the Moel Tryfaen shells were regarded by Edward Hull, a former director of the Irish Geological Survey, as evidence of marine submergence (Davies, 1969, pp. 300–1).

Glacial erosion, unlike glacial deposition, was a much more recent concept to come into vogue. Scientists, in the mid-nineteenth century, found it hard enough to accept a former ice-cover without stretching things so far as to accept that this ice could cause major erosion. When Ramsay, in 1859, suggested 'that the Pleistocene glaciers had played a major role in shaping the Earth's present landscape', and that 'all glaciers must deepen their beds by erosion', he was widely disbelieved. In 1870 Murchison, that influential geologist, thundered: 'Where in any ice tract is there the evidence that any glacier has by its advance excavated a single foot of solid rock? In their advance, glaciers striate and polish, but never excavate rocks' (quoted from Davies, 1969). No doubt Murchison's outburst was a belated response to Ramsay's 1862 paper on the glacial origin of rock basins, but Murchison was not alone. Whilst accepting that many areas had been covered with ice E. J. Garwood and his followers stated, but failed to show, that the ice had protected the underlying rock from experiencing as much erosion as it would have been subject to had it been exposed. Thus, at the beginning of the present century, the theory of glacial protection evolved. Today, without seriously questioning ourselves, we accept that cirques, arêtes, rock basins and the like are due to the erosion of ice, but even in 1900 there were many doubters, although as long ago as January 1838 Robert Mallet had read a paper to the Geological Society of Dublin in which he stated that 'the bed of a glacier is in continual process of degradation, or deepening by the resistless passage of these vast masses of ice and rock over it' (Davies, 1969, p. 302).

Glacial hypotheses, like hypotheses in all worthwhile subjects, are liable to alterations, or even rejection, as new knowledge is gained. Today we are on the verge of startling discoveries about the length of time during which ice covered Wales; perhaps tomorrow we shall discover a completely unknown process of erosion, or deposition, which might cause us to reject many of our most cherished ideas. Nevertheless it appears that the glacial and interglacial history of Wales is one of the most important features in the country's geomorphic evolution.

The most obvious evidence of glaciation is provided by the more spectacular landforms of glacial erosion, such as cirques and arêtes, which are widespread in mountainous areas like Snowdonia. In areas of deposition eskers, terminal moraines and tracts of kettled outwash similarly point to the former existence of glacial conditions. None of these landforms, however, are of any great use in aiding our interpretation of the number of glaciations.

One of the most effective methods of distinguishing periods of glaciation is based on the principle of stratification. This is a basic geological idea founded on the supposition that lower layers of material (in our case, usually glacial drifts), must be older than the layer(s) on top of them. Sometimes, of course, disturbance of the original bedding has occurred, with the result that the lowest is not the oldest deposit, but for the time being we shall overlook that problem. If we see what are obviously two different drifts, one on top of another, then that suggests two periods of glaciation, or two stages within the same glacial period. Obviously there are drawbacks to this method, for how can we be sure that two different drifts are present?

If the stone content of a layer of drift is examined it may well be that at least some of these stones are of different composition from the underlying bedrock. If this is so, then such stones are referred to as *erratics*, since they have been transported from their place of origin to their present position. Naturally, if we can locate their place of origin and if we know how they have been transported (in our case, by ice), then we know the route that was followed by the transporting agent. Furthermore, if there are two layers of drift, the one with erratics of different origin from those of the other, it is obvious that two different stages of ice movement, and possibly even two different glaciations are represented. We cannot, however, be certain that two different periods of glaciation are present unless further evidence is available, for glaciers from two different directions, existing at the same period, but waxing and waning alternately, may deposit layers of drift of different erratic content, one above the other, within the same glaciation.

Glacial periods within the Pleistocene were divided from each other by periods during which warm conditions prevailed. These warm periods, sometimes warmer than our present climate, are known as *interglacials*, and they can be recognised by the remains of plants and animals that lived in them. If between two beds of till, there exist mammalian or floral remains which are *in situ* and which date from an interglacial, then the tills must represent two different glaciations. Normally, where the remains of an interglacial have been discovered, they exist as a layer of peat in which the pollen grains of interglacial plants can be identified. A peat-bed between two tills is therefore of the greatest importance for dating purposes.

Because of the climatic conditions that existed in interglacials, the superficial material that existed on or near the surface in an interglacial was often deeply weathered, just as glacial drifts are being weathered in Wales at the present day. Thus if there are two layers of till, the bottom one of which is deeply weathered, then it is fair to suggest that different glaciations are, once again, represented. Occasionally, instead of finding just a weathered layer, we find remnants of a soil formed under interglacial conditions, perhaps located between tills or on top of a till. Ball (1960) has reported a soil of this type on Worm's Head in Gower. Such a soil is a *palaeosol* and its qualities reflect the nature of the period in which it was formed. Bowen (Chapter 9) has used the distribution of such soils in Glamorgan as a means of differentiating between areas glaciated in the Last Glaciation and those glaciated only in an earlier and larger glaciation, for palaeosols exist, in Glamorgan, only outside the limits of the Last Glaciation.

Another method used to differentiate between glaciations is based on recording the orientation of striae, the scratches left by moving ice on underlying rock surfaces. The alignment of the striae is normally parallel to the direction of ice movement, so that the presence of two or more sets of striae, pointing in different directions, indicates the former existence of different lines of ice-flow. Although differently orientated striae have been used to support other evidence for separate glaciations, as in Farrington's study of the glaciations of the Wicklow mountains (1934), striae alone prove nothing but that ice-streams moving from different directions affected an area. These may equally as well have existed in the same or in different glaciations.

Recently geomorphologists have undertaken stone orientation analyses to determine the direction from which glacial deposits have been derived. These analyses are based on the findings of Holmes (1941), who was able to prove that the direction of orientation of the long axes of the majority of stones in a deposit was parallel to the direction from which that deposit had been derived. Saunders (1968), by a study of stone orientations in the glacial drifts of the Lleyn peninsula, has already shown that the pattern of glaciation in that area is generally as Ball and Whittow describe in Chapter 2. Watson, in particular, has made a minute study of stone orientations to elucidate the origins of material in the Aberystwyth region. He describes his results in Chapter 6, to which he has appended a note on the techniques of such analyses. Of course, as with striae analyses, the fact that stone orientations might indicate ice movement from two or more directions does not necessarily mean that different glaciations are represented. Nevertheless it is a useful aid to our studies.

In some cases evidence of glaciation is afforded, not by the presence of undisputable glacial deposits (such as till, or outwash), but by the existence of scattered erratics. These may vary in size from large boulders, like the 'fifty-ton block of garnetiferous microcline gneiss' (Mitchell, 1960, p. 316) known as the Giant's Rock, which exists at

Porthleven in Cornwall, to microscopic mineral fragments such as occur on the northern side of the Scilly Isles (Kiely, in Mitchell and Orme, 1967). In many cases these erratics are incorporated in beach deposits as Ball and Whittow (Chapter 2) record in north-west Wales, or as Mitchell (1960) has recorded at Courtmacsherry in County Cork. The obvious inference to be drawn from such deposits is that the area was once glaciated, then sea-level rose and the glacial deposits were eroded, during which period erratics from the glacial material were incorporated in the beach formed by the invading sea. Naturally there can be refinements of this argument, for the erratics may have been carried into place by floating pack-ice (as Stephens argues in Chapter 11), although this does not destroy the general tenor of the argument.

Stratigraphically, then, we can define different glaciations on the basis of deposits and their place in the overall sequence. In many areas, however, certain parts of the sequence are missing, possibly because earlier deposits have been eroded by later ice advances. This means that it is often necessary to study sections over wide areas to reconstruct the whole picture, with the result that the evidence is subject to human judgment and hence, possibly, to misinterpretation or to different interpretations by different scientists. Even when the stratigraphic sequence of events has been perfectly reconstructed, it still gives only a relative idea of what has happened. The dates at which events occurred is still unknown, and different methods must be used to date their sequence.

In addition to examining the stratigraphy of glacial deposits, it is occasionally possible to determine the relative chronology of glaciations by examining glaciofluvial deposits, such as the gravels that are washed out of glaciers by meltwater. Under favourable conditions, such as existed along the Rhine valley and along the right-bank tributaries of the Danube, these deposits form a sequence of valley terraces. The most favourable sites for the development of such terraces are along valleys that lead directly away from an ice centre, as the Rhine does from the Swiss Alps. Preferably the valley should never be overwhelmed by ice, for that disturbs the continuity of the terrace sequence and may destroy terraces formed in lesser glaciations. In fact, even a more extensive glaciation following a lesser one would increase the difficulties of interpretation, for its terraces would probably be deposited on top of those of the lesser advance, thereby obscuring them.

The basic assumption of the 'terrace method' is that the first glaciation leads to the deposition of outwash in the valley outside the glacial limit. During the succeeding interglacial (or lesser, inter-stadial phase) the river dissects the outwash, thereby forming two terraces, one on either side of the valley. The next glaciation then leads to the further deposition of outwash material on the valley floor but at a lower level than that of the previous glaciation. (This assumes that the earlier glaciation was larger than the later.) Naturally, in the following interglacial this outwash is dissected by the river and a second, lower, set of terraces is formed. Thus in the tributaries of the Danube, as Penck and Brückner (1909) have shown, there are four sets of terraces, each formed of the glaciofluvial outwash of a particular glaciation. On a lesser scale, if a glacier retreats up-valley from one halt to another, dwindling in size, it will first deposit outwash material on the valley floor, then, as it retreats, its meltwater will erode the centre of the valley, so that the previous deposit forms a pair of terraces, one on either side of the stream. At the next halt of the glacier, when it is smaller than formerly, a new sheet of outwash material will be deposited along the valley centre. During the succeeding retreat of the ice this outwash will be terraced. The overall result of these two periods of deposition and erosion is the formation of two sets of terraces, such as Lewis describes in the Wye gorge (Chapter 7).

Naturally there may be complications in interpreting terrace deposits. Often terrace remains, particularly those of older terraces which have been subject to prolonged erosion, are only fragmentary, possibly being separated from each other by many miles. During these distances the longitudinal profile of the depositing stream will have decreased in altitude, so that a terrace of altitude $x$ in one location is by no means the same age as that at the same altitude either further upstream or downstream. Furthermore, it is possible that certain terraces owe their initiation to fluvial, rather than to fluvioglacial deposition, as happens in the Severn valley (Wills, 1938). To complicate the matter even more, how are we to determine whether terraces were formed during separate glaciations, or during halts in the recession of the same glaciation? Finally, as in the lower Severn valley, base-level may have altered since the formation of important terraces, possibly causing fresh deposition at higher levels than before, thereby covering older and lower terraces.

Elucidation both of terrace and of drift sequences is sometimes aided by the analysis of fossil plant and animal (including human) remains that have been incorporated in them. In North Wales, for example, one of the most recent advances of Irish Sea ice led to the deposition of glacial till over the mouths of a number of caves, thereby sealing them. On archaeological examination of these caves it has been found that they contain implements and evidence of habitation by man of Aurignacian age. It is therefore possible to ascribe a maximum age for the glaciation, based upon the archaeological evidence, for the glaciation must have succeeded the Aurignacian cultural period. Thompson and Worsley (1966) have used this evidence to date glacial deposits in north-east Wales. Evidence of much earlier human activity is recorded by Stephens (in Chapters 5 and 11) and used to

TABLE 1.1. *The terraces of the Severn and Avon* (*Wills*, 1938)

| Severn | Avon | Glacial event |
|---|---|---|
| 1. Woolridge Terrace | No equivalent | Great Welsh glaciation (1) |
| 2. Bushley Green Terrace | No. 5 | Great Welsh glaciation (2) |
| 3. Kidderminster Terrace | No. 4 | Hoxne interglacial |
| 4. Main Terrace | { ?No. 3 <br> { No. 2 | Irish Sea glaciation, ice advancing to Shrewsbury area |
| 5. Worcester Terrace | No. 1 | Little Welsh Advance to Shrewsbury |
| 6. Power House or Forest Period Terrace | No equivalent | Postglacial |

provide a lower (i.e. oldest possible) date for gravel terraces in the Axe valley in Somerset and Devon. Stephens notes the presence of hand axes of Acheulian type incorporated in the Axe terraces, and suggests that these axes have been moved from their original site and redeposited by the great volume of water that initiated the terraces. Similarly Bowen (Chapter 9) uses archaeological evidence provided by cave deposits in Gower, such as those at Paviland, to date events in South Wales.

Evidence of such early human remains, dating from interstadial (see p. 13) or even interglacial times, is rare. Far more common (but still rather scanty) are plant and animal remains which have been incorporated in deposits. Of these, certain trees, or certain animals, are thought to have existed only during limited periods of time. Therefore, if we find such an 'indicator' (perhaps a woolly rhinoceros), we can date the deposit in which it exists. Wills (1938) has used this method to date the terraces in the Severn valley and to correlate them with terraces in the Avon valley. He defined six terraces in the Severn valley and five in the Avon, as shown in Table 1.1.

1*

In addition to using fossil remains, Wills was able to use erratic material, such as pebbles of Ailsa Craig micro-granite which are incorporated only in the Main Terrace, to date the terrace sequence. Since his work was published in 1938 more has been discovered of the glacial history of the upper Bristol Channel area, and this has led Stephens (in Chapter 5) to reinterpret Wills's evidence.

But the discovery of fossil 'indicator' material does not, of itself, provide an absolute date for a deposit, it only allows the deposit to be placed in its general position in the sequence of events. It may, for example, allow us to say that the deposit dates from the most recent glaciation, but it does not tell us the absolute age of that event. Nevertheless it is often possible to date organic or ossiferous material by using radiocarbon techniques.

All living bodies, animal or plant, contain a known amount of radioactive carbon, better known as their carbon 14 content. When these bodies die the amount of carbon 14 contained in their remains (e.g. bones, shells) declines at a known and measurable rate. By measuring the quantity of carbon 14 in a sample (such as a bone deposited in a glacial outwash terrace), and by knowing the rate of decline of carbon 14 and the amount that was originally present in the sample (which was the same as is in the atmosphere today), we can divine the date at which the sample was deposited. (Here it is assumed that the bone was deposited as soon as its owner died.) Unfortunately the amount of carbon 14 present in even a living body is so small that an extremely accurate instrument is needed to measure it. As Shotton (1967) has pointed out, a relatively minor error in the determination of the amount of carbon 14 present can give a completely false age for a body, especially when it is of considerable antiquity and has therefore shed most of its carbon 14. Furthermore, some materials are easier to date than others, the best being: wood, charcoal, hair and dried flesh; whilst the least easy to handle are molluscs, foraminifera, ostracods, birds' egg shells, bones, peat and organic mud. It is perhaps unfortunate that many of the carbon 14 dates that have been obtained for Wales are derived from examination of those very deposits that are least easy to measure. In addition to the physical problem of measuring infinitesimal quantities of carbon 14 it is also possible that the samples have been contaminated before reaching the laboratory, perhaps through water percolating through the deposit in which they lay, or through penetration by roots. A 2 per cent contamination of a deposit which is really 30,000 years old will cause the laboratory determination of its age to err by almost 5000 years, giving a reading of 25,315 years. Even under ideal conditions radiocarbon dating can only be used reasonably for deposits less than 70,000 years old, a time-span that covers only the last, or Weichsel/Würm glacial era. When reading of the 'absolute ages' of deposits, such as those quoted by John in Chapter 10, it is as well to remember the drawbacks of radiocarbon dating techniques. Naturally, the less old the sample, the less is the possibility of its being notably falsely dated, and vice versa.

Over periods of time in excess of 70,000 years, it becomes necessary to use techniques other than carbon 14 for dating. These include potassium–argon (K–Ar) dating, which can only be made on igneous rocks, and uranium decay dating, undertaken on sea-floor sediments which have not been disturbed since their formation. Neither of these methods has yet been used for Pleistocene deposits in Wales, and the latter method is obviously invalid, even for the seas around Wales, for the sediments on the floors of these seas have been disturbed many times, both by glaciation and by submarine currents and turbulence.

Over geologically short periods of time, such as that between the waning of the most recent ice caps in Wales and the present, it is sometimes possible to date events using the techniques of pollen analysis. Virtually all plants release pollen at some time of the year, as sufferers from hay fever know, and this is deposited over the ground surface. In some cases the pollen grains fall into a lake which, through the process of time, becomes choked with partially decayed vegetation and is transformed into a peat bog. Such a bog contains layer upon layer of pollen, each layer representing an individual year. Pollen grains are exceedingly resistant to decomposition, and it is possible to identify individual grains many thousands of years after their deposition. By counting the number and type of pollen grains in a deposit at any one level it is possible to get some idea of the composition of the vegetation from which that pollen was derived. Taking matters even further, by examining different stratigraphic layers of pollen it is possible to reconstruct the development of the surrounding vegetation from the beginning of pollen deposition to the present. Research has shown that, from the Late Glacial period to the present day there has been a definite pattern to vegetation development with the existence of a number of marked vegetation periods. These periods are called 'pollen zones', and the types and quantities of pollen differ markedly from zone to zone, thereby enabling one to ascribe a particular pollen deposit to one or other of the zones. In many cases the age of these zones has been determined through radiocarbon methods, and through examination of archaeological material found in the same layer of the deposit. Therefore if, through pollen analysis, one can relate a deposit to a particular zone, as Whittow and Ball (Chapter 2), Lewis (Chapter 7) and Bowen (Chapter 9) have done, then the age of the deposit is known with considerable accuracy.

The following table indicates the zones that have existed since about 14,000 B.P. As later chapters will show, Zone Ib deposits have been recorded in Snowdonia and in the Brecon Beacons, and they

TABLE 1.2. *Late and post glacial pollen zones*

| Pollen zone | Name | Age (B.P.) |
|---|---|---|
| *Postglacial* | | |
| VIII | Sub-Atlantic | 2,500 to the present |
| VIIʙ | Sub-Boreal | 5,000 to 2,500 |
| VIIa | Atlantic | 7,000 to 5,000 |
| VI ⎫ | Boreal | 9,000 to 7,000 |
| V ⎭ | | 9,600 to 9,000 |
| IV | Pre-Boreal | 10,300 to 9,600 |
| *Late glacial* | | |
| III | Younger Dryas | 10,800 to 10,300 |
| II | Alleröd | 12,000 to 10,800 |
| Ic | Older Dryas | 13,000 to 12,000 |
| Ib | Bölling | 14,000 to 13,000 |
| Ia | | Older than 14,000 |

document the end of the regional ice-caps in those areas. Zone III represents a reversion to colder conditions after the amelioration of Zone II, and the last glaciers in Wales, located in some of the major cirques, existed at that period. The time scale which is also shown in Table 1.2 is only meant to give a relative idea of the length of each zone. No doubt, as further information on absolute dating becomes available, the time scales will be altered, although basically they will remain much as they are given here.

Research in the Alpine areas of Europe has shown that, during the Pleistocene, four periods of glaciation have existed. The extent of ice in each glacial period, however, often varied considerably and in some glacial periods there were two, or even three, main periods of ice advance. These were divided from each other by interstadials, periods of time in which the climate ameliorated to such an extent that conditions resembled those of modern tundra regions. Naturally, in an interstadial the ice caps shrank in size, although they certainly did not melt completely. In the following worsening of climate the ice area increased, so that glacial periods as a whole were times of alternately waxing and waning ice masses. Each glacial period was divided from its successor by an interglacial, a time in which the climate ameliorated to such an extent that it could, at least, be classed as Temperate (in contrast to the tundra, or Boreal, classification of an interstadial).

Whilst the Alpine areas experienced fourfold glaciation, a separate ice mass developed over Scandinavia and spread south and west to cover much of the North German Plain and the area that now lies under the North Sea. This ice body apparently existed in three separate glaciations during each of which, as in the Alps, it alternately waxed and waned. Although the pulsations of the Alpine and north European glaciers did not necessarily coincide in time, it appears that they were broadly coeval (Flint, 1957).

In the British Isles, as Mitchell (1960) suggests, evidence is forthcoming for no more than three periods of glaciation, and it is by no means certain that these stages equate with the glaciations of northern Europe or of the Alps. If correlation has to be made between Britain and Europe it is, perhaps, better to correlate between Britain and the north European glaciations, both of which spread south from an area of limited relief, than between Britain and the Alps, with their high altitudes and more southerly location. Nevertheless the glaciations of the Alps, tabulated as long ago as 1909 by Penck and Brückner, are generally better known than those of northern Europe and both Alpine terminology and that of northern Europe will be found in the following chapters. Table 1.3, based on that of Flint (1957) and incorporating evidence published by Mitchell (1960) and placed in its European context by the present writer, is intended to help the reader follow the arguments of succeeding chapters. It should be remembered that, although events are shown as coeval in this table, they may not, in fact, be so, and further research may prove that the deposits of England, in particular, may be of a different age from that to which they are here ascribed.

Table 1.3 indicates only glacials and interglacials, but as has already been stated, it is known that the glacials were subdivided by a number of less severe interstadial phases. The Günz, Mindel, Riss and Würm glaciations are all thought to have been divided into two glacial maxima separated by an interstadial. Table 1.4 is a simplifica-

TABLE 1.3. *The correlation of Pleistocene events in the northern Alps, northern Germany and England**

| Event | Northern Alps | Northern Germany | England |
|---|---|---|---|
| Postglacial | Recent | Recent | Recent |
| Glacial | Würm | Weichsel | Smestow |
| Interglacial | Riss/Würm | Eemian | Ipswichian |
| Glacial | Riss | Saale | Gipping |
| Interglacial | Mindel/Riss | Holstein | Hoxnian |
| Glacial | Mindel | Elster | Lowestoft |
| Interglacial | Gunz/Mindel | Preglacial | Preglacial |
| Glacial | Gunz | — | — |
| — | Preglacial | — | — |

Pliocene
(warm)

* In 1968 the Geological Society of London suggested that the Quaternary period as it affected England should be divided into the following stages (most recent given first): FLANDRIAN, DEVENSIAN, IPSWICHIAN, WOLSTONIAN, HOXNIAN, ANGLIAN, CROMERIAN, BEESTONIAN, PASTONIAN, BAVENTIAN, ANTIAN, THURNIAN, LUDHAMIAN, WALTONIAN. Although all these stages lie within the geological period of the Quaternary, there is no evidence of glaciation before the ANGLIAN stage.

tion of the sequence of events in central Europe as shown by Gross (1964) and dated, very approximately, by Shotton (1967).

Neither the Early Würm nor the Würm Maximum were altogether periods of complete glaciation. The ice advanced and retreated on a

TABLE 1.4. *Würm events in central Europe*

| Event | Name | Date in years B.P. |
|---|---|---|
| | Postglacial | 10,000 |
| Glacial | Würm Maximum | 23,000 |
| Interstadial | Interpleniglacial | 50,000 |
| Glacial | Early Würm | 70,000 |
| Interglacial | Riss/Würm interglacial | |

number of occasions in both divisions, but at all times there was sufficient present for us to think of them as dominated by ice. During the Early Würm, following the initial advance of the glaciers, a retreat termed the Amersfoort occurred. This was succeeded by a further ice advance, followed by the Brörup retreat, before the glaciers swelled to their Early Würm maximum. The Interpleniglacial, an interstadial lasting over 20,000 years, was interrupted about 30,000 years ago by an ice advance which gave forwarning of the Würm Maximum that was to come. But between this advance and the Maximum the ice retreated to give an interstadial complex.

Coope and Sands (1966) have shown that the same pattern of events was not exactly reproduced in the English Midlands in the glacial period that they term the Weichselian. Figure 1.1 is, in fact, a reproduction of their work, showing that the Weichsel glaciation there comprised two glacial maxima, separated by a marked interstadial in which there were several climatic oscillations. The nature of these oscillations in the English Midlands has been indicated through

TABLE 1.5. *The age of glacial events*

| Event | Age in years before the present |
|---|---|
| Riss–Würm Interglacial | 105,000 |
| Riss Glaciation | 140,000 |
| Mindel–Riss Interglacial | 195,000 |
| Mindel Glaciation | 275,000 |
| Gunz–Mindel Interglacial | 325,000 |

*Source:* Shotton (1967), p. 379.

examination of faunal deposits from various localities. Each locality where fossil faunas have been found and examined in detail is named on the left-hand side of the temperature curve on Fig. 1.1.

Methods of absolute dating have made it possible to ascribe tentative ages to the various glaciations, as well as to interstadial divisions incorporated in them. By using the uranium decay series Shotton (1967) has been prepared to suggest the time-scale for glacial events (Table 1.5).

As will be shown in later chapters, the glaciations of Wales, as far as is known, were no more than threefold, and even the most ambitious author is not prepared to postulate any Welsh glaciation earlier than that of Mindel/Elster times. The evidence for this relatively early event is far from unambiguous and, as Synge shows in the concluding chapter, the stratigraphy and dating of Pleistocene events in Wales still leaves much to be desired.

During the present century several attempts have been made both to correlate the glaciations of Wales with those of mainland Europe, and to delimit the areas covered by ice in each glaciation. In 1929

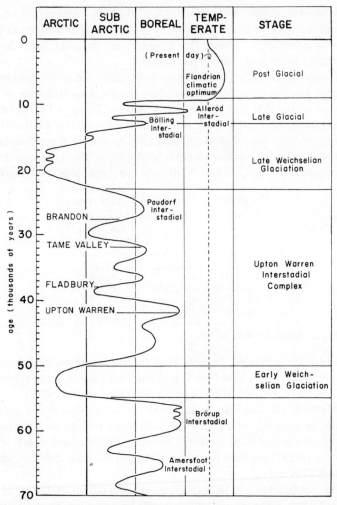

Fig. 1.1. The Weichselian of the English Midlands, after Coope and Sands, 1966. *Proc. Roy. Soc. B*

Charlesworth published his 'South Wales End Moraine' map, where-on he attempted to depict the southern limits of Welsh ice in the last glaciation. In 1956 he again published a map, incorporating his previous limits, but extending them to apply to the whole of Wales. At the same time he indicated the main retreat stages of the ice of the Last Glaciation. This latter publication had been preceded in 1953 by

Wirtz's map, which suggested that the Cardigan Bay coastlands, except for a small area around Cardigan itself, were not affected by ice of the Last Glaciation. In 1960 Mitchell suggested that none of Wales south of the Lleyn peninsula was affected by Irish Sea ice, and

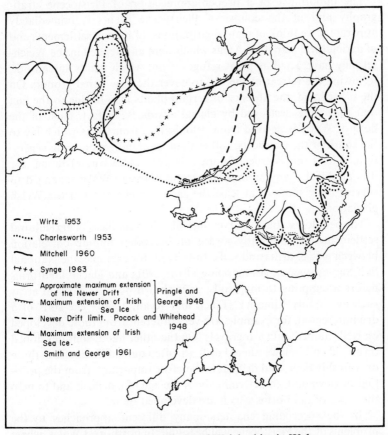

Fig. 1.2. Proposed limits of Weichsel ice in Wales

that large areas of the interior were unglaciated in the Smestow (i.e. Weichsel) glaciation. Finally Synge (1963) suggested an even less extensive Weichsel glaciation for the Welsh coastlands. Quite apart from all these are the limits suggested in the Regional Memoirs of the Geological Survey, where limits for North Wales, South Wales and the Welsh Borderland are depicted. All these various limits are alluded to in the following chapters and they are shown on Fig. 1.2.

Naturally most of these limits are debatable and, in the concluding
chapter, Synge describes the various major glacial limits as shown by
recent research in Wales, the south-west peninsula of England and
along the south coast of Ireland.

The importance of a proper appreciation of Pleistocene stratigraphy and of the course of Pleistocene events is immediately
apparent if we think of plant distributions, of human settlements and
of agricultural activities. Areas which were unglaciated in the Weichsel may have acted as plant refuges, from which plants spread out,
following the retreat of the ice, to cover the now bared ground. On
the other hand, the absence of certain plants may be due to the fact
that they were destroyed by glacial advances. In fact, most of the
flora of Wales is derived from colonisation from areas which lay to
the south of the ice sheets, rather than from indigenous relic vegetation. Whilst this is true of flora it is even more true of fauna, for
it is doubtful if any animals at present native to Wales managed to
survive the rigours of at least the earlier (and larger) of the Welsh
glaciations.

The pedologist, busily surveying soils (the very basis of agricultural
settlement), is constantly aware of the often minute correlation
between glacial drifts and soils. In Ireland, for example, the National
Soil Survey encourages mapping of the drifts and glacial landforms
before attempting to map soils, as is shown by their work in areas
such as County Limerick (Finch and Ryan, 1966). The soils of a
drumlin region, for example, are very different from those developed
on well-drained outwash gravels. On the other hand, soils developed
on an old and well weathered drift are often quite different from those
on a freshly deposited drift. It is therefore important, from the pedo-
logical viewpoint, to determine both the age of a deposit and to map
the types of landforms which are developed on it.

In the succeeding chapters many different approaches to the
problems of the Pleistocene will be found. In north-west Wales, where
there was a complex battle, on a number of different occasions, between Welsh and Irish Sea ice, it has been necessary to concentrate on
the stratigraphy of the area. By comparison, in mid-Wales no deposits
other than those of the most recent glaciation have been found, and
here the emphasis is on the fluctuations of that glaciation and on the
landforms it has left. In other areas, such as Somerset, considerable
attention is paid to archaeological evidence to unravel the history of
the area.

Of course, many details of the Pleistocene in Wales are still uncertain, as the authors of this book readily acknowledge. The reader must therefore realise that he is being presented with current ideas about the glaciations of Wales and that it will probably be many years hence before all the intricacies of the Welsh Pleistocene are fully understood.

## REFERENCES

AGASSIZ, L. (1842) 'On glaciers and the evidence of their having once existed in Scotland, Ireland and England', *Proc. geol. Soc.* **3**, 327–32. (Publication of his paper read in 1840.)

BALL, D. F. (1960) 'Relic soil on limestone in south Wales', *Nature, Lond.* **187** (4736), 497–8.

CHARLESWORTH, J. K. (1929) 'The South Wales end-moraine', *Q. Jl geol. Soc. Lond.* **85**, 335–55.

CHARLESWORTH, J. K. (1956) 'Late glacial history of the Highlands and Islands of Scotland', *Trans. Roy. Soc. Edin.* **62**, 769–928.

COOPE, G. R. and SANDS, C. H. S. (1966) 'Insect faunas of the last glaciation from the Tame Valley, Warwickshire', *Proc. Roy. Soc.* B, **165**, 389–412.

DAVIES, G. L. (1968) 'The tour of the British Isles made by Louis Agassiz in 1840', *Ann. Sci.* **24**, 131–46.

DAVIES, G. L. (1969) *The Earth in Decay*. Macdonald.

FARRINGTON, A. (1934) 'The glaciation of the Wicklow Mountains', *Proc. Roy. Ir. Acad.* **42**B, 173–209.

FINCH, T. F. and RYAN, P. (1966) *Soils of County Limerick*. An Foras Taluntais, Dublin.

FLINT, R. F. (1957) *Glacial and Pleistocene Geology*. New York, Wiley.

GARWOOD, E. J. (1910) 'Features of Alpine scenery due to glacial protection', *Geogr. Jl.* **36**, 310–39.

GEIKIE, A. (1863) 'On the phenomena of the glacial drift of Scotland', *Trans. geol. Soc. Glasg.* **1**, 1–190.

SAXO GRAMMATICUS (c. 1200) *Gesta Danorum*, cited by Thorarinsson (1960).

GROSS, H. (1964) 'Das Mittelwürm in Mitteleuropa und angrengenden Gebieten', *Eiszeitalter Gegenw.* **15**, 187–98.

GROVE, J. M. (1966) 'The Little Ice Age in the massif of Mont Blanc', *Trans. Inst. Br. Geogr.* **40**, 129–43.

HOLMES, C. D. (1941) 'Till fabrics', *Bull. geol. Soc. Amer.* **52**, 1299.

JAMEISON, T. F. (1862) 'On the ice-worn rocks of Scotland', *Q. Jl geol. Soc. Lond.* **18**, 164–84.

MITCHELL, G. F. (1960) 'The Pleistocene history of the Irish Sea', *Advmt Sci. Lond.* **17**, 313–25.

MITCHELL, G. F. and ORME, A. R. (1967) 'The Pleistocene deposits of the Isles of Scilly', *Q. Jl geol. Soc. Lond.* **123**, 59–92.

PENCK, A. and BRÜCKNER, E. (1909) *Die Alpen im Eiszeitalter*.

POCOCK, R. W. and WHITEHEAD, T. H. (1948) *The Welsh Borderland* (British Regional Geology). H.M.S.O.

PRINGLE, J. and GEORGE, T. N. (1948) *South Wales* (British Regional Geology). H.M.S.O.

RAMSAY, A. C. (1860) *The old glaciers of Switzerland and North Wales*, London. (This contains the most accessible statement of his 1859 ideas.)

RAMSAY, A. C. (1864) 'On the erosion of lakes and valleys', *Phil. Mag.* 4, 293–311. (This is essentially the paper which he read in 1862.)

SAUNDERS, G. E. (1968) 'A fabric analysis of the ground moraine deposits of the Lleyn Peninsula of southwest Caernarvonshire', *Geol. Jl.* 6, 105–18.

SHOTTON, F. W. (1967) 'The problems and contributions of methods of absolute dating within the Pleistocene period', *Q. Jl geol. Soc. Lond.* 122, 357–84.

SMITH, B. and GEORGE, T. N. (1961) *North Wales* (British Regional Geology). H.M.S.O.

SYNGE, F. M. (1963) 'Correlation between the drifts of South-East Ireland and those of West Wales', *Ir. Geogr.* 4, 360–6.

THOMPSON, D. B. and WORSLEY, P. (1966) 'A Late Pleistocene marine molluscan fauna from the drifts of the Cheshire Plain', *Geol. Jl.* 5, 197–207.

THORARINSSON, S. (1960) 'Glaciological knowledge in Iceland before 1800', *Jökull.* 10, 1–18.

WILLS, L. J. (1938) 'The Pleistocene development of the Severn from Bridgnorth to the sea', *Q. Jl geol. Soc. Lond.* 94, 161–242.

WIRTZ, D. (1953) 'Zur Stratigraphie des Pleistocäns im Westen der Britischen Inseln', *Neues Jahrb. für Geol. und Paläont.* 96, 267.

# CHAPTER TWO

# North-west Wales

J. B. Whittow, B.A., Ph.D., F.G.S. and
D. F. Ball, M.A., Ph.D.

Louis Agassiz (1842), who was one of the first to describe the effects of glaciation in Britain, included the valleys of Snowdonia among the localities in which he saw evidence of former glaciers, and Cwm Idwal was the first Welsh glacial site to be described in detail (Darwin, 1842). The interest this aroused reached its first and in many ways still its most significant peak in north-west Wales in the work of A. C. Ramsay (1860 and 1881). Later, Reade (1893) described the section of Moel Tryfaen and those of the north Lleyn coast; in 1909 Jehu further described the glaciation of Lleyn; and in 1919 Greenly gave a detailed account of the glacial features of Anglesey. Many other relevant papers have appeared during and since this period (see Bassett, 1961, 1963), but no comprehensive recent attempt has been made to interpret the glacial history of the region.

It has been generally accepted that Anglesey and western Caernarvonshire were crossed twice by ice from northern Britain (Northern or Irish Sea ice). Upper and Lower boulder clays were described as separated by Intermediate sands and gravels which formed during interglacial melting of the Lower boulder clay. The apparent absence of a similar association of meltwater deposits from the Upper boulder clay was not explained, and it was assumed that similar advances and intervening meltwater deposits were produced by the Welsh ice. Recent studies have shown as inadequate this simple account of two ice advances which covered the whole region. Drift stratigraphy in north-west Wales results from a complex ebb and flow of Northern and Welsh ice. Although current interpretation of glacial history is necessarily incomplete it is now possible to make a preliminary synthesis with some confidence.

## THE GLACIAL HISTORY OF NORTH-WEST WALES

Table 2.1 summarises what is at present known of glacial events in the region; the detailed evidence on which this table is based is given as an Appendix at the end of the chapter.

TABLE 2.1. *North-west Wales chronology*

| Classification | Morphology and stratigraphy | Suggested dating | |
|---|---|---|---|
| | | Zones IV–VIII | Postglacial |
| 8 | Postglacial deposits | Zones IV–VIII | Postglacial |
| 7 | Inner corrie moraines—Snowdonia / Solifluction, periglacial scree | Zone III | LATE GLACIAL |
| 6 | Organic deposits / Nant Ffrancon, Llyn Dwythwch, Cors Geuallt, (Caernarvonshire), Cors Goch (Anglesey), Glanllynau (Lleyn) | Zone II Allerød | LATE GLACIAL |
| 5 | Outer corrie moraines—Snowdonia / Solifluction, periglacial scree | Zone Ic | |
| 4 | Organic deposits / Cors Geuallt | ? Zone Ib Bolling | |
| 3 | a. Liverpool Bay Phase deposits — b. Welsh Intermediate Moraines Phase deposits / Yr Eifl corrie moraines / Cryoturbation, Upper head formation | ? Second phase | Second stage (Stadium) — GLACIAL |
| 2–3 | ?– ? ? ? | ? Interval | |
| 2 | a. Main Anglesey Advance deposits — b. Arvon Advance deposits / Tor modification and summit blockfields in Lleyn and Snowdonia | ? First phase | |
| 1–2 | Zone of weathering and erosion | | INTERSTADIAL |
| 1 | Cryoturbation / a. Irish Sea Advance deposits — b. Criccieth Advance deposits / Lower heads formed | First stage (Stadium) | GLACIAL |
| d | Raised beach deposits, Porth Oer, Red Wharf Bay | | INTERGLACIAL |
| c | Lowest head formed, Red Wharf Bay | COLD PERIOD (? GLACIAL) | |
| b | Raised marine platform, Lleyn, Anglesey | | INTERGLACIAL |
| a | Deep weathering in bedrock Porth Wen, Pen-y-bryn | | INTERGLACIAL |

*The earliest evidence*

Relics of interglacial or even preglacial weathering may be present, as depths of up to 15 m (50 ft) of rotted rock occur in Anglesey at Porth Swtan (310894) and Porth Wen (402947) and also near Caernarvon as oxidised shale beneath till at Pen-y-Bryn brickworks (490614). The most widespread Pleistocene feature which antedates the boulder clay however, is a raised platform of marine abrasion found in Anglesey and western Lleyn at approximately 7·5 m (25 ft) O.D. (Whittow, 1965). At Porth Oer (167301) a beach deposit on this platform contains a few erratic pebbles, perhaps from an otherwise unrecorded earlier glaciation, although at Red Wharf Bay (532816) the raised beach is composed entirely of local (Carboniferous) rocks.

In Anglesey, owing to the absence of unequivocal Irish Sea Advance deposits it is only possible to show that the raised marine platform and the deeply weathered bedrock are older than the Main Anglesey Advance. But since the raised beach and platform at Porth Oer are older than the Irish Sea Advance it seems probable that the Red Wharf Bay beach and platform also are older than the Irish Sea Advance. It is possible to demonstrate at Red Wharf Bay that the raised beach was deposited after the formation of the lowest head which in turn buries the platform. It is suggested, therefore, that the marine platform was fashioned during a period of high sea-level prior to that during which the raised beach was formed.

The evidence of earlier cold periods is very fragmentary and the earliest glaciation that can be described in any detail is that of the Irish Sea Advance (1a).

*The first recognised glacial stage* (1a–1b)

The earliest glacial drifts appear to be the relic high-level patches of Snowdonian till and the shelly sands on Moel Tryfaen. These shelly sands must have been dredged from a pre-existing sea floor and therefore represent the onset of a glaciation following a substantial interglacial episode. Preservation of such an early deposit in isolation is difficult to explain and while we tentatively correlate it with the Irish Sea Advance (1a), responsible for the thick shelly tills of western Lleyn, it is impossible to feel entirely convinced of this correlation. Similarly, if the Welsh till which underlies these Northern sands on Moel Tryfaen is correlated with the lowest Welsh till of the lowlands (the Criccieth Advance (1b) deposits), then the Welsh ice must have been in retreat to allow northern ice to penetrate the Snowdonian foothills to a height of 400 m (1300 ft) O.D. If this is so, it is difficult to understand how Criccieth Advance till is seen at a number of lowland localities in eastern Lleyn without a comparable association of Northern drifts. The position of Moel Tryfaen in the chronology

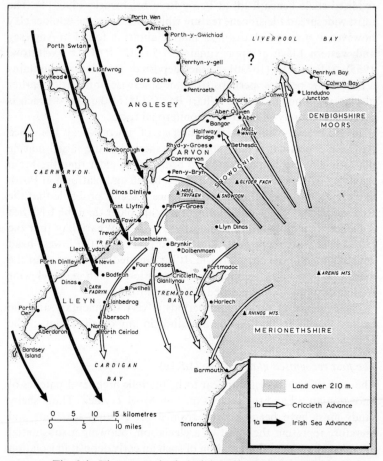

Fig. 2.1. First recognised glacial stage, north-west Wales

remains obscure but on balance its Northern and Welsh drifts could represent 1a and 1b deposits respectively. Certainly any attempt to suggest a more recent age creates many difficulties.

The unequivocal exposures of the Irish Sea Advance (1a) deposits are confined to western Lleyn, principally west of a line Abersoch–

Sarn–Nevin. It seems likely that this ice advance must have crossed western Anglesey before reaching Lleyn although there is a total absence of similar shelly till and sand in Anglesey (a problem remarked upon and unsolved by Greenly). It is possible that deposits were formerly present but have been removed by later ice advances, although the drift thicknesses remaining in Lleyn make this unlikely. With less probability it could be suggested that a single ice-sheet which had previously crossed Anglesey collected all its shelly material entirely from the now submerged area between Anglesey and Lleyn.

All the striae and stone orientations of Bardsey Island and south-western Lleyn show an ice movement from NNW to SSE which suggests that the Irish Sea Advance moved down the central channel of the Irish Sea, perhaps impinging only on western Anglesey and western Lleyn before passing southwards towards Pembrokeshire. It is unlikely that any nunataks survived in Lleyn.

In central Lleyn a contemporaneous Welsh ice-sheet, the Criccieth Advance (1b), appears to have been in contact with Irish Sea ice in the area from Nevin to St Tudwal's peninsula and their basal tills are mutually exclusive. As in Anglesey, no deposits of till 1a have been recognised anywhere in Arvon north of Dinas Dinlle due in the latter area to the magnitude of the Criccieth Advance Welsh ice-cap which appears to have radiated from a centre over the mountains south-east of Snowdon. The major through-valleys of Snowdonia were over-deepened and straightened at this time by lowering of the watersheds by ice moving radially outwards from a centre near Llyn Tegid (Bala Lake). No end-moraines of either till sheet are found on the present land surface. The general absence of Snowdonian till in Anglesey, even on the Menai Straits shore, is notable. Any passage of Criccieth Advance till on to the island has been subsequently removed, but this or earlier Welsh ice may have been responsible for the erratics recorded in Anglesey by Greenly (1919) a few kilometres inland from the Menai Straits.

As the combined ice-sheets melted a series of spectacular meltwater channels were produced along the preglacial divide of the Lleyn Peninsula. The Carn Fadryn hills in central Lleyn emerged first from beneath the down-wasting ice and the subglacial gorges of Nant Llaniestyn (*c.* 265344) and Nant Horon (*c.* 287330) were formed. Submarginal chutes are present, for example at Llangian (296290) and Nant Pig (309250) where the channel is buried beneath subsequent solifluction deposits at Porth Ceiriad. The Soch gorge at

Abersoch (*c.* 310281) probably originated at this stage as a sub-glacial channel, and was clearly not formed by the Afon Soch.

## An interstadial episode (1–2)

A retreat of Criccieth Advance and Irish Sea Advance ice was follow-ed by a passage through periglacial conditions to a nonglacial climate in which the upper surface of the basal tills was deeply weathered. Solifluction processes affected the upper layers of both Northern and Welsh tills at this period. Iron pan formations and indurated layers in

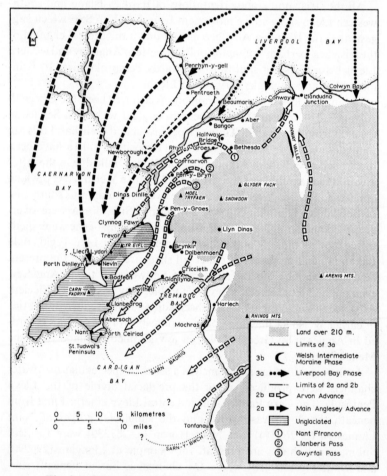

Fig. 2.2. Second recognised glacial stage, North-west Wales

gravels may indicate the length of the weathering period, as do the frequency and scale of the channels cut during interstadial erosion and infilled by solifluction deposits. No organic remains have yet been found in the area associated with this episode to allow absolute dating.

*The second glacial stage—1st phase (2a–2b) (Fig. 2.2)*

The next ice-sheets to appear in North Wales are those of the Main Anglesey Advance (2a) and its associated Welsh ice, the Arvon Advance (2b). The greater part of Anglesey has drift deposits very different from the shelly till of the Irish Sea Advance and the characteristic feature of the Main Anglesey Advance drift is the prevalence of predominantly local material and the relatively low proportion of far-travelled erratics (Smithson, 1953). The direction of ice movement produced north-east to south-west aligned boulder trains and drumlin swarms (Greenly, 1919). Subsequent periglacial solifluction and erosion of this till, which must always have been thinner than the Lleyn till of the Irish Sea Advance has left large areas of the island substantially drift-free and made it difficult to observe any contact between Anglesey Advance and Arvon tills. That this contact zone must have been more or less in the present position of the Menai Straits is shown by the virtual absence of Snowdonian stones in the tills near the Straits on the Anglesey side, and a corresponding absence of characteristic Anglesey rocks in the weathered tills remaining in pockets on the Bangor ridge immediately on the Arvon shore of the Menai Straits.

The Arvon ice-sheet had its northward and north-westward moving piedmont glaciers (e.g. Llanberis and Nant Ffrancon) deflected south-westwards by the Northern ice. Welsh ice was also probably thinner at this time than in the Criccieth Advance since a glacial trimline can be seen on the valley sides. In Nant Ffrancon, Llanberis Pass and Nant-y-Betws, ice apparently reached no higher than about 610 m (2000 ft O.D.), for above this height the peaks emerge littered with thick sheets of periglacial debris. It has been suggested that a soil containing gibbsite ($Al_2O_3.3H_2O$) on weathered granite on Llymllwyd, Nant Ffrancon, at 747 m (2450 ft) (631609) may be a preserved relic of interglacial weathering (Ball, 1964). Earlier Welsh till (1b) is locally present on higher ground but was swept out of the valley floors which were probably further deepened at this time. Glacial breaching of secondary divides may also be attributed to the period of the Arvon Advance, for example at Llyn Cowlyd (720615), Bwlch Ehediad (665524),

Bwlch Cwm Llan (605520), Drws-y-Coed (545535), (Dury, 1955) and Cwm Trwsgl (552495). The high Snowdonian range acted as a barrier to the free passage of Welsh ice much as the peripheral mountains of Greenland impede the Greenland ice today.

Above the suite of glacial cwms (corries) which were further excavated at this period in Snowdonia, and on Yr Eifl in Lleyn, tor-like forms were first exhumed and ultimately refashioned by peri-glaciation on the nunataks. These and associated blockfields, which continued to be added to in Snowdonia during the remainder of the Pleistocene, are present on Tre'r Ceiri (375446), Carn Ddu (402468), Bwlch Mawr (426478) and Mynydd Craig Goch (498485) in northern Lleyn, while summits such as Carn Fadryn (280352) and Carreg Lefain, Mynydd Rhiw (226275) are severely frost-shattered. Glyder Fach has the most dramatic summit tors of the high peaks, but simi-lar features are widespread in the Carneddau, e.g. Foel Grach (689659) and Garnedd-Uchaf (688670).

Meltwaters near the junction of Welsh and Northern ice along the Menai Straits removed any morainic evidence and the limits of this ice sheet in north Caernarvonshire have been destroyed by later advances of Welsh ice or lie beyond the present coastline. In Lleyn a series of moraines can be identified. Arvon ice encroached on the northern slopes of the Gurn Ddu–Bwlch Mawr massif to a height of 90 m (300 ft) and its limit is marked by the Clynnog moraine and a large marginal channel around the southern flank of Y Foel (450505). An ice lobe passed south through the Pant Glas gap and deposited gravels and sands around Brynkir (480445) and southwards to the southern Lleyn shore. It is in these gravels that kettle holes are formed at Glanllynau. Reinforcement of the ice from northern Snowdonia came from the western valley glaciers, especially from the Llyfni, Pennant and Glaslyn valleys, from which ice must have extended into Cardigan Bay. It is probable that at this period the Merionethshire ice-sheets of the Harlech Dome were active and their glaciers crossed the present coastline to leave their lateral moraines as submarine sarns (ridges) in Cardigan Bay, for example, Sarn Badrig, traceable seawards from the moraine of Mochras 'Island' (555265). Further south still it is probable that at this period the Mawddach glacier reached the Ton-fannau moraine and its seaward extension, Sarn-y-Bwch.

The combined Cwm Pennant–Glaslyn–Ffestiniog ice stretched as a piedmont lobe across Tremadoc Bay to reach St Tudwal's peninsula in Lleyn where the ice margin can be mapped quite accurately. An ice-tongue crossed the low valley from Sarn Bach (305625) to Nant, Porth Neigwl (289247), but failed to reach Porth Ceiriad (Fig. 2.2). Striations of an east–west trend have been mapped on St Tudwal's peninsula by Nicholas (1915), who also collected Tremadoc Slate erratics from Llanengan (293270). The Afon Soch which formerly flowed out near Sarn Bach or at Abersoch (the present mouth of the river) was temporarily diverted to Porth Neigwl. The diversion channel is seen cut into the till cliffs west of Tai Morfa (285265) but it is now blocked by postglacial dunes. Marginal channels associated with the piedmont ice can be traced northwards from Abersoch via Oerddwr (310303) and Llanbedrog (323317) to central Lleyn and the now marshy infilling of Cors Geirch (315365). The latter is flanked by remarkably flat-topped terrace features. Although formerly interpreted as proglacial lake terraces (Matley, 1936), the sections show no evidence of lacustrine deposition and are kame terraces formed on the flanks of a dead ice tongue which stagnated in the Cors Geirch hollow. From Penrhos airfield (335335) outwash sands and gravels at least 48 m (160 ft) in thickness can be traced for about 4 km (2½ miles) northwards. These terraces are lobate in plan and flat-topped in profile. The highest terrace reaches about 48 m (160 ft) O.D. near Rhydyclafdy (328549), the terrace at Penrhos airfield being much lower at 22·5 m (75 ft). Mineralogical analysis (Dr D. A. Jenkins, personal communication) showed characteristic Northern minerals to be important in the sands but their contribution apparently faded in the higher horizons. Near Traian (328365) a large fossil ice-wedge (drawn to our attention by Dr B. Seddon) cuts the deposits in a small quarry, and near Maesoglan (302379) the fluvioglacial cross-bedded sands and gravels contain lumps of shelly till. The latter section occurs on the flanks of a zone of morainic relief which, with its kames and kettle holes (most marked between Tan-y-Graig (313387) and Ceidio (285383) can be traced north-westwards towards Porth Dinllaen (275415). The marginal channel west of Garn Bodvan (313393), the disrupted drainage pattern near Ceidio, and the deflection of the Afon Geirch from Porth Dinlleyn to Aber Geirch (265405) (Cope, 1902), all support the association of this moraine with a lobe of Main Anglesey Advance ice (2a) which just impinged on north-central Lleyn.

The fresh ice-contact slopes (15°–20°) and re-entrant features noted on the inner flanks of the Cors Geirch terraces are present also on their eastern side. It is believed that Arvon Advance ice from northern Snowdonia reached eastern Lleyn through the Pant Glas gap, swung south-westwards along the southern slopes of Yr Eifl and reached the area of Bodfean (325378). The principal streams of eastern Lleyn (Rhyd-Hir, Erch and Wen) are still deflected westwards and a large marginal channel is followed into Pwllheli by the A499 road, supporting the concept that Arvon ice and the Ffestiniog–Glaslyn–Pennant lobe substantially occupied eastern Lleyn, with the river deflections marking the successive melt phases of the Arvon ice in this area. With the presence of a lobe of the Main Anglesey Advance ice in north-central Lleyn south of Nevin, only the south-west of the peninsula remained ice-free at this period (Fig. 2.2), and the thick till sheets of the earlier Irish Sea Advance must have been subject to severe solifluction and mass movement during the periglacial conditions of the time.

*The second glacial stage—2nd phase (3a–3b) (Fig. 2.2)*
*Northern ice (3a).* It is difficult to decide whether the next glacial episode in the area represents a separate cold stage or a still-stand in the gradual retreat of 2a–2b ice from the region. Let us first consider the possibility that the Main Anglesey Advance (2a) and the Liverpool Bay Phase (3a) were separate maxima.

In the first place there is a very marked lithological and mineralogical difference between the deposits attributed to the Main Anglesey Advance and those which impinge only marginally on the Welsh shores of Liverpool Bay and which become progressively more extensive when traced eastwards into Cheshire. The Main Anglesey Advance till (2a) is dominated by local rocks and generally devoid of, or weak in characteristic Northern minerals (Smithson, 1953) whereas the so-called Liverpool Bay Phase till (3a) is a heavy-textured and generally relatively stoneless reddish-brown clay with characteristic marine shells and a Northern mineral assemblage. Such a complete change of facies and the general uniformity of Liverpool Bay till from its most north-westerly occurrence at Llaneilian, Anglesey (470932) to its extensive outcrops in the Vale of Clwyd and north-west Cheshire supports the possible recognition of Liverpool Bay ice as a distinct late Northern Advance into the region. As the latter till is traced eastwards its elevation becomes progressively greater which suggests that in contrast to the earlier Irish Sea Advance (1a) this ice-sheet was not at its maximum thickness in the central Irish Sea Basin, but instead passed southwards essentially to the east of the Isle of Man. The approximate elevations are: 42 m (140 ft) at Llaneilian and 85 m (280 ft) near Llangoed, Anglesey, 122–152 m (400–500 ft) in the Creuddyn peninsula, 170 m (560 ft) at Colwyn Bay golf course (Reade, 1885), and 183 m (600 ft) on the northern Clwydian hills, Flintshire.

Although it may be considered preferable to call this the Liverpool Bay *Advance*, the term *Phase* is retained because of an alternative interpretation, that the Main Anglesey Advance (2a) and Liverpool Bay (3a) tills represent deposits from different zones of the same ice-sheet. The basal layers of such an ice-sheet would be heavily charged with shelly material dredged from the sea-floor of Liverpool Bay whilst the upper layers remained relatively 'clean'. Blocking of the lower layers against the orographical change now marked by the present coastline or against Welsh ice would cause deposition of the burden from the lower layers whilst the 'cleaner' upper ice, when not so checked by Welsh ice, as in Anglesey, moved forward beyond the basal till limits, transporting only local material. In this reconstruction the absence of a change in till lithology in the Vale of Clwyd would be ascribed to the presence of Triassic rocks in the vale similar to those postulated as present on the floor of Liverpool Bay. The almost total absence of evidence showing a weathered horizon separating the 'Lower' (2a) from the 'Upper' (3a) tills might be regarded as significant when contrasted with the presence of such a horizon between the Criccieth Advance (1b) and Arvon Advance (2b) tills. But there remains the possibility that evidence of the hypothetical interval (2–3) may have been destroyed or is still to be discovered. Furthermore, whether there were two ice advances or only one, there does appear to have been a stillstand of downwasting Northern ice coincident with the limits of the Liverpool Bay red tills, judging by the outwash features at Pentraeth and Beaumaris in Anglesey. Nevertheless the latter are obviously products of stagnating ice and there is no evidence of push moraines at the margins of 3a ice in North Wales. Thus it might be preferable to regard the Liverpool Bay Phase as a prolonged stillstand in the withdrawal of Main Anglesey Advance ice from north-west Wales, with a possibility of spasmodic rejuvenation as the Welsh ice downwasted and withdrew into Snowdonia.

The detailed surface distribution of Liverpool Bay till is currently best obtained from maps of the Soil Survey for Anglesey, Beaumaris and Denbigh. In Anglesey tongues of the 3a till penetrate short distances into the eastern valleys, as at Dulas and Lligwy, but a deeper penetration is suggested in the Pentraeth–Malltraeth depression. This is shown by the presence of reddish-brown clay (including local Carboniferous rocks but with a high proportion of Northern minerals) found beneath sand in Newborough Warren (*c.* 399650). This may be the furthest southern limit of the Liverpool Bay deposits for the main stillstand is marked by the outwash sands and gravels near Pentraeth, described as eskers and kames by Embleton (1964) who thought that the features were of subglacial origin. However, the linking of the meltwater channel south-west of Pentraeth with the height of the esker crests by the latter author is not conclusive evidence of an englacial water-table.

On the Arvon mainland east of Bangor it is difficult to distinguish between the Main Anglesey Advance (2a) and the Arvon Advance (2b) tills because of the extremely local character of their incorporated materials, and because this was an area marked by ice-margin fluctuation. Near Penmaenbach (738781) a section shows Liverpool Bay red till (3a) (with gravel lenses) to a depth of 4 m (12½ ft) overlying a grey stony till which includes abundant shale, sub-angular to rounded grits, rhyolites and quartzites. This lower till contrasts in general lithology with the basal (1b) tills at Llandudno Junction, Castle View, Conway, and at Llanfairfechan and may therefore belong to either the Arvon Advance (2b) or its coeval, Main Anglesey Advance (2a). Elsewhere in the Arvon coastal fringe the local till is intermingled with the Liverpool Bay till, suggesting that during the 3a Phase Welsh and Northern ice-sheets remained longest in contact in north-east Caernarvonshire. A significant contrast is apparent between the Conway valley and the Vale of Clwyd at this time. In the former Welsh ice held off Northern ice at or about the present estuary, but the Vale of Clwyd was deeply penetrated by Liverpool Bay ice because of the lack of an effective barrier of Welsh valley ice resulting from the small catchment of the Denbighshire Moors during this late phase.

*Welsh ice.* Turning now to a consideration of Welsh ice limits during the later glacial phase (3b) one is again faced with the difficulty of distinguishing in the stratigraphic record evidence of a weathering interval separating the Arvon Advance (2b) deposits from those hereafter referred to as the Welsh Intermediate Moraines (3b). Evidence provided by the relief suggests that there was a very marked stillstand

1 South west Lleyn, Caernarvonshire. A view from Nant, looking northwards across the plain of Porth Neigwl towards Carn Fadryn (1217 ft).The gentle drift slopes and the rolling relief reflect the intense degree of solifluction that occurred in this region which remained unglaciated during the last ice advance. In the foreground, cliffs of boulder clay (Irish Sea Advance) are overlain by solifluction deposits of Arvon Advance age and blown sand which now forms dunes in an earlier outlet of the Afon Soch. Raised marine platforms of Lower Pleistocene age form the distant plateaux above which rises Carn Fadryn with its summit blockfield.

2 Eastern Lleyn, Caernarvonshire. A view looking north-eastwards from Blaen-y-Cae farm (487453) across Cors Graianog towards Mynydd Craig-goch (1996 ft), one of the western foothills of Snowdonia. Note the Kame/Kettle surface in the foreground which is part of the Welsh Intermediate Moraines sequence. Contrast this with the subdued relief of the older glacial drifts in Plate 1. Tors can be seen on the summit of Mynydd Craig-goch.

3 Cwm Tregalan and upper Cwm y Llan, Snowdon. Soliflucted drift slopes mantle the fore-ground behind the slate spoil. Beyond a rock ridge in the middle distance can be seen the fresh moraine features of Cwm Tregalan. Behind these, scree mantles the foot of the rocky cliffs leading to Y Wyddfa (3560 ft O.D.), the highest peak of the Snowdon group.

4 Large-scale sorted stone-stripes on the south-west slopes of Rhinog Fawr at the head of Cwm Nantcol, Harlech Dome, Merionethshire. The stripes are clearly exposed on the ground over an area of some 10 hectares. The exposed boulder zones consist of Cambrian grits and conglomerates of a general size range between 60 and 150 cm in length, the intervening vegetated zones being partly of peat-covered boulders and partly of relatively stoneless fine earth. The repeat distance of the pattern averages 5 to 8 m. (Ball and Goodier, 1968).

of the local ice as it fell back from its piedmont form (Arvon Advance, 2b) into the foothills and valleys of Snowdonia, the Welsh Inter- mediate Moraines Phase (3b).

During the formation of outwash (2b″) at Glanllynau in which kettle holes were subsequently formed, ice from Cwm Pennant must have passed across the 244 m (800 ft) high col at Llwyd Mawr (508458). As the Welsh ice downwasted and withdrew from Tre- madoc Bay the ice between Brynkir (480445) and Cors Graianog (495453) would have become isolated from the Cwm Pennant glacier by the appearance of the Llwyd Mawr ridge above the thinning ice- cap surface. The melting of this dead ice produced the kame and kettle surface at Blaen-y-Cae (Simpkins, 1968) during the Welsh Intermediate Moraines Phase (3b). Thus the Brynkir kames need not be regarded as the regional end-moraine of the Weichsel glaciation as previously claimed (Synge, 1964) but are merely local retreat pheno- mena. Further supporting evidence is found in the nearby Llyfni valley where the glacier formerly crossed a col (485495) north-west of Cwm Dulyn. Its retreat phase (3b) is marked by a massive moraine at Cwm Bran (475482) (Simpkins, *op. cit.*) while the main valley glacier produced the kames between Pen-y-groes and Craig-y-Dinas (450520). The Lleyn peninsula appears to have been ice-free during this retreat phase although thick deposits of head accumulated locally over the existing tills and small glaciers remained in the corries of Yr Eifl. Evi- dence for this is supplied by the large but subdued moraines at 180 m (600 ft) on the corrie floors, features which contrast sharply with the fresh, steep slopes (often between 20° and 30°) which characterise the Late Glacial moraines (5 and 7) of the higher Snowdonian corries. The Gwyrfai valley, like the Pennant valley, has no evidence of kame moraines on the valley floor to mark this retreat phase, but it is pos- sible that subsequent river erosion has destroyed such remnants at low elevations. The same may be true at the mouth of the Llanberis valley, but on the interfluves between these former glacier routeways large areas of kame and kettle morphology have survived below 150 m (500 ft) on the northern flanks of Snowdonia. Thus the 3b retreat phase is marked by a kame moraine (with numerous kettles) between Hafodlas (535617) and Prysgol (515617).

Between Caernarvon and Bangor there is evidence that much of the Arvon coastal plain was already ice-free during the Welsh Intermediate Moraines Phase (3b). The shores of the Menai Straits are essentially driftless, suggesting that the gorge was formed as a subglacial

2

channel (Embleton, 1964) during the preceding glacial maximum (2a–2b) and was only functioning as a spillway for Liverpool Bay ice meltwaters during this late phase (3a–3b). Further evidence is provided by a comparison between the two pre-Cambrian ridges of north Caernarvonshire. The Bangor ridge, extending from Bangor to Caernarvon, has only pockets of weathered Arvon drift (2b) remaining among outcrops of deeply frost-shattered, clay-veined sedimentary and igneous rocks, while a Tertiary dolerite dyke near Bangor (553694) is very deeply weathered (Benayas and Ball, 1962). The Padarn ridge, passing between Llanrug (530632) and Waenfawr (530592) in the foothill zone, is, in contrast, heavily ice-scoured, leaving grooved and striated but unweathered and non-fractured rocks. This would suggest late ice-passage across only the innermost ridge as far as the Hafodlas–Prysgol kame moraines described above, leaving the coastal ridge exposed to deep periglacial weathering. The pattern is repeated on the northern foothills between the Llanberis and Ogwen valleys. In the Pentir area, between Rhydau (560655) and Rhyd-y-Groes (579672) a kame-kettle complex, previously ascribed to melting of a Northern ice-sheet (Embleton, 1964) is in fact composed almost entirely of Snowdonian material (largely from the Cambrian slate belt). Paleocurrent analysis has shown that these fluvio-glacial sediments were brought into the area by melt water streams emanating from the south east. (D. Helm, personal communication). Thus it is preferable to regard the line of kames around Pentir as marking the Welsh Intermediate Moraines Phase (3b) with ice still emerging from the large corries of Marchlyn Mawr (616620) and Marchlyn Bach (608624) and crossing the col near Pen-y-bwlch (590652). Excavations at Rhydau have disclosed that thick peat layers have formed on the outwash gravels associated with this line of kames. Similar outwash from the Ogwen glacier formed a 'sandur' or outwash plain which can be traced northwards as terraces flanking the Ogwen river below Halfway Bridge (603691) before spreading laterally into the Cegin valley near Llandegai, where extensive sections were revealed at Bangor cricket field (594710) during archaeological excavations in 1967. The bedded sands and gravels may also be seen in a river cliff east of Lon-isaf (599694) and these are obviously related to a partly destroyed moraine at Halfway Bridge which may be ascribed to the Welsh Intermediate Moraines Phase (3b).

Eastwards from the Ogwen valley the drift stratigraphy suggests that the Liverpool Bay ice (3a) and the Welsh Intermediate Moraines ice (3b) remained in contact here longer than elsewhere in North Wales. Interdigitation of Northern and Welsh tills around Conway points to a complex fluctuation of the respective ice fronts. The deglaciation of the Conway valley has been dealt with in some detail (Embleton, 1961) and it appears that the retreat from the 3a–3b maximum was a continuous process punctuated by periods of more or less stable stands of ice at specific valley locations marked by lateral moraines at Merchlyn (769750) and Cefn (765702). At Tal-y-cafn (787718) a moraine-like feature is exposed on the valley floor and there is a suggestion that outwash terraces extend downstream towards Glan Conway (802761). In its upper reaches the Conway valley becomes narrower and steeper-sided. More intense periglacial and solifluction processes on the valley sides coupled with alluvial fill of the valley floor have removed any clear evidence of the Conway main glacier. The same is true of the upper reaches of the Ogwen in the Nant Ffrancon, and in the Llanberis Pass. In Nant Gwynant, however, two possible stillstands of the valley glacier after the 3b Phase maximum are seen in the morainic mound of sand and gravel at Dinas Emrys (608492) and in a cross-valley ridge of similar material north of Llyn Dinas (622501). At Pont-Rhyd-goch (679604) there is a line of morainic material which prevents the Llugwy flowing into Llyn Ogwen and which now acts as the regional watershed in the through-valley of the Llugwy–Ogwen.

*Late glacial period* (4–8)

Seddon (1957) has given an account of the distribution of Late Glacial cwm (corrie) glaciers in Snowdonia and has also discussed (Seddon, 1958, 1962) their dating from pollen analysis and stratigraphy of sediments in representative lake-basins. He has shown that several cwms have inner and outer moraines. Only postglacial deposits occur within the inner moraines but, outside these, Late Glacial deposits are found. He has therefore dated the final inner cwm glaciers (e.g. at Cwm Idwal) (*c.* 645593) to the post-Allerød climatic recession (i.e. Pollen Zone III) and the outer moraines to the pre-Allerød cold period. Pollen data showed the Allerød climatic amelioration to be represented in the peat of Nant Ffrancon, above Bethesda. Recent work in Cwm Dyli, Cwm Merch and Cwm-y-Llan, three valleys of the Snowdon massif, has traced the relative stages of down-wasting of the valley glacier period and the extent of the Late Glacial ice advance. The conclusions are that in the valley with the smallest catchment (Cwm Merch) solifluction processes were intense, resulting in extensive areas of bare rock with only till remnants. In the southerly facing but larger Cwm-y-Llan, which would have supported a larger valley glacier, smoothly sloping soliflucted till occurs in the lower reaches but fresh morainic relief (perhaps post-Allerød) occurs in its upper reaches in Cwm Tregalan. Similar fresh morainic features cover the full length of the floor of Cwm Dyli from Llyn Llydaw to its descent to Nant Gwynant (Ball, Mew and Macphee, 1969). This is in agreement with the general dominance of a north-easterly aspect for Late Glacial cwm moraines found by Seddon (1957).

Apart from the palynological data by Godwin (1955) and Seddon (1962) there is little evidence for absolute dating of any glacial deposits in Snowdonia. The oldest known date has been obtained from Cors Geuallt (730596) near Capel Curig by Crabtree (1966), who found evidence in the lake sediment stratigraphy for deposits both of the Allerød phase and of a minor warm phase earlier than the Allerød which it can be suggested may represent the Bølling Oscillation (Zone 1b). This site lies centrally in Snowdonia and would confirm that the Welsh Intermediate Moraines are, as interpreted here, clearly older than Zone 1b of the Late Glacial period. In the Glan-llynau kettle hole infillings in eastern Lleyn, Simpkins (1968) has found pollen evidence of the pre-Allerød cold phase, the Allerød amelioration and the post-Allerød recession. Again then we have evidence that the Arvon Advance also antedates Late Zone I although otherwise the available internal evidence in the area for dating is entirely relative.

Away from the high level corries soliflual processes were intense. Periglacial processes have left their mark in Anglesey and Lleyn from the earlier episodes when glacier ice was present in Snowdonia, but it is possible that frost-shattering and congelifluction continued on the higher ground in these regions in the Late Glacial period. The periglacial block-field above 150 m (500 ft) O.D. on Holyhead Mountain (summit 219 m, 720 ft O.D.) which was described by Greenly (1919) as Holyhead Mountain Moraine, may have remained in active formation at this period, as also may the accumulation of local head material on the upper slopes of Yr Eifl in Lleyn. The significance of periglacial processes in glaciated upland Britain has become increasingly realised only in recent years and much active work in this field is in progress.

Many hill slopes in upland Wales, including Snowdonia, are mantled by thick periglacial scree (Ball, 1966), especially where the rock type is most suitable, for example, slate. Examples of such screes remaining under a soil and vegetation cover are seen at Coed Camlyn (655397), Gwydyr (800595) and partially exposed by postglacial erosion, near the Aber Falls (671703). Near Eigiau (723647) the upper metre (3 ft) of a scree is affected by disturbance due to mass movement. A large block scree or head of similar origin covers the drift seen in the Llyn Dinas section where the deeply downwasted valley drift (? 2b or 3b) was seen to have an eroded surface marked by a large boulder pavement. Comparable pavements of smaller stones have been found in a number of soil profiles in Caernarvonshire (Ball, 1967) at a surface separating underlying drift from the upper soil horizons. In a number of drift sections (e.g. Figs. 2.3, 2.6, 2.8) it is seen that stone pavements can mark discontinuities at erosion surfaces in complex till and solifluction sequences and it has been suggested that such surfaces in soil profiles also result from congelifluction, erosion and deposition under permafrost conditions in the Late Glacial period. Stone stripes, another feature of periglaciation, are excellently developed on the slopes of Rhinog Fawr, in the Harlech Dome. That periglacial processes remain active today on the highest peaks can be seen in the areas of small active frost polygons on the ridge between Carnedd Llewelyn and Foel Grach at about 945 m (3100 ft) O.D. (Tallis and Kershaw, 1959). These polygons and other frost action features are currently the subject of further study.

Thus glacial problems and their influence on the region are seen to cover a range in time and scale from regional effects of the distribution of major till sheets and the creation of the present mountain scenery to minor features seen in soil profiles and to contemporary periglacial sorted structures. We are only on the fringe of understanding the glacial history of north-west Wales and many aspects of our theme are virtually virgin fields for further study.

### APPENDIX

*The regional stratigraphy of North-west Wales*

#### The Lleyn peninsula

Drift stratigraphy is complex, especially in central Lleyn which was a transition zone between Northern and Welsh ice on two separate occasions. Till sheets overlain by or interbedded with fluvioglacial deposits are rarely found in a simple stratigraphic succession. Early deposits have been subjected to periods of weathering and periglacial disturbance and deformed or destroyed by later ice advances, while finally solifluction and landslipping have created further difficulties for correlation and interpretation. The important sections which are believed to hold the key to unravelling this complexity are given below, commencing on the south coast and moving from east to west. Symbols given in brackets for each section refer to Table 2.1 and to the key for the respective diagrams. Some of these sections have been published by Synge (1964) but our interpretations are not always in agreement.

*Section* 1. Criccieth (Merllyn) (507380) (Fig. 2.3)

Beyond the eastern end of Criccieth promenade a low drift cliff is undergoing marine erosion which is maintaining relatively fresh

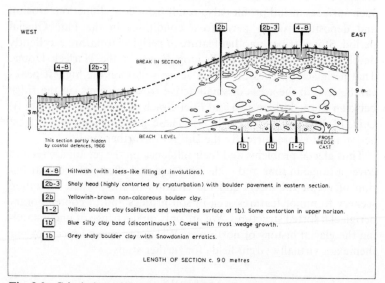

Fig. 2.3. Criccieth section (see text and Table 2.1 for the explanation of the numbers on this and the following six figures)

sections (briefly summarised in Ball, 1961), so that a stratigraphic sequence of the following type is usually to be seen.

(4–8)   Hillwash, with loessic material in pockets formed by cryoturbation.

(2b–3)   Head of angular shale, 3 m (10 ft) thick in the west, thinning eastwards to 1 m (3–4 ft) severely disturbed by cryoturbation and with a solifluction boulder-pavement separating the horizon from:

(2b)   Light yellow-brown* (10YR6/4) non-calcareous boulder clay.

(1–2)   A yellow soliflucted till, interpreted as the weathered and slumped surface of the underlying basal till. It can be seen to infill fossil ice wedges which penetrate the grey basal till.

(1b′)   Blue stoneless silty band.

(1b)   Grey (2·5Y6/0) shaly boulder clay. Both (2b) and (1b) are crowded with small shale fragments comparable to those contained in the overlying head but include also larger boulders, several feet in diameter, of Snowdonian origin, including shale, rhyolite and dolerite. The exposed thickness of this lower till above the modern beach level varies from 0·3 to 2 m (1–6 ft) and shows tectonic features in its uppermost horizons which may have formed as slump structures during the weathering phase, or as drag structures (from a subsequent ice advance).

\* Munsell colours.

*Section* 2. Criccieth Castle (East) (500389)

A high drift cliff rises from the beach immediately east of the Castle rock, with the following sequence.

(4–8)   Hillwash.

(2b–3)   An upper head.

(2b)   Brown boulder clay, 24 m (80 ft) thick, containing much Snowdonian material. Lower horizons were previously exposed, but are now covered by coastal defences built in 1966:

(1–2)   A sandy clay, highly indurated in places, probably a weathering zone.

(1″)   Fine textured head with involutions.

(1b)   Blue-grey basal Welsh till interdigitating against the Castle Rock with:

(1′)   Coarse blocky head.

West of Criccieth the sections are masked by slumping or coastal defences as far as Afonwen.

*Section* 3. Glanllynau Farm, Afon Wen (456372) (Fig. 2.4)

Between the farmhouse and the sea, several marshy hollows are partly infilled kettle holes. Two of these are breached by marine erosion giving one of the most significant sections in Lleyn. Within the kettle hole the sequence is:

(8b)   Hillwash.

(8a)   Postglacial peat.

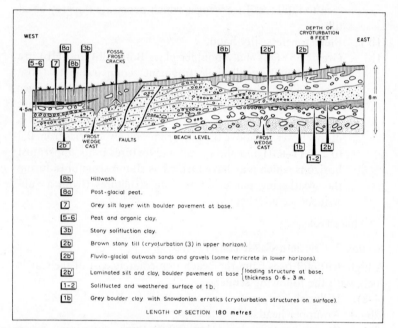

Fig. 2.4. Glanllynau section

(7)   Grey silt layer with boulder pavement (Zone III).

(6)   Organic pond clay (Zone II, Allerød).

(5)   Lake mud (Part of Zone I) (Simpkins, 1968).

(3b)   Stony solifluction clay (pre-Zone I).

(2b″)   The kettle holes are formed in fluvioglacial outwash gravels, which are overlain eastwards by:

(2b)   Brown (10YR5/3) very stony Welsh till. Stone orientation in the lower part suggests deposition from a north-easterly direction while upper horizons are strongly disturbed by cryoturbation.

(2b′)    Laminated stoneless clays, silts, and sands with minor contortions perhaps due to loading.

(1–2)    Yellow-brown (2·5Y4/2) weathered surface of

(1b)    Basal grey (5Y5/1) Welsh till, the surface disturbed by fossil frost wedges.

Fluvioglacial sands and gravels associated with Welsh drifts (2b″). with strong cryoturbation in upper horizons, are seen in most coastal sections west of Afon Wen to Abersoch, especially at Butlin's Holiday Camp (436363) where the Welsh basal till (1b) is also present, and at Carreg-y-Defaid (342326) where the gravel is associated with stony-brown Welsh till (2b), which rests on the most easterly exposure in Lleyn of the raised marine platform.

*Section* 4. Porth Ceiriad (315248) (Fig. 2.5)

Here the 30 m (100 ft) high cliff exposes both Welsh and Northern till but their relationship is obscure.

(4–8)    Hillwash.

(2b′–3)    Head with cryoturbation structures.

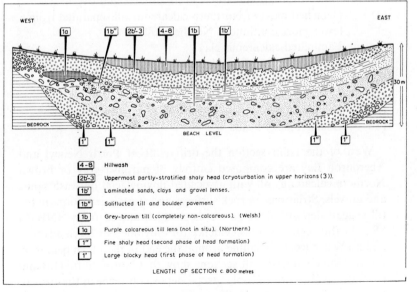

Fig. 2.5. Porth Ceiriad section

(1a)     Lens of Northern calcareous till included in a solifluction deposit (1b″–1b′) which may represent the eroded surface of

(1b?)    Indurated grey-brown non-calcareous Welsh till with dominantly shale stones, comparable in erratic content to the lower Welsh tills of Criccieth and Glanllynau.

(1″ and 1′)  Head deposits with evidence of several phases of deposition.

Although Northern drift is not *in situ* it appears to be slumped from a nearby deposit, suggesting that the St Tudwal's peninsula was a transition zone between Northern and Welsh ice. This suggestion is supported by the section at Nant, Porth Neigwl, given below.

*Section* 5. Nant, Porth Neigwl (290256)

(4–8)    Hillwash.

(3)      Boulder pavement indicating a solifluction and erosion episode.

(2b)     Brown Welsh very stony till comparable to the upper Welsh till at Glanllynau. This till, which with its associated gravels (2b″) appears to have been moved downslope by solifluction, has cryoturbation features in its surface, and overlies:

(1)      Detached mass of Northern calcareous till separated from a lower more continuous Northern till by

(1a′)    Laminated calcareous clay.

(1a″)    Lens of calcareous sands and gravels.

In the east beyond a rocky headland, the section shows:

(2b)     Upper Welsh till.

(1b)     Soliflucted Welsh till incorporating a fine head (1″).

(1′)     Coarse blocky head.

West of the Nant section the drift cliffs of Porth Neigwl and Aberdaron Bay are composed entirely of purplish grey or brown Northern calcareous till with associated fluvioglacial outwash sands and gravels. Striations on rock surfaces and stone orientations in the till suggest deposition by an ice-sheet which moved from NNW to SSE. In this region there is evidence for only one Northern advance and no Welsh ice. Continuing around the end of the Lleyn peninsula for 25·7 km (16 miles) drift sections show that Northern till (1a) and its outwash (1a′) are the only glacial deposits present in the bays of this generally rocky coast.

*Section* 6. Porth Oer (167301) has a raised beach deposit between the marine platform and the overlying Northern calcareous till (1a).

Inland the relief of south-western Lleyn is extremely subdued with gently undulating slopes running seawards for several miles from the isolated residuals of the interior (Whittow, 1957). However, north of Aber Geirch (265405) there is a change in relief and in superficial stratigraphy. The raised marine platforms present at Penrhyn Nevin and at Porth Dinlleyn which have glacial striae of NNE–SSW orientation are overlain by moraine-like drift mounds (Whittow, 1960). Coastal sections are usually obscured by slumping but a cliff fall in 1965 exposed the following sequence:

*Section* 7. Porth Nevin (300407)

(2a″) Fluvioglacial sands and gravels.

(2a) Non-calcareous till containing coal and other erratics from Anglesey or from the seafloor between Anglesey and Lleyn.

(1a″) Pockets of calcareous fluvioglacial sands and gravels associated with:

(1a) Purple-grey massive calcareous till, with its upper surface thrust into fold and drag structures by a subsequent ice-sheet moving from a direction 5°E of North.

Coastal sections between Nevin and Trevor are also largely obscured by slumping, though the purple-grey Northern till (1a) is present at most localities. In places, as at Llech Lydan (332435), there is inter-digitation with a local Welsh till (either 1b or 2b) and adjacent to rock cliffs there is mixing of head and solifluction horizons with the till. It seems that the Llech Lydan–Pistyll–Trevor area represents the northern equivalent of the transition zone between Welsh and Northern ice which was located in St Tudwal's peninsula in the south. Thus drifts having the characteristic heavy texture of Northern till but a high proportion of slate and other Snowdonian stones can be found, for example, in the stony clay loam in roadside exposures leading to the quarry at Carreg Llam (338437). Upper Northern till (2a) is thought to be absent here, its stratigraphic position being occupied by very local glacial and periglacial material from Yr Eifl massif (2b–2b′).

Near to Trevor the drift stratigraphy becomes less complex again and close to a prominent marginal channel the following section is seen at Trwyn-y-Tal, where Ramsay (1881) reported striae of 43° orientation.

*Section* 8. Trwyn-y-Tal (363468) (Fig. 2.6)

(4–8)  Hillwash.

(3)  Fine shaly head with cryoturbation structures, containing occasional till-derived rounded stones.

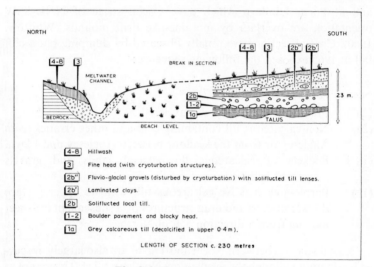

Fig. 2.6. Trwyn-y-Tal section

(2b″)  Fluvioglacial gravels with soliflucted and eroded till also disturbed by cryoturbation.

(2b′)  Undisturbed laminated clays.

(2b)  Local till with much micro-granite derived from the corries on the north face of Yr Eifl.

(1–2)  Solifluction layer and boulder-pavement.

(1a)  Grey (10YR4/1) calcareous heavy textured Northern till with thin weathered surface.

Moving north-eastwards along the coast from Trevor the lower Northern till (1a) which was noted by Reade (1893) as 38 m (125 ft) thick in the bay east of Port Trevor (375473) gradually disappears, its most northerly exposure being as pushed material in the base of the complex high drift mound at Dinas Dinlle (436563). It is superseded by thick layers of stony Welsh till (2b) and associated outwash (2b″). These latter deposits form a moraine-like feature of hummocky relief at Clynnog Fawr (415497) which was termed the Clynnog end-moraine by Synge (1964). From Clynnog Fawr to Dinas Dinlle, the coastal drifts contain largely Snowdonian, Arvonian and eastern Anglesey stones (the latter probably incorporated from earlier Northern advances), and are seen, for example, in the pale brown (10YR6/3) gravelly till (2b) at Aberdesach (420509). These drifts are considered to have been deposited by an ice sheet moving south-westwards from the Menai Straits/Arvon area.

The described sections are dominantly coastal. There are few inland sections of any depth in Lleyn to show stratigraphic relation-ships but the present surface distribution of lower Northern till (1a), Welsh tills (2b) and the extensive central Lleyn belt of fluvioglacial sands and gravels (Bodvel sands of the Soil Survey) can be obtained with considerable accuracy by consideration of soil parent materials on the Pwllheli 1:63,360 sheet of the Soil Survey of England and Wales.

### The Arvon lowlands

The hill fort at Dinas Dinlle which was sited on a coastal drift mound has been cut into by marine erosion to expose a complex drift sequence.

### Section 9. Dinas Dinlle (435563) (Fig. 2.7)

Glacial tectonics complicate the interpretation of this section which is partly obscured by slumping, and has thrust-faults and over-folds from a direction 80° east of north cutting the drifts.

(4–8)  Hillwash.
(3)    Cryoturbation features in upper horizons of Welsh till.
(2b″)  Fluvioglacial outwash sands and gravels in thrust lenses and wedges, occasionally of reddish-brown tinge.
(2b′)  Sands and silts.
(2b)   Stony grey-brown Welsh till.
(1a)   Isolated patches of lower Northern purple-grey calcareous till at beach level.

No comparable coastal sections occur elsewhere in Arvon but quarry workings provide some inland sections.

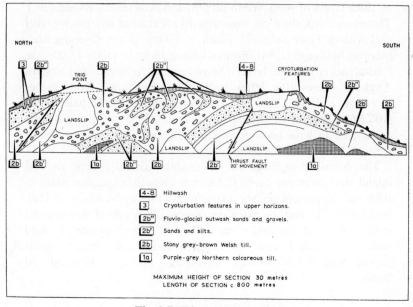

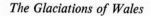

Fig. 2.7. Dinas Dinlle section

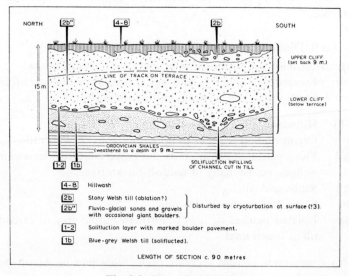

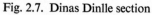

Fig. 2.8. Pen-y-Bryn section

*Section* 10. Pen-y-Bryn Brickworks (490614) (Fig. 2.8)

This pit near Caernarvon shows the following succession, allowing for probable correlation between different portions of the very large quarry.

(4–8)   Hillwash overlying and infilling cryoturbation structures in:
(2b)    A thin layer of stony Welsh till.
(2b″)   Thick layer of fluvioglacial Snowdonian gravels and sands containing huge (ice-rafted?) Snowdonian boulders 4·5–6 m (15–20 ft) in length.
(1–2)   Soliflucted material with a marked boulder pavement infills a channel cut into:
(1b)    Soliflucted blue-grey Welsh till composed mainly of Ordovician shale overlying shale bedrock.

North-east of Caernarvon the Arvon lowlands are blanketed by thick deposits of meltwater sands and gravels most of which are of Snowdonian derivation, for example Cambrian slates and grits and Ordovician shale and igneous rocks. The majority of the deposits belong to the Arvon Advance (2b) but locally kame and kettle moraines on the flanks of the foothills mark an important retreat phase (Welsh Intermediate Moraines 3b). At Rhyd-y-Groes, Pentir (579672) a thick deposit of these kame gravels is locally covered by ablation till and the gravel surface is disturbed by cryoturbation.

Northern till is absent at the surface in Caernarvonshire between Dinas Dinlle and Bangor, nor has it been seen in sections. Reddish-brown stony loamy drift exposed about Caernarvon (e.g. *c.* 491643) is the result of incorporation of local Carboniferous red measures in Welsh drift. East of Bangor the surface distribution of Northern till (3a) can be traced mainly as Aber series soils on the Soil Survey sheets for Bangor and Beaumaris, as an intermittent outcrop along the coastal plain east of the river Ogwen. The north Caernarvonshire coastal lowland was a transition zone between Northern and Welsh ice but around Aber Ogwen coastal sections in a drift mound exhibit entirely Welsh drift.

*Section* 11. Aber Ogwen (612723)

Grey-brown massive Welsh till with abundant shale, weathered to a depth of 3·5 m (12 ft). Some fossil frost-cracking and a solifluction layer with a discontinuous boulder pavement marks the surface.

There is no overlying Northern till nor any other stratigraphic break in the exposed section of 6 m (20 ft).

Dating of this deposit is difficult. It is probably (2b) but might be (1b). The similar sources and tracks followed by the Welsh erratics in all the episodes make isolated occurrences difficult to date with certainty.

Continuing eastwards, road reconstruction of the A55 in 1967 produced a shallow (about 1·5 m (*c.* 5 ft)) trench running for about 2·5 km (1½ miles) from Tai'r Meibion (633718) to the village of Aber. Locally the exposure showed outcrops of relatively stoneless reddish-brown calcareous clay (3a) as small mounds (e.g. at 650725) in sectors which tended to be protected from valleys in the hills behind the coastal plain. However, the general material at the base of the trench beneath colluvial hillwash was a reddish-brown stony till incorporating textural characteristics of Northern till and stone types of Welsh till. The North Caernarvonshire coastal plain is considered to have been a zone of fluctuating advance and retreat of Northern and Welsh ice which were generally in contact with each other. The red-brown stony till is now disturbed by fossil frost-cracking and small fossil ice-wedges. These small wedge casts varied in width at their upper surface from 2·5–38 cm (1–15 inches), and occurred at spacings which were fairly regular over short stretches, e.g. in one instance twelve such cracks occurred in 12 m (40 ft). The reddish-brown frost-cracked till was levelled by erosion so that the wedges were truncated to different depths in different sectors. It is now generally covered south of the A55 by 0·5–1 m (1·5–3·5 ft) of a grey-brown stony solifluction drift dominantly of local shale derivation (?4–7). The ebb and flow of Welsh and Northern ice is seen by contrasting the coastal site of the Aber Ogwen Welsh drift (Section 12) with a patch of Northern (3a) till exposed in a mound beneath Snowdonian till (3b) in the Aber valley (663719), this being the most southerly exposure yet seen of Northern till in Arvon.

Other temporary exposures adjacent to steeper slopes, such as at the Grove, Llanfairfechan (*c.* 675743) showed no Northern till in the exposed depth of 3 m (10 ft), but the drift was an admixture of Welsh till with angular local rock, and interbedded stoneless solifluction layers (Plate Xb, Ball, 1963). In Llanfairfechan (683751) a recent excavation showed a 0·5–1 m (1·5–3·5 ft) long lens of dark-grey (5Y4/1) heavy till (?1b) incorporated within indurated brown (10YR4/3) sandy clay (?2b) containing rounded Snowdonian boulders, to a depth of 6 m (20 ft) capped by a soliflucted till and head (?3b).

An important section was exposed in 1965–6 during construction of the Dyson Wilkinson depot at Llandudno Junction.

*Section* 12. Dyson Wilkinson Depot, Llandudno Junction (805777) (Fig. 2.9)

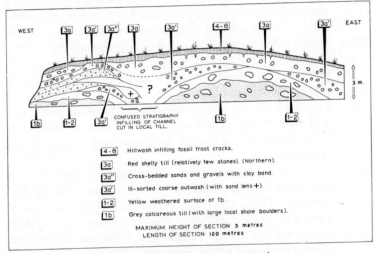

Fig. 2.9. Llandudno Junction section

(4–8)   Hillwash locally filling fossil frost-cracks.
(3a)    Red shelly Northern almost stoneless till.
(3a″)   Cross-bedded sands and gravels.
(3a′)   Poorly sorted coarse gravels and sand.
(1–2)   Pale yellow-brown weathered till, representing the surface of
(1b)    Grey (2·5Y5/0) till containing heavily striated shale boulders and stones (Silurian). This till was locally cut by a channel into which the upper series of red gravels and sands were deposited. There would appear at this section to be a substantial discontinuity above (1–2) since Welsh tills of (2b) and (3b) type with lithologically varied Snowdonian stone content are absent.

In contrast to this section an exposure in 1967 west of the Conway and rather nearer the source of general Snowdonian till, at Castle

View Housing Estate, Gyffin, Conway, showed Welsh till (3b) which included within it lenses of (3a) Northern till. A composite section derived from exposures over a length of several hundred metres of road and drain construction is as follows:

*Section* 13. Castle View, Gyffin, Conway (782773–780770)

(4–8)   Solifluction head of local Silurian shale with included rounded stones of glacial derivation. This in part overlies a shale outcrop, but in hollows either overlies or is replaced by solifluction clay with a discontinuous stone pavement over:

(3b)   Grey-brown Snowdonian till. Active transport of this till is shown by the presence of wedges of stony loam comparable to this drift, driven into fractures in the cleaved shale which in places directly underlies this till.

(3a)   Lenses of red-brown Northern till are included within (3b) and were clearly caught up in it and pushed northwards following an earlier slightly more extensive advance of the Northern ice.

(1b)   A deeper hollow in the rock contained blue-grey heavy till with large (Silurian?) striated stones beneath (3a) till. This material is identical to (1b) at Section 12 and similar to that at Criccieth. At the Llandudno Junction section the till (1b) is calcareous but it is non-calcareous at this site and at Criccieth. The present pH of buried tills of course reflects in part their original composition and subsequent degree of leaching and the composition of overlying horizons. At Llandudno Junction the overlying material is calcareous Northern drift (3a) but at Criccieth and Gyffin is mainly or entirely non-calcareous Welsh till.

Welsh till from the Conway valley (3b), similar to that of Section 13, was also seen near Bryn Estyn, Deganwy (*c.* 791785). The dominant till here was a light grey-brown (2·5Y8/2) stony compact silty loam (3b) with mainly shale stones. Where any superimposition occurred, the grey-brown till overlay reddish-brown (2·5YR5/4) massive clay loam (3a) containing some local shale, but generally Northern till was present as lenses within the Welsh till, as at Section 13. Similar Welsh till is present alone at sea level at West Shore, Llandudno (*c.* 772816) but in contrast Northern till (3a) is present, at altitudes above 90 m (300 ft) O.D., on the Great Orme and on the Gloddaeth–Little Orme ridges near Llandudno. The complex relationships between Welsh tills 2b and 3b and the Northern till 3a continue eastwards into Denbighshire until on the coastal plain east of Llandulas, uniform reddish-brown almost stoneless till occurs unchanged to 6 m (20 ft) depth in the excavations for the Abergele bypass near Hen Wrych, Abergele (830781), and the junction zone between Northern and Welsh ice has moved inland to the foothills above the old limestone cliff line.

Sections which may have been comparable to those of the Dyson Wilkinson Site (Section 12) were described by Reade (1885) from Colwyn Bay and by Hall (1868) and Kendall (1893) from the east side of the Little Orme. Both these sections apparently showed Welsh un-weathered and weathered tills (1b) and (1b′) succeeded by Northern sands and till (3a″) and (3a). Red sands (3a″) were more recently exposed near Bryn Euryn, Colwyn Bay (834796).

*Anglesey*

Sections on the eastern coast of the island were described by Greenly (1919). He identified an upper red shelly Northern till separated by 'intermediate sands and gravels' from a blue-grey or brown lower till of local (Anglesey) derivation. The sections as described by Greenly are summarised below, with some modified interpretations and cor-relations with the stratigraphy now proposed.

*Section* 14. Porth-yr-Ysgaw (477930). Reddish Northern till over local blue till.

*Section* 15. Porth-y-Gwichiad (488915)

| | |
|---|---|
| (2a–3a) | Red Northern till mixed with soliflucted local material. |
| (?c?1) | 6 m (20 ft) of bedded gravel (Greenly, 1919) (in reality a head deposit of schistose rocks with a large boulder pavement at its base). |
| (?Pre-1a) | 1 m (3 ft) of rubble (Greenly, 1919), composed largely of quartz pebbles which may represent a heavily weathered till. In an identical position at the south end of the section is a sandy clay. The entire succession overlies the raised beach platform. |

At the following localities: Penrhyn (493883), Porth-y-Forllwyd (494873), Nant Bychan (515856), Penrhyn-y-Gell (517853), Trwyn

Dwlban (531820), Greenly described red boulder clay overlying blue till on rock platforms with glacial striae of two dominant directions, one series from 40°E of North and another series from a few degrees South of East. Correlation of these striae with two tills is tempting but speculative. The present authors have identified the lower till with certainty only at Nant Bychan and Penrhyn-y-Gell.

*Section* 16. Red Wharf Bay (532816)

(4–8) Hillwash (disfigured by quarry spoil).

(?3a) Red-brown till remnants (mixed with quarry spoil). The relationship of the till to the limestone head is obscured by slumping in this section but in nearby sections the red till clearly post-dates the head.

(*d*) Raised beach deposit mainly of Carboniferous limestone pebbles, cemented by secondary calcite. The deposit occasionally stands directly on the platform but for the most part it occupies a wave-cut notch in the head, some of which is incorporated within the beach.

(*c*) Indurated limestone head deposit up to 7 m (23 ft) in thickness. Head formation may have continued after the deposition of the raised beach.

(*b*) Raised marine platform in massive well-jointed Carboniferous sandstone, partly buried by the lowest head. Elevation approximately 3 m (10 ft) above high water mark.

Between Red Wharf Bay and Beaumaris, coastal drift sections show reddish-brown Northern till (3a) associated with thick outwash sands and gravels. Those at Llienawg (621791) contain coal and other erratics and are sometimes cemented by secondary calcite. Near Pentraeth (521785) sand and gravel outwash dominantly of Carboniferous limestone is exposed in quarries and was interpreted as eskers by Greenly (1919). The distribution of these fresh gravel and sand features which Embleton (1964) considers were formed as infilling of subglacial channels, is mapped as the outcrop of the Hendre soil series on the Soil Survey map of Anglesey, and the distribution of Northern till is shown on the same maps by the outcrops of the Flint, Cottam and Fedw soils. It seems clear that the distribution of this till and its outwash represents an important stillstand of the Northern ice and that there was on Anglesey at this time no contact with Welsh ice to give the complex interfingering of two tills seen in Arvon.

Elsewhere in Anglesey, interior and coastal sections show areas of a single grey-brown till, the stones of which are dominantly of Anglesey derivation. There is a close relationship between the stone-content and the underlying solid geology (Roberts, 1958; Ball, 1961). Smithson (1953), however, has shown a general north-east to south-west trend in transport of marker minerals and rocks which together with the alignment of drumlin features in northern Anglesey and the directions of striae on rock surfaces indicate a Northern origin for this till. Its relationship to other Northern tills (1a and 3a) is obscure as it differs from them texturally, lithologically and chemically. It apparently predates 3a as it has been subjected to a lengthy period of erosion and periglacial weathering, so much so that over much of the island drift is absent and the soils are formed directly on underlying rock. In a quarry at Gwyndy, near Gwalchmai (395791), drift remnants with fossil frost-cracking containing much angular country rock overlie frost-shattered rock. The Pentraeth valley has on its southern slope a transition upslope from striated pavements of pre-Cambrian schist overlain by red till (3a) to frost-shattered schist overlain by about 1 m (3–4 ft) of schist-dominated soil (Ball, 1963, pp. 13–14). Near Llanbedrgoch on the northern side of the valley a section (501800) discussed by Ball and Mew (1968) consists of a deep sedentary kaolinitic soil overlying Carboniferous rocks.

A clear conclusion is that Anglesey as a whole has been free of Snowdonian ice except in a possible very early glaciation which is believed to have left mainland erratics westwards to a line from Llanddwyn (385625) to Penmon (641812) (Greenly, 1919) and for a later very small extension of Welsh ice (2b?) to Foel Ferry (475647) opposite Caernarvon (Roberts, 1958).

## Snowdonia

Sections showing a stratigraphic succession are infrequent and the similarity of glacier routes and available source materials makes distinctions of tills of different ages difficult when seen in isolated sections. The essentially uniform petrology of stones in the basal grey Welsh till (1b) compared with the variety of rock types of later Snowdonian tills is an unsolved problem. Variety is what would be expected of material from Welsh mountain ice, and yet till (1b) found in several sites peripheral to the mountain land is thick, and contains large, well-rounded, striated boulders dominantly of one rock type yet suggestive of appreciable transport. No unequivocal exposures of till (1b) in the uplands of north-west Wales have yet been recorded although a section at the Alexandra Slate Quarry, Moel Tryfaen (520560) has posed difficulties since its original description by Trimmer in 1831 and later study by Reade (1893). Sands with marine shells of presumably Northern origin were recorded from this site at an altitude of 396 m (1300 ft). Such a height for a Northern ice-sheet might have been possible during the 1a glacial but later Snowdonian ice (?2b) must subsequently have swept the 1a Northern till from the foothills and the Arvon and eastern Lleyn lowlands. The described section suggests that the sands were interbedded between two Welsh tills over what from Reade's section may have been slate head. Today the exposure is greatly obscured but at the nearby Bettws Garmon quarry (527569) two different Welsh tills may be seen, the lower (?1b) apparently transported from the south-east and the upper solifluced downslope from the west (?2b) from the slopes of Moel Tryfaen. Shell-bearing sands were also recorded at about 400 m (1300 ft) on the east of the Gwyrfai valley near Maen Bras, Snowdon (c. 585545?) but have not, we believe, been relocated.

We cannot attempt to correlate scattered exposures of high level Welsh till in Snowdonia. Most of the clearcut glacial features and tills in the central mountains (e.g. Cwm Idwal and Snowdon) can be ascribed to the final retreat stages from valley to corrie glaciation and the ultimate Late Glacial disappearance of ice from the area. Earlier tills at high levels have been largely removed from the glacially over-steepened slopes by intense periglacial solifluction but the presence of ice-smoothed surfaces, roches moutonnées, perched blocks and relict patches of drift up to at least 900 m (3000 ft) on the eastern slopes of Carneddau testify to the thickness of ice-cover at earlier stages, probably 1b and earlier in the present chronology. With such thick Welsh ice it is unlikely that the Northern sands at Moel Tryfaen would have been preserved unless they post-date the high level Snowdonian drifts.

In the main Snowdonian valleys a number of sections and relief features can be attributed to Welsh ice and its retreat stages, but these have already been discussed in the earlier part of our chapter.

Some features of the Gwynant valley which is likely to have supplied some of the Welsh till to eastern Lleyn can be mentioned briefly. From the watershed at Pen-y-Gwryd (660558), thick bouldery till (probably 3b) occupies the trough's end. Below this for some 4 km

(3 miles) past Llyn Gwynant the steep valley sides carry only relict soliflucted drift patches but at the northern end of Llyn Dinas (622501) current-bedded sands and gravels were exposed in a small transverse mound which may mark a valley glacier retreat stage. Slightly further down the valley a small quarry opposite Llyn Dinas showed after fresh excavation in 1958 the following tentatively correlated sequence.

*Section* 17. Llyn Dinas (619499)

(7–8)  Head incorporating large angular blocks, with evidence of several stages of activity shown by interstratified thin organic horizons (?6) (?5). Pavement of large, rounded, drift-derived boulders marking an intense erosion surface.

(?3b′)  Weathered brown stony loam, 0·5 m (18 inches) thick.

(?3b)  Grey-brown stony, sandy clay-till of variable thickness up to 3 m (10 ft) thick.

(?2b)  Weathered yellow-brown stony till (to base of section).

About 1 km further down the valley a moraine-like bedded sand and gravel feature abuts against the hill at Dinas Emrys (605493) and this could well mark an earlier standstill in the ice retreat from the Welsh Intermediate Moraines Phase (3b).

*The Harlech Dome*

Our study area includes this region but there is very little published work pertaining to it. Ramsay (1881) reported east–west striae on the passes in the central Rhinog mountains, e.g. Bwlch Ddrws Ardudwy (*c.* 665283) and across the summit of Graig Ddwrg (558 m, 1800 ft O.D., 657334). Reconnaissance soil survey by the Nature Conservancy has shown that little high-level drift remains on the hills, stone contents of soils correlating well with the underlying solid rocks. Glacial till (probably soliflucted) occupies the subsidiary valleys in the Trawsfynydd area to the east and can be traced out on to the narrow coastal lowland in the west. A coastal section at Mochras (*c.* 558272) shows drift cliffs up to 6 m (20 ft) high of a brown (10YR6/3) stony boulder clay, the majority of the stones being recognisable as Rhinog grits and shales. Although it is impossible to give a relative date for this till, it seems probable that periglacial and solifluction processes operated for a longer continuous period on the Rhinogs than in north Snowdonia, perhaps because of their lower precipitation and smaller bulk which may have resulted in smaller glaciers at the Intermediate Moraines Phase (3b). A possible consequence of this is seen on the south-western flanks of Rhinog Fawr at about 457 m (1500 ft), where large stone-stripes were recently recognised. These, developed under periglacial conditions, have coarse zones of gritstone boulders up to 2 m (6½ ft) in length, with intervening fine material zones, the repeat distance of the pattern being some 6 m (20 ft) (Ball and Goodier, 1968).

## REFERENCES

AGASSIZ, L. (1842) 'On glaciers and the evidence of their having once existed in Scotland, Ireland and England', *Proc. geol. Soc. Lond.* **3**, 327–32.

BALL, D. F. (1961) *Glacial and periglacial deposits as soil parent material in Wales.* Rept. No. 2 Welsh Soils Discussion Group, 3–10.

BALL, D. F. (1963) *The Soils and Land Use of the District around Bangor and Beaumaris.* Memoirs of the Soil Survey of Great Britain. H.M.S.O.

BALL, D. F. (1964) 'Gibbsite in altered granitic rock in North Wales', *Nature, Lond.* **204**, 673–4.

BALL, D. F. (1966) 'Late-glacial scree in Wales', *Biul. Peryglac.* **15**, 151–63.

BALL, D. F. (1967) 'Stone pavements in soils of Caernarvonshire, North Wales', *Jl. Soil Sci.* **18**, 103–8.

BALL, D. F. and MEW, G. (1968) 'Sedentary Kaolinitic soil on the Carboniferous Limestone series in Anglesey', *Geol. Jl.* **6**, 1–6.

BALL, D. F. and GOODIER, R. (1968) 'Large sorted stone stripes in the Rhinog mountains, North Wales', *Geogr. Annlr.* **50**, A.1, 54–9.

BALL, D. F., MEW, G. and MACPHEE, W. S. G. (1969) 'Soils of Snowdon', *Fld Stud.* **3**, 69–107.

BASSETT, D. A. (1961) *Bibliography and Index of Geology and Allied Sciences for Wales and the Welsh Border, 1897–1958.* Nat. Mus. of Wales, Cardiff.

BASSETT, D. A. (1963) *Bibliography and Index of Geology and Allied Sciences for Wales and the Welsh Border, 1536–1896.* Nat. Mus. of Wales, Cardiff.

BENAYAS, J. and BALL, D. F. (1962) 'Magnetita secundaria en un dique de dolerita de la formacion Arvoniense', *An. Edafol. Agrobiol.* Madrid, **22**, 259–71 (Spanish with English summary).

COPE, T. H. (1902) 'Notes on the titaniferous iron-sand of Porth Dinlleyn', *Proc. Lpool geol. Soc.* **9**, 2, 208–19.

CRABTREE, K. (1966) 'Late Quaternary deposits near Capel Curig, North Wales', unpublished Ph.D. thesis, Univ. Bristol.

DARWIN, C. (1842) 'Notes on the effects produced by the ancient glaciers of Caernarvonshire and on the boulders, transported by floating ice', *Phil. Mag.*, S3, **21**, 180–8.

DURY, G. H. (1955) 'Diversion of drainage by ice', *Sci. News* **38**, 48–71.

EMBLETON, C. (1961) 'The geomorphology of the Vale of Conway, North Wales, with particular reference to its deglaciation', *Trans. Inst. Br. Geogr.* **29**, 47–70.

EMBLETON, C. (1964) 'The deglaciation of Arfon and southern Anglesey, and the origin of the Menai Straits', *Proc. geol. Ass. Lond.* **75**, 407–30.

GODWIN, H. (1955) 'Vegetational history at Cwm Idwal: A Welsh plant refuge', *Svensk bot. Tidskr.* **49**, 35–43.

GREENLY, E. (1919) *The Geology of Anglesey*. Mem. Geol. Surv. London.

HALL, H. F. (1868) 'On the geology of the district of Creuddyn', *Proc. Lpool geol. Soc.* (Session 9), 34–8.

JEHU, T. J. (1909) 'The glacial deposits of western Caernarvonshire', *Trans. Roy. Soc. Edinb.* **47**, 17–56.

KENDALL, P. F. (1893) 'British Drifts', in *Man and the Glacial Period*, ed. G. F. Wright.

MATLEY, C. A. (1936) 'A 50-foot coastal terrace and other Late-glacial phenomena in the Lleyn peninsula', *Proc. geol. Ass. Lond.* **43**, 222–33.

NICHOLAS, T. C. (1915) 'The geology of the St Tudwal's peninsula Caernarvonshire', *Q. Jl. geol. Soc. Lond.* **71**, 83–143.

RAMSAY, A. C. (1860) *The Old Glaciers of Switzerland and North Wales*. London.

RAMSAY, A. C. (1881) *The Geology of North Wales*, 2nd edn. Mem. Geol. Surv. Gt Brit. H.M.S.O. London.

READE, T. M. (1885) 'The drift deposits of Colwyn Bay', *Q. Jl. Geol. Soc. Lond.* **41**, 102–7.

READE, T. M. (1893) 'The drift beds of the Moel Tryfaen area of the North Wales coast', *Proc. Lpool geol. Soc.* **7**(1), 36–79.

ROBERTS, E. (1958) *The County of Anglesey: Soils and Agriculture*. Mem. Soil Surv. Gt Brit. H.M.S.O. London.

SEDDON, B. (1957) 'Late-glacial cwm glaciers in Wales', *Jl. Glaciol.* **3**, 94–9.

SEDDON, B. (1958) 'Geology and vegetation of the Late Quaternary period in North Wales', unpublished Ph.D. thesis, Univ. Cambridge.

SEDDON, B. (1962) 'Late-glacial deposits at Llyn Dwythwch and Nant Ffrancon, Caernarvonshire', *Phil. Trans. R. Soc. Lond.* B, **244**, 459–81.

SIMPKINS, K. (1968) 'Aspects of the Quaternary history of Central Caernarvonshire', unpublished Ph.D. thesis, Univ. Reading.

SMITHSON, F. (1953) 'The micro-mineralogy of North Wales soils', *Jl. Soil Sci.* **4**, 194–209.

SYNGE, F. M. (1964) 'The glacial succession in west Caernarvonshire', *Proc. geol. Ass. Lond.* **75**, 431–44.

TALLIS, J. H. and KERSHAW, K. A. (1959) 'Stability of stone polygons in North Wales', *Nature, Lond.* **183**, 485–6.

TRIMMER, J. (1831) 'On the diluvial deposits of Caernarvonshire between the Snowdon chain of hills and the Menai Straits', *Proc. geol. Soc. Lond.* **1**, 331–2.

WHITTOW, J. B. (1957) 'The Lleyn Peninsula, North Wales. A Geomorphological study', unpublished Ph.D. thesis, Univ. Reading.

WHITTOW, J. B. (1960) 'Some comments on the raised beach platform of South-west Caernarvonshire and on an unrecorded raised beach at Porth Neigwl, North Wales', *Proc. geol. Ass. Lond.* **71**(1), 31–9.

WHITTOW, J. B. (1965) 'The interglacial and post-glacial strandlines of North Wales', in *Essays in Geography for Austin Miller*, ed. J. B. Whittow and P. D. Wood. Univ. of Reading, pp. 94–117.

# CHAPTER THREE

# North-eastern Wales

Clifford Embleton, M.A., Ph.D.

In contrast to the areas of Snowdonia dealt with in the previous chapter, north-eastern Wales lacks any of the spectacular scenery usually associated with glacial erosion, except along the western margins of the Vale of Conway. The relatively gentle country of Denbighshire and Flintshire, nowhere exceeding 670 m (2200 ft) in height and usually below 520 m (1700 ft), is nevertheless of great interest to the glacial geomorphologist, for it represents a part of Wales into which, at times, Northern or Irish Sea ice was able to penetrate as much as 18 km (11 miles) and, in the last main glaciation, it includes extensive areas that were probably ice-free. Glacial deposits are widespread throughout the region, but the complexity of the sequences and paucity of exposures have meant that the problem of elucidating the succession is far from being solved, even though the deposits have been studied for over a hundred years. The inter-digitation of the Welsh and Northern tills here may be of crucial significance in linking the glacial succession in Wales to that of a wider area of western Britain.

The principal source regions of Welsh ice lay to the west in Snowdonia and to the south-west in the Arenigs, while the Dee valley sector was also affected by Berwyn ice. It is likely that even the upper parts of the Denbighshire moors, the Clwydian range, or Llantysilio mountain were insufficiently high to nourish more than minor ice masses or snow patches at times when they were not themselves sub-merged by ice from outside the region. The movements of the Welsh ice (Fig. 3.1) are recorded by occasional striated surfaces, though the Silurian mudstones and shales outcropping over so much of the uplands are unsuited to the preservation of such glacial markings. Where found, striations indicate a general north-easterly movement of Welsh ice (Bryn-Davies, 1936; Strahan, 1885, 1886; Morton, 1876; *Mem. Geol. Surv.* 1923, 1927). On Mynydd-y-Cwm (075765), for example, striae point between NE and E 10° N; on Mynydd Cricor

(149505) E 10° N; at Cefn rocks (017716), the direction is ENE; and near Cyrn-y-Brain (220513), E 15° N. Departures from the general north-easterly course have resulted from the local influence of relief. The most important examples of this are the northerly ice movement down the Vale of Conway, and the eastward flow of the glacier occupying the middle Dee valley. The dispersion of erratics has also helped to reconstruct ice movements. Snowdonian igneous rocks, for instance, were transported northward to the Great Orme by the Conway glacier (Hall, 1870) and north-north-east to Colwyn Bay (Reade, 1885). Arenig igneous rocks were carried north-east to the Clwydian range, crossing it by means of cols such as Bwlch-y-Llech (354 m; 1160 ft). Some such erratics were carried on to the Carboniferous Limestone uplands beyond (for example at Erryrys, 343 m; 1125 ft), while farther south, some were lodged near the summit of Cyrn-y-Brain (562 m; 1844 ft) to which they travelled in an east-north-easterly direction (Mackintosh, 1874, 1879). High-level occurrences of glacially transported boulders show that even the highest parts of the area were ice-covered at some stage, and that the Vale of Clwyd must have then been filled with 600 m (2000 ft) or more of Welsh ice.

The second main ice-sheet to affect the area came from the Irish Sea basin to the north, bearing erratics of Scottish and Lake District origin. Impinging on the north Welsh coastlands, it pressed far up the Vale of Clwyd (Fig. 3.1), where the most southerly occurrence of Scottish or Lake District erratics is 1·6 km (1 mile) south of Denbigh. Farther west, the Irish Sea ice may have penetrated only a short distance up the narrow Dulas and Gele valleys through the Carboniferous Limestone uplands, while in the lower Conway valley, Snowdonian ice prevented the Northern ice from extending any farther south than Llandudno Junction. On the east of the area, Irish Sea ice in the Dee estuary rode up on the flanks of Halkyn mountain and filled much of the Lower Alyn valley. Its limit lies 1·6 km (1 mile) west of Mold, and wraps around Leeswood mountain. It is of considerable interest also to note the vertical extent of Irish Sea ice: on Halkyn mountain at Moel-y-Crio, it left marine shell fragments at 294 m (965 ft) (Strahan, 1886), but in northern Denbighshire, it is rare to find Northern till above 150 m (500 ft). Just outside the area to the west, however, Northern erratics rest on Moel Wnion at 581 m (1905 ft) (Mackintosh, 1882). If the maximum level attained by Irish Sea ice is of the order of 600 m (2000 ft) in Conway Bay, then it is hard

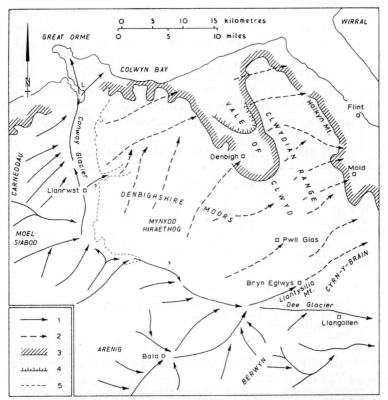

Fig. 3.1. The glaciation of north-east Wales. (1) Ice movement (Welsh Ice) of the last main glaciation (extent uncertain in the south of the area). (2) Ice movement (Welsh Ice) of probably Older Würm glaciation. (3) Maximum known penetration of Irish Sea Ice. (4) Readvance of Irish Sea Ice to the Trefnant moraine. (5) 240 m (800 ft) contour on the east of the Vale of Conway

to believe that the level off Denbighshire and Flintshire can have been significantly less. Thus to prevent Irish Sea ice gaining farther access inland, it is necessary to postulate an equal or greater mass of Welsh ice at such times. It will later be suggested that the unusual degree of low-level penetration of the Vale of Clwyd is related to a late phase of glaciation when Welsh ice was not present here to ward off its competitor. The precise limits of the Irish Sea ice cannot be laid down, because the deposits are interleaved with those of the Welsh ice over a zone of junction, and it is evident that there were several minor advances and retreats. Adequate sections in critical places too, are often lacking, while proglacial deposits further obscure the issue. The line shown on Fig. 3.1 should therefore be regarded as provisional.

### CHRONOLOGY AND THE DRIFT SEQUENCE

No absolute dating of any Pleistocene deposits in north-eastern Wales has yet been undertaken, owing to the lack of suitable materials. The tentative chronology established is therefore based on correla-

TABLE 3.1. *Drift sections in north Denbighshire*

| Observer | Date | Area | Sequence |
|---|---|---|---|
| H. F. Hall | 1870 | Little Orme | N on W on H |
| A. Strahan | 1885 | Rhos | N on S on W |
| A. Strahan | 1885 | North Colwyn | W(?H) on N on S on W |
| A. Strahan | 1885 | Colwyn Station | W(?H) on S on N on W |
| T. McK. Hughes | 1887 | Ddol (992728) | N on W |
| T. M. Reade | 1898 | Colwyn Bay Golf Course | N on W |

N—Northern (Irish Sea) drift, often red-brown in colour.
W—Welsh drift, often a blue-grey till.
S—Sands and gravels.
H—head.

tions with adjacent areas and on some limited archaeological evidence in the Vale of Clwyd.

Throughout the coastlands, the Welsh border, and in the Vale of Clwyd, compound drift sections have long been recorded. Table 3.1 summarises some early observations in north Denbighshire. Many of these sections are now obscure or have vanished. On the Welsh border, compound drift sections have never been very clear (Wedd and King, 1923). Peake (1961) in 1938 noted a road section near Bryn Bellan (214650) in which red Northern boulder clay overlies a small thickness of local Welsh till, with evidence of plucking of the latter by the overriding Irish Sea ice. Borings in the Conway valley and neighbouring areas have provided some additional information. At Mochdre (825785), both Welsh and Irish Sea drift is interbedded; at Conway Bridge (790778), up to 21 m (68 ft) of Irish Sea drift probably overlies 16 m (52 ft) of Welsh drift (Embleton, 1961).

The general picture that emerges for the North Wales coastlands is that of a blue-grey Welsh till lying on head deposits or sometimes directly on solid rock, and overlain by red boulder clay, sands and gravels, containing Northern erratics and sometimes marine shells. The lower Welsh till is found as far from its source in the uplands as the Great Orme and Little Orme, indicating an early glaciation of some magnitude in which Welsh ice spread rapidly to cover the coastlands before the arrival of Irish Sea ice. The overlying red Northern drift may represent a subsequent glaciation, possibly one of less intensity when the initial accumulation of Welsh ice was insufficient to prevent Irish Sea ice from invading the coastlands. However, it was probably later during this phase that Irish Sea ice left the high-level drifts of Moel Wnion and Moel-y-Crio. The fact that the red Northern drift is often separated into two by sands and gravels has often been commented on, but the significance of these interbedded fluvioglacial deposits is not clear, nor has their distribution been adequately studied.

In these early glacial phases, whose age is not known (though judging from the appearance of the high-level shelly drifts, their age may not be greater than Early Würm), the ice cover over the area must have been complete up to 600 m (2000 ft) at least. There is no evidence of any older glaciations yet available. However, there are several features which suggest that the area was affected by a younger and more restricted glaciation. In the Vale of Conway, fresh-looking morainic features and meltwater channels are not found above 180 m (600 ft) in the north, or 250 m (800 ft) in the Llanrwst area. In the eastern parts of Flintshire, fresh depositional features and meltwater channels related to the last Irish Sea ice here are also confined to areas below about 210 m (700 ft) (Embleton, 1956, 1964a; Peake, 1961), and in northern Denbighshire, the uppermost Northern drift does not exceed 180 m (600 ft) (for example, south of Colwyn Bay, or in the lower Elwy valley).

In the Vale of Clwyd, Rowlands (1955) has suggested that in the last glaciation, a lobe of Irish Sea ice extended up the vale approximately as far as Bodfari (Fig. 3.2). A series of low drift hills runs across the vale at this point towards Trefnant, the present River Clwyd cutting through them in a relatively narrow gap at Ty-coch. It is possible that the line of hills represents a terminal moraine of the Irish Sea ice lobe; supporting this view is the evidence of a proglacial lake south of the moraine, standing at levels between 107 and 67 m (350–220 ft) as indicated by deltas around Bodfari and at the points where the Ystrad, Nant Mawr, and Clywedog entered the lake on the west. The Bodfari–Trefnant moraine was breached by the escaping

waters of the lake once the Irish Sea ice to the north had thinned
sufficiently. The moraine probably marks the last stand of Irish Sea
ice in the vale and not its maximum extent, which was some 6·4 km
(4 miles) farther south at a slightly earlier date. There does not appear
to be any evidence of higher proglacial lake levels in the vale (in spite

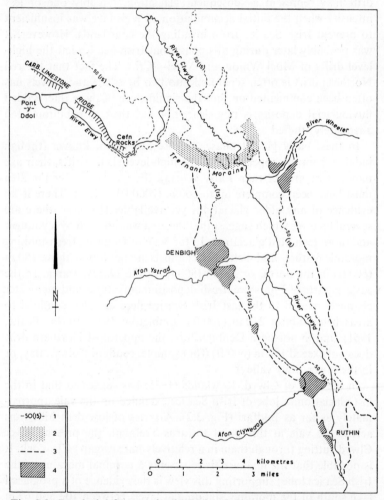

Fig. 3.2. Proglacial features in the Vale of Clwyd. (1) Contour at 50 m
(150 ft). (2) Supposed moraine of Irish Sea Ice readvance. (3) Shoreline of
proglacial Lake Clwyd at 67 m (220 ft). (4) Deltaic deposits. (After B. M.
Rowlands, 1955)

5 Cirque forms on the eastern side of the Clwydian Range. The highest point of the Range, Moel Fammau, rises to 555 m (1820 ft). The cirques may have developed by nivation rather than glacial erosion.

6 The Alyn gorge. The River Alyn at this point breaks through the Carboniferous Limestone ridge east of Cilcain. The gorge was initiated in the Pleistocene by meltwater, and is now up to 75 m (250 ft) deep.

7 Water-bedded sands and gravels of kame ridges in the Wheeler valley, deposited during decay of stagnant ice.

8 A kame mound in the upper Alyn valley, possibly a perforation deposit.

of Rowlands's suggestion of a lake at 165 m (540 ft) escaping east by the Wheeler valley), and the moraine nowhere exceeds 49 m (160 ft), being 30 m (100 ft) above the Clwyd at this point. It may be argued that the last Irish Sea ice to enter the vale was a tongue of an ice-sheet which, on the north Welsh coastlands, did not attain a level more than perhaps 180–210 m (600–700 ft).

Altogether there is some basis for believing that the last glaciation in north-eastern Wales was of a restricted nature, with Irish Sea ice on the coastlands up to 180–210 m (600–700 ft) and sending a tongue part way up the Vale of Clwyd. In the Vale of Conway, a local Welsh glacier appears to have joined concordantly in height with the Irish Sea ice, preventing the latter from spreading south here. There is no clear evidence of the extent of Welsh ice at this time in the Denbigh-shire uplands, but if the Snowdonian mountains failed to nourish more than a restricted Conway glacier, then it seems improbable that the moderately elevated plateaux of Denbighshire could have sup-ported more than a limited ice and snow cover. It is important to note that the eastern side of the Conway valley possesses only one gap whose floor might have been below the level of the surface of the Conway glacier. This is the gap at Farmyard (830640), 249 m (816 ft), leading into the head of the Elwy system. However, an ice-level approaching 290 m (950 ft) at least would seem necessary for any significant diffluence of the Conway glacier to take place here and send a tongue into the Elwy valley, and there is little evidence that this occurred. In other words, the Denbighshire uplands, which lack any fresh glacial forms, appear to have been free from any penetra-tion by Snowdonian ice in the last glaciation, and at the same time to have been incapable of nourishing more than a local ice cap on the highest parts. If so, then the Vale of Clwyd farther east must have also been free of Welsh ice at this stage, and Irish Sea ice would have had no obstacle to prevent it pushing inland to Denbigh.

To sum up: the sequence of glacial events tentatively proposed is shown in Table 3.2. The only evidence so far known in the area which bears on the relative ages of these events is that from the fossiliferous deposits of the caves near Tremeirchion in the Vale of Clwyd. Excavations here were first carried out by Hicks (1886) and the sites have been revisited many times since (see, for example, Garrod, 1926). Boswell (1932) summed up the evidence thus: 'In North Wales, a boulder-clay from the Irish Sea ice sealed up the mouths of

3

TABLE 3.2. *Glacial chronology in north-east Wales*

III.  *Last glaciation:* Irish Sea ice up to 180 m (700 ft) and sending a tongue up the Vale of Clwyd; a restricted Conway glacier.

II.  *Main Irish Sea glaciation:* Irish Sea ice up to 600 m (2000 ft), leaving high-level shelly drift. Welsh ice covering most of the interior.

I.  *An early Welsh glaciation:* Welsh ice extending into the coastal regions, covering all the interior. Irish Sea ice unable to penetrate inland.

several well-known caves in the Carboniferous Limestone, containing floors of Middle Aurignacian implements. . . . although for some time difference of opinion was held as to whether or not the undisturbed boulder-clay of the Vale of Clwyd actually sealed up the caves, the consensus of opinion was finally in favour of that view' (p. 77). The boulder clay which sealed the caves is that of phase III above, which indicates a post-Aurignacian age, probably 20–30,000 years B.P., for this last glaciation. The Main Irish Sea glaciation then falls in the Older Würm period.

The Tremeirchion cave deposits are now too much disturbed for further investigations to be of any value. No other interglacial or interstadial deposits are known in north-east Wales. The chronology suggested above may have to be revised if and when suitable materials for radiometric age determinations are found in the area. Meanwhile, it seems possible that the glaciation of phase III corresponds to the Main Anglesey glaciation described in Chapter 2. On the eastern border of the region, Peake (1961) found evidence for two separate Irish Sea ice advances in the Late Würm for which the suggested ages are 28,000 and 20,000 years B.P. Possibly it was the first of these (the 'Ellesmere Readvance') that sealed the Vale of Clwyd caves, while the second, so far only traced on the Welsh border (Peake—'Llay Re-advance'), may be the equivalent of the 'Liverpool Bay Phase', which has been recorded in Anglesey (Chapter 2). The tentative nature of these proposals should be stressed.

GLACIAL EROSION

It has been established that at certain earlier glacial stages, the whole area was submerged beneath ice fanning out from Snowdonia, the Arenigs and the Berwyns. The maximum height attained by the ice surface then is a matter for conjecture, but it must have lain at least 600 m (2000 ft) high over Cyrn-y-Brain in the east of the region, and a level in excess of 900 m (3000 ft) for Snowdonia appears likely. Over the Denbighshire uplands, a level of 750 m (2500 ft) may thus be inferred, giving ice at least 240 m (800 ft) thick over even the highest parts of these uplands (Mwdwl Eithin and Gwylfa Hiraethog). Yet, surprisingly, obvious signs of glacial erosion over the area east of the Vale of Conway are few. The explanation for this anomaly is three-fold. First, at these stages of maximum glaciation, the ice was not confined to movement along existing valleys and hollows, but was able to spread freely in a general north-easterly direction. Secondly, its rate of motion was probably always sluggish, for its eastward and north-eastward surface slope may have been no more than 5 m/km (25 ft per mile), and it must have been impeded in its escape on the north and east by an equally massive barrier of Irish Sea ice. Furthermore, its principal direction of movement lay transverse to some major relief barriers: the Clwydian range, for example, must have prevented any significant movement in the lowest 300 m (1000 ft) or so of ice in the Vale of Clwyd. Thirdly, at times of more restricted glacierisation, the form, height, and location of the preglacial relief features were not conducive to the formation of any important cirque or valley glaciers.

In parts of Britain and elsewhere, attempts have been made to recognise zones of intensity of glacial erosion (Linton and Clayton, 1964). In north-eastern Wales, it is useful also to distinguish such zones, even though their limits cannot be precisely defined and the scheme is no more than a tentative proposal:

A. Areas of little or no modification by glacial erosion:
  (i) The Denbighshire uplands. A few small-scale ice-moulded features. No rock basins. Valleys represent fluvial action almost entirely.
  (ii) The Vale of Clwyd. No evidence of glacial erosion.
  (iii) The Carboniferous area of Flintshire and eastern Denbighshire. No evidence of glacial erosion apart from local striation and smoothing of rock surfaces.

B. Areas of slight modification by glacial erosion:
  (i) The Clwydian range. Immature cirques, U-shaped valleys and cols. No rock basins.
  (ii) The middle Dee valley. Immature cirques at tributary valley heads; some widening and deepening of the main valley; partial demolition of spurs in meander loops.

C. The Vale of Conway and its western border. Partial modification
   of the preglacial surface by glacial erosion. Cirques, rock basins,
   breached watersheds, discordant tributary valley junctions.

   The extent of glacial modification in zone C is not easy to assess.
There is some indisputable evidence of locally intense glacial erosion.
Llyn Cowlyd (68 m; 222 ft deep), for example, occupies a magnificent
glacial trough gouged out across a former watershed by transfluent
ice. There are undoubtedly glacially eroded rock basins, such as the
small but unusually deep (58 m; 189 ft) example of Llyn Dulyn
(700667). Llyn Crafnant (750610) occupies a rock basin of unknown
depth (the lake is 22 m; 71 ft deep), and there may be a shallow one in
the case of Llyn Geirionydd (763610). In the main valley it is likely
that a glacially eroded basin lies upstream from Caerhun, for there is
a very conspicuous narrowing of the valley as rocky outcrops approach the river at Tal-y-Cafn bridge (786717). However, post-glacial
alluviation has obscured the situation and no deep borings in the
critical area are available. Solid rock does not appear in the bed of the
Conway until just below Waterloo Bridge (799558). Supporters of the
hypothesis that glacial erosion has extensively modified the Vale of
Conway may also point to the extremely abrupt nature of its western
wall as far north as Llanbedr (755695) and again at Llangelynin
(760735), and to the hanging junctions of the left-bank tributaries
such as the Afon Ddu and the Afon Crafnant. It can also be argued
(Embleton, 1961, 1962) that the dominant factor in this situation is
the resistance of the Ordovician tuffs and rhyolites along the western
edge of the vale, into which the tributary valleys have been unable to
cut down as fast as the main Conway flowing on fault-shattered
Silurian sediments. The eastern side of the Vale of Conway only
shows abrupt slopes in two places (Plas Maenan, 788667, and Tan-
yr-Allt, 783700) formed in the Denbigh Grit Series, and tributary
junctions are accordant.

   In zone B, evidence of glacial erosion is even more controversial.
The rounded heads of some valleys, such as those tributary to the
middle Dee, or on the east face of the Clwydian range, may represent
immaturely developed cirques. Viewed from Cefn Mawr (200635),
the three hollows below the summit of Moel Fammau (161627) look
quite unlike normal fluvial valley heads. Yet whether such features
are the work of minor cirque glaciers or snow patches is problematical (see p. 80). The recesses do not apparently contain rock

basins, and rock outcrops on the smooth walls are infrequent. The transverse cols and valleys of the Clwydian range are perhaps more convincing evidence of glacial modification, with their parabolic cross profiles and the established fact that ice poured through them in a north-easterly direction, constricted by the intervening summits. In the middle Dee valley, an eastward moving ice stream has been established. The former course of the river is clearly marked by abandoned meander loops. The abandonment of these has been attributed to erosion of new channels by meltwater or by the Dee at times when ice still partially choked the valley (Wedd *et al.*, 1927; Wilkinson and Gregory, 1956). It is also likely that glacial erosion of the spurs and promontories within the meander loops, as the ice attempted to carve itself a less sinuous course along the gorge, initiated the process of diversion, and was partly responsible for the dry gaps at Pen-y-Bryn (162424) and above Plas Berwyn (182435) which partly sever the respective spurs. The tops of the spurs are in places grooved and ice-moulded, supporting this contention, while the steep sides of the middle Dee valley as a whole may also be partly the result of glacial erosion.

Zone A, possessing little or no sign of glacial erosion, requires only brief comment. Areas in this zone were free from any substantial ice accumulations in the last glaciation, according to arguments already presented, and any features resulting from earlier glacial erosion may have been modified by subsequent periglacial action. The deeply cut valleys of the Denbighshire uplands show no evidence of glacial modification, except possibly for some widening and deepening of the Dyffryn Gallen (850645) and the Cledwen above Gwytherin (875615). On the plateaux and ridges, there are occasional signs of small-scale ice moulding, such as the crag-and-tail feature at Bryn-cnap (950658). The ineffectiveness of glacial erosion in this zone is substantiated by the way in which features of preglacial origin, such as E. H. Brown's peneplain surfaces (1960) or the sequences of valleyside benches and river knick points (Embleton, 1957, 1960), have survived glaciation and still permit of altitudinal correlation, suggesting that their forms have hardly been altered.

### GLACIAL DEPOSITION

With the exception of areas covered by postglacial alluvium and peat, few parts of the area are completely devoid of drift if this term is taken in its broadest sense. Great thicknesses of glacial deposits occur in the lowlands—at least 37 m (120 ft) under the Conway estuary, 29 m (95 ft) at Sandycroft beside the Dee estuary beneath 13 m (42 ft) of alluvium. Still greater depths to bedrock are encountered in Cheshire (Chapter 4), and reveal a minimum postglacial submergence of nearly 90 m (300 ft), though Gresswell (1964) has suggested that glacial erosion below sea-level may partly account for the abnormally low level of bedrock beneath the Dee estuary in places.

Detailed drift maps of the Geological Survey are available for the Clwydian range and areas to the east; over the Denbighshire uplands, the distribution of drift has been mapped by Boswell (1949, Fig. 1.5); but drift mapping in the Vale of Clwyd is still incomplete. The Soil Survey map of north Denbighshire provides further relevant information, the Eriviat soil series indicating areas of relatively thick Welsh drift, and the Powys series areas of thin or absent Welsh drift.

It is difficult to make useful generalisations about the types of drift, though one important distinction between Northern (mainly red-brown or red-yellow) and Welsh (mainly blue-grey or yellow-grey, often very shaly on the Silurian areas) tills has already been made. The Welsh drift is often very full of local rock fragments, the ice having very effectively quarried material from the Silurian shale and Carboniferous limestone outcrops. Occasionally the Welsh drift becomes a stiff boulder clay, but a rubbly character is more usual. Large ice-scratched boulders are relatively common. The Irish Sea drift is usually a hard tenacious clay, with a variable stone content; its reddish colour represents the incorporation of Triassic debris from the sea floor to the north. It is important to exercise caution, however, when using colour to distinguish tills of northern and local origin, for the Welsh till often becomes reddish-brown in the neighbourhood of the outcrop of the Carboniferous Basement Beds (as noted by Montag, 1928–9) and red Triassic material was carried by Welsh ice high up the western slopes of the Clwydian range. The red Triassic debris of the Wheeler and Alyn valleys will be considered later (p. 74).

### LANDFORMS OF GLACIAL DEPOSITION

*Drumlins*

Apart from some scattered examples in the Denbighshire uplands and along the north-eastern flank of Halkyn mountain, the only drumlin field lies in the Afon Morwynion valley, extending north towards Llanelidan (110505). In an area of some 15 km² (6 square miles), there are about forty drumlins, aligned along north-east axes in accordance with the last ice movement across this area. The largest attain heights of over 30 m (100 ft) and lengths of 320 m (350 yd). Irregularity of shape and alignment of some may result from bedrock cores. In several places, the Afon Morwynion has been forced to meander around drumlins (for instance, above Pont-swil, 113454), and it is a series of drumlins south-east of Moel-Fodig (090455) that has blocked the former course of the Morwynion and turned it south to join the Dee at Carrog.

*Moraines*

Moraines are nowhere a conspicuous feature of the area. The Irish Sea ice, fed from far distant sources and moving only sluggishly across the relatively flat sea-floor, was insufficiently active to produce any distinctive terminal accumulations, except for the series of low hills crossing the Vale of Clwyd from Trefnant to Bodfari. The Welsh ice at stages of maximum glacierisation was confluent with the Irish Sea ice, so that any debris carried towards the coastlands was received by the Irish Sea ice and, in part, transported outside the region. This is the main reason why no terminal moraines are to be seen in the Vale of Conway; another contributory factor must have been the later stagnation of the Conway glacier as a result of thinning during deglaciation (Embleton, 1961). The earlier activity of the Conway glacier has, however, produced some notable lateral moraines. One of these runs for 1 km (0·6 mile) along the western edge of the Afon Henrhyd valley, consisting of a chain of hills (e.g., 764740) rising 30–43 m (100–140 ft) above the floor of the latter and separated from the steep rock side of the Vale of Conway by a marginal drainage channel. Another example is at Coed Maenan (795653), reaching 90 m (300 ft) above the Conway valley floor and locally diverting a stream to flow northward for 1 km (0·6 mile) before it succeeds in entering the vale.

Small terminal moraines have been claimed to mark brief halts in the retreat of some other Welsh glaciers. Kendall and Lomas (1895–6) thus interpreted the moundy coarse gravel deposits containing boulders up to 0·6 m (2 ft) in diameter, near Pwll-glas (115543) where the Clwyd breaks through the limestone ridge to enter the Vale of Clwyd. To the south of the mounds lies a peat-filled hollow, while finer outwash gravel appears to the north. Further investigation of the deposits is needed but they may well represent no more than an ice stagnation deposit. The group of mounds in the Morwynion valley at Bryneglwys (144475) is of similar form: whether it is of morainic origin is still uncertain, but some support for this view is given by the evidence of a former small lake on the north-east of Bryneglwys, which may have been ponded up by an ice tongue in the valley terminating at the mounds. There are three of these mounds forming a crescent concave towards the north-east. Exposures show sand and gravel, with both rounded and angular stones.

## Fluvioglacial deposits

The most striking and complex features of glacial deposition are associated with meltwater. The features may be separated for convenience into three groups—proglacial, marginal, and subglacial/englacial—though one may merge into another.

Proglacial features include gravel terraces representing former phases of aggradation by torrents of meltwater. The Elwy valley from Llangerniew downstream possesses many examples at heights ranging from 88 m (290 ft) to 171 m (560 ft) (Embleton, 1960, Fig. 6), the lowest set falling on a thalweg passing Pont-y-Ddol at about 94 m (310 ft) and apparently grading to the lowest level of proglacial Lake Clwyd (67 m; 220 ft). There was a similar situation in the Vale of Conway, where terraces have been found to slope gently northward from Caerhun to Llandudno Junction (Embleton, 1961). West of Tal-y-Cafn, the terrace is as much as 0·8 km (0·5 miles) broad; an exposure here reveals 10 m (30 ft) of gently bedded sand, silt and gravel, with some rounded boulders up to 0·6 m (2 ft) in size. Meltwater from decaying ice to the south swept these materials northward to the Llandudno Junction area where there was a shallow lake hemmed in by Irish Sea ice, the lake level standing first at about 31 m (100 ft), and later falling to about 21 m (70 ft) and 15 m (50 ft), as shown by deltas built out by meltwater pouring in from the Gloddaeth channel to the north.

Proglacial deltaic deposits occur in many other areas of north-eastern Wales, reflecting the existence of numerous temporary ice-dammed lakes. It is impossible to mention all. A beautiful set of delta terraces, the highest at 159–162 m (520–30 ft) now much excavated for gravel, stands at the exit of the Bellan gorge (213654) in Flintshire (Fig. 3.4b). Lower terraces here stand at 140–143 m (460–470 ft) and 122–125 m (400–410 ft), all three levels being remarkably even. They were formed in a narrow proglacial lake impounded between rising ground to the west and Irish Sea ice to the east (Peake, 1961; Embleton, 1964a). Other examples of proglacial lake deltas in the lower Alyn basin have been examined by Peake, including the largest example of all, the Wrexham delta terrace. In the latter, first described in detail by G. W. Lamplugh (in Wedd *et al.*, 1927), the sands and gravels attain a maximum thickness of 38 m (123 ft), and the surface slopes gently eastward from 100 m to 60 m (300–200 ft) in 5 km (3 miles) ending abruptly in a steep face at Marford. The deposits were built out progressively as the lake level fell, the lake draining first to the south but later to the north as the Irish Sea ice barrier decayed.

Marginal (kame) terraces and other less well-defined marginal gravel accumulations are widespread in those areas affected by ice in the last glaciation, particularly along the coastal zone where the Irish Sea ice margin lay. They are typically moundy gravel platforms, sometimes with kettle holes. The platform bordering Colwyn Bay at levels between 15 m and 60 m (50–200 ft) consists in part of gravels laid down at the melting fringe of Irish Sea ice occupying the bay, the meltwater draining off through the Bryn Euryn gorge (833795) and later around the north of Bryn Euryn hill. Roughly terraced spreads of gravel also lie along Halkyn mountain from Holywell southeastward, especially at about the 60 m (200 ft) level, also deposited against the edge of Irish Sea ice in the Dee estuary.

Subglacial and englacial forms include kames and short esker-like ridges. The kame-and-kettle complex at Padeswood (275620) in the lower Alyn valley, although interpreted by Peake (1961) as morainic and the result of a late readvance of Irish Sea ice, seems more likely to represent deposition of subglacial and englacial debris as stagnant ice in the valley slowly decayed. The accordant summits of the kames (98–99 m; 320–325 ft) may relate to an englacial water-table at the time of deposition; the feature as a whole owes its preservation to the diversion of the River Alyn here into the Pontblyddyn gorge (88 m: 290 ft). The kettle holes are up to 50 m (150 ft) in diameter and up to 10 m (30 ft) deep.

Kame mounds also occur singly or in small groups in many districts, and vary greatly in size. One approaching 60 m (200 ft) in height stands at a point (775747) where the Conway and Henrhyd valleys approach one another before once again diverging. Another stands in the Alyn valley at Bryn-y-castell, while a particularly conspicuous example is that at Knowl Hill (288642), 15 m (50 ft) high

3*

and reaching an altitude of 172 m (563 ft), the highest point for several kilometres around. It is not clear what glaciological features controlled the location of such isolated mounds.

True eskers are only rarely encountered, though the term 'esker' has unfortunately been applied to many kames by the geological surveyors. The largest true esker is that near the margin of the Wrexham delta-terrace, forming a steep-sided ridge 12–18 m (40–60 ft) high and about 1 km (0·6 mile) in length. Short linear ridges, some of which perhaps just fall in the category of eskers rather than kames, are abundant in the Wheeler valley. This valley, running from Bodfari (095700) to Nannerch (167695) contains one of the most intriguing and impressive assemblages of fluvioglacial deposits in the whole area (Fig. 3.3). Along its length, numerous banks and short kamiform ridges of sand and gravel have been deposited by meltwater. The sand is bright red in colour in the Bodfari gap nearest to the red Bunter sandstone of the Vale of Clwyd from which it is derived, but as one passes up the valley, the red colour gradually fades, though it is still noticeable at the Wheeler head and is traceable through the Hendre gorge into the lower Alyn valley approximately as far as Mold. The form and bedding of the Wheeler sand deposits are suggestive of an ice-contact or subglacial origin. Since the red sand has been lifted from levels of 45–140 m (150–450 ft) in the Bodfari gap to 150–180 m (500–600 ft) at the Wheeler head, it is suggested that it was carried up the valley by subglacial or englacial streams confined beneath immobile ice. The kames and kame ridges on this hypothesis represent the infillings of tunnels or cavities roofed with ice. Many of the ridges reach a height of 150–170 m (500–550 ft) (for example, Swan Wood, 150709, and Coed Maes-mynam, 118724), as well as the kame terraces around Nannerch with their unusually large kettle holes (up to 20 m: 70 ft deep). The level of 150–170 m (500–550 ft) may represent the englacial water-table, itself controlled by the col at Star Crossing (155 m: 510 ft) through which the meltwater passed into the Hendre gorge. With further decay of the ice in the Wheeler valley, the meltwater began draining west in a normal manner, carrying back with it a proportion of the red sand and gravel. A right-bank tributary of the Wheeler, the Afon Pant-Gwyn, also possesses abundant hillocks and kamiform ridges, which probably also formed subglacially as meltwater from Irish Sea ice poured southward through the col at its head (201 m: 660 ft) and passed into or beneath Welsh ice lower down the valley.

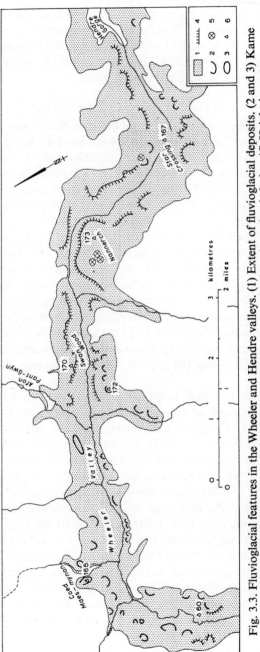

Fig. 3.3. Fluvioglacial features in the Wheeler and Hendre valleys. (1) Extent of fluvioglacial deposits. (2 and 3) Kame and banks of sand and gravel. (4) Margins of kame terraces. (5) Prominent kettle holes. (6) Height in metres

The extensive occurrence of kames, kame terraces, and kettles in Flintshire and parts of Denbighshire points clearly to widespread stagnation of the associated ice masses during deglaciation. Only in the Vale of Conway are there signs that ice activity was maintained somewhat longer in the last deglacial phase.

### GLACIAL DRAINAGE CHANNELS

In deglacial phases, meltwater from the Irish Sea ice was frequently trapped against the rising ground and local ice of the Welsh mainland and forced to escape along the coastlands. Generally, it moved to the east and south-east along northern Denbighshire from the area of Colwyn Bay towards the Vale of Clwyd. The Dulas basin, for example, impounded meltwater which escaped by way of the Dinorben channel (973746; maximum floor-level 146 m: 480 ft) to the Elwy valley and thence into the Clwyd. Meltwater in the Vale of Clwyd passed eastward up the Wheeler valley when ice conditions were favourable; it also utilised the Dyserth–Llanasa channel farther north (085810; another subglacial escape route, with a maximum floor level of 142 m: 466 ft) as well as later, with thinning ice, travelling around the limestone headland south of Prestatyn. Along the Cheshire-facing slopes of Halkyn mountain and the hills to the south, Irish Sea ice directed meltwater mainly southward, the eventual outlet doubtless being to the Severn system after the opening of Ironbridge gorge, discussed in Chapter 4.

West of Colwyn Bay, meltwater on the coastlands was directed westward and south-westward, for instance through the Bryn Euryn gorge (833795), the Gloddaeth channel (800801), and possibly over the Sychnant Pass (749772). In the latter case, the meltwater came from the Conway glacier, and its escape route, when Welsh ice supplied part of the channel sides, can be traced from Gwern Engan (757766; 167 m; 548 ft) to the top of a dry waterfall at 162 m (531 ft) (749772), where the meltwater plunged beneath the Irish Sea ice. South-westward from here, meltwater followed the Penmaenmawr coast towards Anglesey and Arfon, its latest course here being now drowned to form the Menai Straits (Embleton, 1964b).

Thus, an important divide in the direction of meltwater flow along the coastlands appears to be situated in the Colwyn Bay region, just as the movement of Irish Sea ice appears to have been cleft by the Great Orme nearby to pass south-east and south-west around the Welsh massif.

Directions of meltwater flow associated with the Welsh ice are less susceptible of generalisation owing to the greater variety in the directions of ice slope and ice movement, and the controlling effects of local relief. In the Vale of Conway, meltwater moved northward, as evidenced by numerous marginal and submarginal channels; in the middle Dee valley, it travelled eastward.

The distribution of meltwater channels shows the greatest numbers in the areas affected by ice in the last glaciation. There is a noticeable concentration in areas where drainage was impeded by the Irish Sea ice and also in the Vale of Conway. In the Denbighshire uplands, the Vale of Clwyd, and the Clwydian range, meltwater channels are infrequent, for these areas may have been partly or largely ice-free in the last glaciation, and meltwater drainage in earlier phases appears to have mainly used existing valley systems. As already mentioned, the vertical distribution of meltwater channels shows that most are concentrated below 200 m (700 ft) in coastal areas.

There is a great variety of size and form of meltwater channel. They range from rocky gorges such as that of the Alyn (195658; 60 m: 200 ft deep), Bryn Euryn (833795; up to 120 m: 400 ft deep), and the Ffrith gorge (285555; about 90 m: 300 ft deep), to shallow depressions cut in drift or partly filled with drift. Some represent, in part at least, erosion by overflowing lake waters, such as the gorge of the Dell (799718) which drained a shallow lake upstream in the Eglwys–Fach valley. Others were formed along gently sloping ice margins, or were the result of local superimposition of supraglacial (or englacial) streams. By far the larger number, however, appear to be of the submarginal or subglacial type. It is clear that many of the deeper ones were re-used in successive glaciations, not only because of their great size but because of the drift found in them. In such cases it is usually impossible even to guess at the way in which the channel was initiated; one can only study the latest phases in its functioning and evolution.

Details of meltwater channels in two areas will be presented briefly: first, the Glan Conway area, and secondly, the south-east Halkyn complex. In the Glan Conway area (Fig. 3.4a), several irregularly spaced parallel channels run along the hillside, descending northward at gradients between 1 in 20 and 1 in 50. In some cases, channels turn abruptly west to fall steeply towards the Conway, or in one case to drop into the Mochdre valley to the north. It is possible that the most gently graded channels were formed marginally to the

Conway glacier, in part at least, and the steeper members must in that case be of subglacial origin. The channel descending to Plas Isaf with an average gradient of 1 in 15 is probably submarginal. It is difficult to determine former ice-marginal gradients because of the irregularity of the channel floors, and no attempt has been made to mark ice-margin positions on Fig. 3.4a. It is, in fact, quite possible that all the

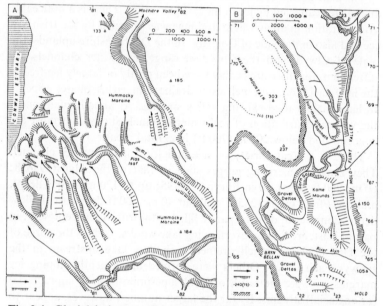

Fig. 3.4. Glacial drainage and associated features in: (A) The Glan Conway area. (1) Direction of meltwater flow; minor meltwater channels. (2) Steep sides of larger meltwater channels and related features. (B) The area south and east of Halkyn Mountain. (1 and 2) As for A. (3) Contour at 240 m (800 ft). (4) Possible maximum limit of Irish Sea Ice in the last glaciation

channels in this area developed subglacially or submarginally; if so, then nothing can be inferred about former ice surface slopes or stages in downwasting.

The channel system south-east of Halkyn mountain (Fig. 3.4b) is even more complex. It has been discussed by Peake (1961), Derbyshire (1963) and Embleton (1964a). In the last of these, four categories of channel are differentiated: (1) marginal, suggesting ice-margin gradients declining from north to south at 1:100 to 1:200, which seem reasonable for the Irish Sea ice here; (2) submarginal, with gradients averaging 1:50; (3) subglacial chutes (gradients 1:8 to 1:25); and (4) the Sarn Galed channel, of subglacial origin but separated from group (3) because of its great size, winding course, and humped long profile. Meltwater was forced south and south-west along it under pressure, in accordance with the direction of slope of the Irish Sea ice. The largest channel of all, a compound feature, is the Mold-Flint trench-valley, 60 m (200 ft) deep in places and up to 900 m (100 yd) wide. It incorporates sections of preglacial valleys. It may have originated subglacially, as meltwater was forced southward over the divide from the Dee lowlands to the Alyn valley. Later, it may have existed as an open ice-walled channel before finally being modified by marginal and extra-marginal drainage.

## GLACIAL DIVERSION OF DRAINAGE

The depth of erosion by meltwater in some instances, and the plugging of former courses by drift, led to the adoption of new courses for some rivers. The most radical alteration of all was the diversion of the Alyn (Embleton, 1957), which preglacially flowed from its headwaters northward through the Cilcain gap (180660) to become the headstream of the Wheeler. Blocking of the Cilcain gap by ice and some drift caused diversion of the Alyn in an early glaciation. The col through which the meltwater passed and which was subsequently deepened to form the Alyn gorge, stood at a height of about 207 m (680 ft), so that a minimum thickness of 30 m (100 ft) of ice was needed in the Cilcain gap to effect diversion.

The lower course of the Elwy was also changed in an early glaciation (Embleton, 1960). Formerly flowing east from Pont-y-Ddol across the limestone ridge at about 131 m (430 ft) into the Vale of Clwyd (Fig. 3.2), Irish Sea ice forced it southward for 3 km (2 miles) as a marginal stream to the point (Cefn Rocks) where it now enters the vale. This diversion clearly occurred before the last glaciation which deposited thick drift in the valley between Ddol and Cefn rocks.

Numerous large and small diversions of drainage have taken place in the Vale of Conway (Embleton, 1961) as a result of glaciation. The most significant are the abandonment of the Mochdre valley by the main river, which now reaches the sea to the west of the Llandudno peninsula, and the more complex diversion of the Afon Roe below Ro Wen, where the river unexpectedly turns south in a direction opposite to that in which the main flow of meltwater occurred. Although the routeway was first opened by glacial meltwater (flowing north), postglacial aggradation was the final factor turning the stream southward here.

Other instances of glacial diversion of drainage (in the middle Dee, and the Morwynion valleys, and in many places in east Flintshire) have already been referred to elsewhere in this chapter.

### PERIGLACIAL FEATURES

These have not yet been studied systematically in the area and only brief comments are therefore possible. Further investigations may throw useful light on the precise extent of the area not covered by ice in the last glaciation.

Deposits of frost-shattered and soliflual debris are undoubtedly widespread, though sections are few. Probably a great deal of the rounded appearance of the landscape on the Silurian shales and mudstones is related to the susceptibility of these materials to frost shattering and to solifluction. Ball (1961) suggests that the Aber and Abergele soil series of north Denbighshire and adjacent areas are developed on local fine-textured material laid down during a late cold phase. Older 'head' deposits have been recorded underlying early Welsh drift in some localities (such as Colwyn Bay), and the thin Welsh 'till' reported as overlying the last Irish Sea ice till in several localities may well also be a solifluction deposit.

Ancient scree deposits are known from a few places, such as Glyn-Dyfrdwy in the middle Dee valley, where 13 m (40 ft) of scree underlies glacial drift. Many 'modern' screes, such as those in the Sychnant Pass or at Creigiau Eglwyseg near Llangollen, may have originated substantially under periglacial rather than presentday conditions. The Eglwyseg screes have been examined by Tinkler (1966); they attain slopes of up to 36 degrees and many parts appear to be still active today. However, the occurrence in a few places of breccia beneath till of the last glaciation indicates that scree formation dates back to at least the last interstadial.

Dry valleys in the Carboniferous limestone may also relate to a cold climate and freezing of water in the joints. In areas that were ice-free in the last glaciation, snow patches may have been locally thick and extensive, and it has already been suggested that some of the cirque-like hollows in the Clwydian range and the Denbighshire uplands may be the work of nivation rather than erosion by glacier ice, and comparable to the 'nivation cirques' studied by Watson (1966) in central Wales (Chapter 6).

## REFERENCES

BALL, D. F. (1961) 'Glacial and periglacial deposits as soil parent materials in Wales', *Rep. No. 2, Welsh Soils Discussion Group*, Aberystwyth.

BOSWELL, P. G. H. (1932) 'The Ice Age and early man in Britain', *Rep. Br. Ass. Advmt Sci.* (York), 57–88.

BOSWELL, P. G. H. (1949) *The Middle Silurian rocks of North Wales.* E. Arnold.

BROWN, E. H. (1960) *The Relief and Drainage of Wales.* University of Wales Press.

BRYN-DAVIES, D. A. (1936) 'The Ordovician rocks of the Trefriw district', *Q. Jl geol. Soc. Lond.* **92**, 62–87.

DERBYSHIRE, E. (1963) 'Late-glacial drainage in part of north-east Wales: an alternative hypothesis', *Proc. Geol. Ass.* **73**, 327–34.

EMBLETON, C. (1956) 'Late-glacial drainage in part of north-east Wales', *Proc. Geol. Ass.* **67**, 393–404.

EMBLETON, C. (1957) 'Some stages in the drainage evolution of part of north-east Wales', *Trans. Inst. Br. Geogr.* **23**, 19–35.

EMBLETON, C. (1960) 'The Elwy river system, Denbighshire', *Geogrl Jl.* **126**, 318–34.

EMBLETON, C. (1961) 'The geomorphology of the Vale of Conway, North Wales, with particular reference to its deglaciation', *Trans. Inst. Br. Geogr.* **29**, 47–70.

EMBLETON, C. (1962) *Snowdonia.* Geogr. Ass., 28 pp.

EMBLETON, C. (1964a) 'Subglacial drainage and supposed ice-dammed lakes in north-east Wales', *Proc. Geol. Ass.* **75**, 31–8.

EMBLETON, C. (1964b) 'The deglaciation of Arfon and southern Anglesey, and the origin of the Menai Straits', *Proc. Geol. Ass.* **75**, 407–30.

GARROD, D. A. E. (1926) *The Upper Palaeolithic Age in Britain.* Oxford University Press.

GRESSWELL, R. K. (1964) 'The origin of the Mersey and Dee estuaries', *Geol. Jl.* **4**, 77–86.

HALL, H. F. (1870) 'On the glacial and post-glacial deposits in the neighbourhood of Llandudno', *Geol. Mag.* **7**, 509–13.

HICKS, H. (1886) 'Results of recent researches in some bone-caves in North Wales', *Q. Jl geol. Soc. Lond.* **42**, 3–19.

HUGHES, T. MCK. (1887) 'On the drifts of the Vale of Clwyd and their relation to the caves and cave deposits', *Q. Jl geol. Soc. Lond.* **43**, 73–120.

KENDALL, P. F. and LOMAS, J. (1895–6) 'Contributions to the glacial geology of N. Wales, No. 1', *The Glacialists' Mag.* **3**, 171–85.

LINTON, D. L. and CLAYTON, K. M. (1964), 'A qualitative scale of intensity of glacial erosion', Abstracts of papers, *20th int. geogr. Congr.* (London, 1964), Supplement, 18–19.

MACKINTOSH, D. (1874) 'Additional remarks on boulders, with particular reference to a group of very large and far-travelled erratics in Llanarmon parish, Denbighshire', *Q. Jl geol. Soc. Lond.* **30**, 711–21.

MACKINTOSH, D. (1879) 'The erratic blocks of the West of England and the East of Wales', *Q. Jl geol. Soc. Lond.* **35**, 425–52.

MACKINTOSH, D. (1882) 'Additional discoveries of high-level marine drifts in North Wales', *Q. Jl geol. Soc. Lond.* **38**, 193–4.

Mem. Geol. Surv. Gt Br. See STRAHAN *et al.* (1890); WEDD and KING (1923); WEDD *et al.* (1927).

MONTAG, E. (1928–9) 'Erratics from the boulder clay of Pen-y-Coed, Abergele', *Proc. Lpool geol. Soc.* **15**, 144–5.

MORTON, G. H. (1876) 'Records of glacial striae in Denbighshire, Flint-shire and Anglesey', *Proc. Lpool geol. Soc.* **3**, 123–6.

PEAKE, D. S. (1961) 'Glacial changes in the Alyn river system, and their significance in the glaciology of the North Welsh Border', *Q. Jl geol. Soc. Lond.* **117**, 335–66.

READE, T. M. (1885) 'The drift deposits of Colwyn Bay', *Q. Jl geol. Soc. Lond.* **41**, 102–7.

READE, T. M. (1898) 'High-level marine drift at Colwyn Bay', *Q. Jl. geol. Soc. Lond.* **54**, 582–4.

ROWLANDS, B. M. (1955) 'The glacial and post-glacial geomorphological evolution of the landforms of the Vale of Clwyd', unpublished M.A. thesis, Univ. of Liverpool.

STRAHAN, A. (1885) *The geology of the coasts adjoining Rhyl, Abergele, and Colwyn.* Mem. geol. Surv. Gt Br. 32 pp.

STRAHAN, A. (1886) 'On the glaciation of south Lancashire, Cheshire, and the Welsh Border', *Q. Jl geol. Soc. Lond.* **42**, 369–90.

STRAHAN, A. *et al.* (1890) *The geology of the neighbourhoods of Flint, Mold, and Ruthin.* Mem. geol. Surv. Gt Br. 242 pp.

TINKLER, K. J. (1966) 'Slope profiles and scree in the Eglwyseg valley, north Wales', *Geogr. Jl.* **132**, 379–85.

WATSON, E. (1966) 'Two nivation cirques near Aberystwyth, Wales', *Biul. Peryglac.* **15**, 79–101.

WEDD, C. B. and KING, W. B. R. (1923) *The geology of the country around Flint, Hawarden, and Caergwrle,* Mem. geol. Surv. Gt Br. 222 pp.

WEDD, C. B. *et al.* (1927) *The geology of the country around Wrexham. Part II. Coal Measures and Newer Formations,* Mem. geol. Surv. Gt Br. 237 pp.

WILKINSON, H. R. and GREGORY, S. (1956) 'Aspects of the evolution of the drainage pattern of north-east Wales', *Lpool Manchr. geol. Jl.* **1**, 543–58.

# CHAPTER FOUR

# The Cheshire–Shropshire Lowlands

Peter Worsley, B.A., PhD., F.G.S.

An analysis of the pattern of glacial events within the Cheshire–Shropshire area is necessarily concerned with the characteristics of lowland glaciation. A precise definition of the Cheshire–Shropshire Plain area is a somewhat arbitrary procedure, for despite problems related to the marginal zones, appreciable areas of relatively higher relief occur within the plain area. In Fig. 4.1 the major drainage components of the general area are shown together with the more important localities mentioned in the text. The area transgresses parts of four major drainage basins, the Mersey, Dee, Severn and Trent, the divide between the latter two rivers forming part of the major watershed of England. During the Pleistocene the positions of the watersheds changed considerably, largely through ice activity. In addition the watershed lines have distinguished areas of differing modes of ice wastage, especially depositional styles.

The area to be described will be taken as that covered by the glacier or glaciers which in the Upper Pleistocene originated in northern Britain and moved generally southwards to cross Lancashire and the northern Irish Sea basin, finally to occupy the low ground between the Welsh Massif and the Pennines. For the moment it will be assumed that we are dealing with the products of a single glacial advance called the Irish Sea ice-sheet. The deposits of this glacier are distinctive and, as might be expected, they reflect the bedrock materials within the glacier's path. An examination of the geological map shows that above sea-level the major sources or parent lithologies are rocks of Upper Palaeozoic–Lower Mesozoic age, namely the Carboniferous, Permo-Triassic and lowest Jurassic. The rock types are all sedimentary, being generally geomorphically unresistant materials including shales, sandstones, and marls. The Permo-Triassic lithologies in particular are well known for their predominant red coloration and this is also imparted to the glacial deposits. Naturally the thick Keuper rock salt deposits do not directly contribute to the

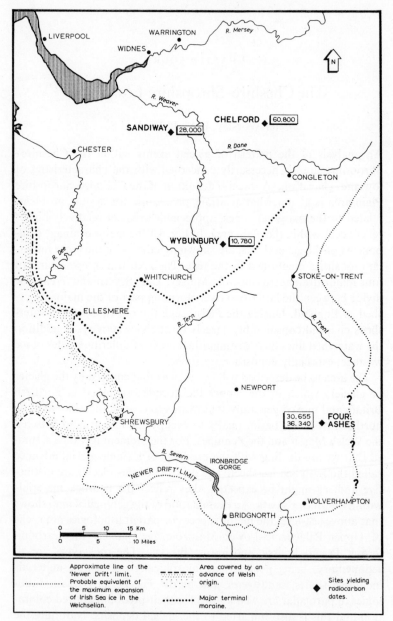

Fig. 4.1. Glacial limits in the Cheshire–Shropshire lowlands

glacial materials. The matrix of the till in hand specimens appears, at first sight, to be composed of clay and silt particles only, but mechanical size analysis invariably shows a sand content within the range of 20–40 per cent. Scattered within the till matrix are larger pieces of local material plus erratics from more distant sources. As might be expected, these erratic materials are more durable than the local debris and are easily distinguished. An assemblage of these foreign or erratic materials fairly accurately reproduces the range of resistant rock types within the source areas, the commonest occurrences being various igneous rocks, particularly types of granite, with greywacké and basic pyroclastic materials. These are easily matched with known outcrops in the Lake District and Southern Uplands. Flint is also commonly found and is likely to be derived from concealed outcrops beneath the Irish Sea.

Today, little if any of the areas formerly covered by the glacier are in an undisturbed state since human activity, agricultural in particular, has greatly modified the land surface. As a consequence the formerly extensive scatter of variously sized surface erratic boulders has been largely eliminated, to be incorporated in buildings and walls. Fortunately some records do exist which give an impression of the original situation. The Victorian 'glacialists', led by Mackintosh, determined the erratic lithologies and attempted to deduce their sources by matching with bedrock outcrops to the north. In addition the erratic distributions were mapped. Later the Boulder Research Committee, set up under the auspices of the British Association, made many reports on this theme including much information from the area. Some localities were particularly prolific in boulder concentrations and a good example is afforded in the Trysull area, near Wolverhampton (850940) where many walls lining the local lanes are built entirely from mainly rounded erratic boulders of northern origin. The best available comprehensive map to show the erratic boulder distribution throughout the area was published by Harmer (1928).

The Irish Sea ice-sheet probably came into contact with Welsh valley ice streaming eastwards from the Welsh Massif, but their relationships have yet to be fully worked out. The type of situation where a major southward moving ice-sheet was in contact with an eastward flowing valley glacier system might be expected to produce complex tectonic and depositionary sequences.

No true glacigenic deposits are known which might have been derived from a local ice centre in the south-eastern Pennine area. As a consequence the eastern margin of the Irish Sea glacier was only obstructed by the local relief. An approximation of the maximum extent of the ice margin may be obtained by plotting the area covered by the surficial northern erratic boulders and the indicator erratic materials within till lithologies. Caution must be exercised so that outwash lithologies in which erratics transported far beyond the maximum ice extent are not included in the distribution mapping. The southern and eastern boundaries of the Irish Sea glacier's maximum extent as determined by the above evidence forms part of a zone which is recognised by many workers as extending across Wales and England separating areas of essentially recent or fresh glacial landforms, in which ice contact slopes may be recognised, from older glaciated terrain to the south in which depositional morphologies are absent or rare. This fundamental contrast of glacial depositional morphology was used as the basis for distinguishing areas of 'older' and 'newer' glacial drift. On this criterion the surface glacial deposits of the Cheshire–Shropshire lowlands are invariably classified as being entirely of 'Newer Drift' age.

The area under examination has been investigated by workers interested in establishing and interpreting the sequence of glacial deposits, for more than a hundred years. Most early observations were made in and around the urban areas since constructional and quarrying activities were concentrated there. In the Manchester area in particular originated the ideas relating to the palaeo-environments and modes of glacial deposition, and from this locality they were then extended and applied to successions exposed in the rural lowlands of Cheshire and Shropshire.

During the mid-nineteenth century two differing schools of thought emerged regarding the interpretation of the sequence of glacial events. The first recognised the deposits of one glaciation only, whilst the other invoked a biglaciation concept, the two phases being separated by a nonglacial period. It should be remembered, however, that the majority of workers at this time, including Binney and Hull, were confirmed supporters of 'the great submergence' which was thought to be the depositionary environment for glacial deposits. The land-ice hypothesis was not fully accepted in north-west England until about 1900, but even after that time the two contrasting modes of interpretation remained.

A basically monoglacial view was initially stimulated by the writings of Binney (1848). He maintained that the glacial successions

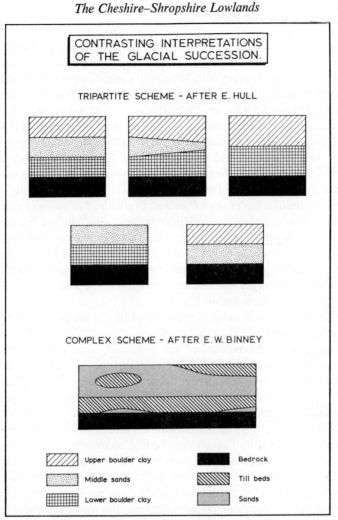

Fig. 4.2. The Cheshire–Shropshire lowlands; contrasting interpretations of the glacial succession

were complex and in particular that sand and till lithologies could apparently replace each other within any given sequence at random. Hence, accepting this interpretation the prediction of the lithological sequence to be encountered in a given area was difficult. In contrast the biglaciation hypothesis, originally proposed by Hull (1864), envisaged a rigid tripartite framework which consisted of a basic till-sand-till sequence when fully developed. The later arrangement was interpreted in terms of an initial glacial advance period resulting in till deposition succeeded by a retreat phase associated with outwash sand deposition, culminating with nonglacial conditions to be followed by a further glaciation which left another till sheet. These three successively deposited rocks in a tripartite sequence were called 'Lower boulder clay', 'Middle sands', and 'Upper boulder clay' respectively. It is important to note that the tripartite stratigraphers did recognise that one or two of the members might be absent through non-deposition or erosion. For example, if only one till sheet was encountered lying on the pre-Pleistocene bedrock in a given locality then two alternative explanations seemed available, the bed was either a 'Lower boulder clay' or an 'Upper boulder clay'. By tracing the till bed laterally and determining its relation to the nearest sand bed that might lie either above or below, its stratigraphical status as a Lower or Upper boulder clay deposit would be indicated. The general application of the tripartite nomenclature to typical sequences encountered in the area is illustrated in Fig. 4.2.

The consensus of opinion in favour of one or the other interpretation has varied with the passage of time. Currently the interpretative framework which recognises the potential complexity of glacial depositional sequences would appear to be in the ascendancy (Boulton and Worsley, 1965; Yates and Moseley, 1967). Such an opinion is well illustrated by the findings of the Geological Survey Burland Borehole (Fig. 4.3) (Poole and Whiteman, 1966, p. 106) in which four till horizons were encountered above an interglacial or interstadial deposit. This sequence could be interpreted in a manner which would assign each till sheet to a glaciation, but this (*a*) refers the organic beds to a period well back into the Pleistocene, which is unlikely, and (*b*) shows that the tripartite scheme is an oversimplification of reality. A more plausible interpretation would recognise that a single glaciation might produce a multi-till sequence and this is substantiated by similar successions found above deposits known to date from early last (Weichsel) glaciation times. However advocates of the tripartite scheme remain (Poole, 1968).

## THE SUBDRIFT SURFACE

The form of the subdrift surface and the processes and the mechanisms responsible for it are both imperfectly known and little comprehended. Obviously in the areas of economic activity more information

is available and this particularly applies to the northern half of
the area, especially the Mersey valley and saltfield areas. The first
detailed maps covering part of the subdrift surface in the area were

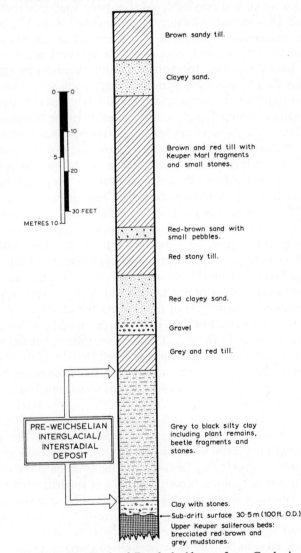

Fig. 4.3. The Burland Borehole (drawn from Geological
Survey data). The borehole was located at SJ 601533 at an
altitude of 68·6 m (225 ft) O.D

produced by Wills (1912) for the lower Dee region. He was able to identify a buried channel extending southwards from the head of the Dee estuary which was thought to connect with a buried former course of the Dee which had previously taken a more southerly route via Chirk and Ellesmere after leaving the confinement of the Llangollen gorge. A similar buried channel became identified beneath the Mersey valley. An unsuspected find associated with the construction of the Mersey Tunnel was that the buried subdrift channel beneath the presentday river cannot have drained directly north-westwards into Liverpool Bay since the minimum threshold height was considerably higher than the depths found to exist upstream in the Warrington–Widnes area. This problem has been discussed by Owen (1950).

A later more comprehensive study by Howell (1965) analysed the form of the subdrift surface of the northern Cheshire–Shropshire lowland. He was able to demonstrate that an integrated buried channel system was extensively developed and appeared to have a general fall to the west. In particular he defined the channel beneath the Mersey and various branches, one of which appeared to trend southwards from Lymm via Knutsford into the working saltfield. Although positive evidence is not as yet available, it is likely that the buried Mersey Channel has a continuous fall westwards but with an outfall to the south-west as a tributary to the master Dee system.

In the more southerly part of the lowland the incidence of thick drift sequences is much less than in the north. In addition borehole records in the areas of thick drift development are not sufficient to make any coherent picture of the buried relief. Deep 'pockets' of drift have been established in the Shrewsbury area, some bottoming below sea-level, and it may be that the upper Severn once formed part of the Dee drainage basin.

The logging data derived from boreholes within the areas of buried rock-cut channels indicate complex relations among the infill materials, which usually consist of poorly sorted sand and gravels with some tills. Thus it may be concluded that they have some relationship to the glaciations. Since the pattern of subdrift relief is imperfectly known, it is not possible to make any firm conclusions on their mode of origin. The few areas of intensive drilling reveal that the channels are well defined and have the attributes normally associated with fluvial erosion. Should it be possible to demonstrate areas of enclosed subdrift relief which are not part of an overall river valley pattern then two alternative explanations may be postulated, subglacial meltwater erosion or pure glacier erosion. The latter mechanism producing ice ways has been advocated by Gresswell (1964) to account for the subdrift relief of the lower Dee and Mersey areas. However, on the data available the interpretation of subdrift contour patterns is necessarily subjective and it may well be that the lower Dee and Mersey were fluvially eroded.

Accepting a fluvial origin for the buried valley system, it follows that erosion must have been related to a sea-level much lower than that at present. The greatest depths of drift below sea-level are found beneath the Dee estuary and these approach 120 m (400 ft). A causal relationship with eustatic lowering of sea-level during times of widespread glaciation is apparent and recent estimates of the amount of lowering are in accord with the local evidence. Although the whole valley system was buried at the maximum of the Weichselian glaciation the area only just lies within the Weichsel limit. Hence the full effects of the lowest sea-levels within the glacier period are likely to have been operative, particularly since the maximum ice margins are unlikely to have been synchronous.

### THE CHELFORD SANDS

In parts of east Cheshire deposits known as the Chelford Sands directly overlie bedrock. These were formerly thought to be glacial (Middle sands), but have been identified as of nonglacial outwash origin overlain unconformably by a series of glacial deposits. The Chelford Sands appear to have undergone sedimentation within a fan complex emanating from the Pennines. Outcrop and known borehole evidence indicates that the Chelford Sands Formation always lies directly on the Permo-Triassic bedrock, without any intervening deposit.

The type locality for the Chelford Sands is the Farm Wood quarry, Chelford (810730), where in a large excavation the deposit is worked as a source of industrial silica. An organic horizon within the sands, exposed in 1957, consisted of felted peats, tree trunks and detritus mud in a bed which varied from 0–3·5 m thick (0–12 ft) (Simpson and West, 1958). Later investigations were able to extract a prolific beetle fauna from the deposit (Coope, 1959).

After a pollen analytical study of the organic bed West was able to demonstrate that the floral evidence could be interpreted in terms of a forest community in the local area at the time of peat accumulation. The dominant tree was found to be *Pinus sylvestris* occurring together with *Betula* sp, *Salix* and *Picea abies*. Of special note was the discovery of *Picea* because this tree has not been indigenous to Britain since the last glaciation, thus the organic bed must be of some antiquity. In addition the occurrence of *Picea* also suggests a continental climate. As the quarry is extended and more of the organic bed is stripped back, occasional tree stumps in the growth position are exposed and both *Picea* and *Pinus* have been identified. There can be little doubt therefore that at least a good deal of the organic materials are essentially *in situ*. Amongst the non-tree pollen types, that from aquatic plants is common, suggesting the occurrence of acid-water pools which probably became progressively infilled. Both West and Coope are in agreement in that their independent palaeo-temperature estimates based upon the floral and faunal assemblages suggest that at the time of deposition the climatic régime was similar to that of south-central Finland today.

On grounds of stratigraphical position and similarities with the vegetation and pollen-analytical evidence associated with the early Weichselian interstadial deposits in western Europe, West concluded that the most probable correlation of the Chelford organic bed was with the Brørup interstadial of Denmark. The possibility that it may be equated with zone i of the Eemian interglacial was raised. Unfortunately the study of fossil beetles is not as yet sufficiently developed to confirm or deny this proposal. A radiometric dating using the controversial carbon 14 method yielded an age of 60,800 years B.P. (G r N 1475), which on face value is very similar to that obtained from the Brørup deposit. It should be noted however that this is one of the oldest absolute radiocarbon age determinations and is well within the range where contamination can produce substantial errors. It should therefore be regarded as tentative.

As the Chelford Sands are almost entirely buried beneath the deposits associated with the Irish Sea glacier they are only well exposed in working quarries. The principal localities are near Chelford, Arclid, Congleton and Brereton Heath. Since they rarely form part of the natural land surface they have virtually no influence on the present-day landforms. By contrast they do give an insight into the local environmental conditions prior to the last glaciation. The Chelford organic bed forms a very insignificant part of the total sediment mass comprising the Chelford Sands and is likely to be the product of a more temperate episode of short duration, within a much longer period of cold sufficient to develop permafrost. This latter statement is largely based upon the recognition of many intraformational fossil frost wedges throughout the sand (Worsley, 1967). However it is necessary to stress the palaeo-climatic importance of these structures. As is well known permafrost forms in areas which experience a mean annual temperature of at least a few degrees below zero. When the temperature rapidly falls at the onset of winter the contractional stresses within the permafrost exceed the tensile strength of the ground and cracking results. The fossil frost structures are genetically related to this phenomenon and they therefore indicate the severity of climate. This suggests that the Chelford Sands probably accumulated over a permafrost table.

Statistical analysis of the directional structures within the sands reveals a basic east–west pattern of sediment transport which, in the local geographical context, indicates movement away from the Pennines. This conclusion is borne out by an examination of the pebble material which is occasionally found within the sands, since the bulk of this can be matched with known outcrops along the west Pennine margin. Further consideration is necessary to seek the origin of the sands. Three sources appear probable: (a) the Permo-Triassic sandstones which form part of the subdrift surface; (b) fluvioglacial sand deposits which have survived from a previous glaciation; and (c) deeply weathered Carboniferous sandstones. As yet the relative importance cannot be assessed but extensive quantities of the latter are known and are quarried, for example at Biddulph.

The Chelford Sands probably originated as a permafrost zone alluvial fan, comparable in some respects to those building out onto the Mackenzie delta area of North-West Territories, Canada (Legget et al., 1967). This analogue is not perfect, however, since the Mackenzie fans contain more fine sized material and are not so extensive as the postulated Cheshire fans. Despite these imperfections the Mackenzie delta is probably the nearest modern environment to the inferred conditions during the time of the Chelford Sands deposition.

### THE GLACIATION

Most north to south traverses across the lowland reveal a basic contrast between the glacial erosional and depositional morphologies on either side of the general latitude of Whitchurch.

In the northern sector 'fresh' glacial depositional terrain is relatively frequent whilst the reverse holds in the south. The zone between the two sectors is defined by a fine terminal moraine complex, which can be traced from near Padeswood (275616) southward to pass through Wrexham as far as Overton-on-Dee trending parallel to the Welsh uplands. After Overton the line of the moraine swings eastward through the Ellesmere district in an arc towards Whitchurch. Another arcuate feature of similar dimensions leads from Whitchurch to north of Market Drayton and then follows the well-defined westward limit of the north Staffordshire uplands north-eastwards from Bar Hill. The southerly limit of this terminal moraine is shown on Fig. 4.1.

This bi-lobate terminal moraine has been recognised for many years but its precise spatial relationship to the area covered by the glacier advance has not been convincingly demonstrated to date. The ideas pertaining to its genesis may be grouped into three classes: (a) an end-moraine *sensu stricto*, that is, being produced ice marginally at the maximum expansion of a major glacial advance; (b) a retreat or stadial moraine resulting from a period of equilibrium or minor re-advance conditions; and (c) overridden pre-existing moraines from a previous glaciation.

The last of these has been strongly advocated by Poole and Whiteman (1961) who have assigned the older moraines to their 'Lower boulder clay' glaciation. Since they attribute only a thin till which mantles extensive areas of fresh morainic relief to their second glaciation their hypothesis has not won the support of many other workers. The end-moraine interpretation was adopted by Boulton and Worsley (1965) following the view of T. I. Pocock (1938) and this hypothesis has subsequently been supported in a detailed study of its eastern end by Yates and Moseley (1967). Both morphologically and structurally an end moraine can be virtually undistinguishable from a stadial moraine and thus the problem is essentially one of nomenclature. Many workers have in the past been impressed by the previously noted contrast in the frequency of occurrence of fresh relief forms on either side of the Whitchurch moraine and this in conjunction with the massive development of the moraine itself has tended to favour hypotheses which recognise a significant age difference between north and south.

In an attempt to establish the validity of this impression Boulton and Worsley (1965) reported the results of leaching determinations made in the two regions of suspected age differences. This method of investigation was possible because the tills are calcareous and therefore with time the carbonate will be progressively leached by percolating rainwater. On the results obtained a systematic variation was found, the northern tills having an average depth of carbonate removal of 1·37 m (53 inches) whilst to the south the measurements indicated an average of 2·64 m (104 inches). This finding would appear to give support to the idea of an age difference but it should be noted that the method is fraught with difficulties especially with regard to the selection of sites which are likely to have similar run-off characteristics.

South of the Whitchurch moraine the area is covered with sediments derived from an Irish Sea glacier which extended to the 'Newer Drift' limit, a postulated glacier advance terminal zone which is not fully defined but which includes the well-documented Bridgnorth–Wolverhampton line. Within this area fresh glacial relief forms are not common although it does include the often cited Newport–Wolverhampton esker chain. This feature was first identified by E. E. L. Dixon while he was engaged in the Geological Survey's primary mapping of the area. Unfortunately no map was subsequently published showing the exact esker (*sensu stricto*) distribution but only a map of the sands and gravels thought to be associated with the esker chain (Whitehead *et al.*, 1928, p. 177). Consequently the outcrop distributions do not show the sinuous patterns normally associated with eskers but rather irregularly shaped sand areas with a vague linearity. Occasional areas of sand and gravel called 'kames' by Dixon, were described as being bounded by steep northern slopes and these were interpreted as ice-contact slopes, the products of ice margin deposition. By projecting the trends of these marginal features Dixon was able to postulate 'ice-stand' positions which were generally concordant with an ENE–WSW alignment, although he admitted that they were to some extent conjectural. Indeed some workers have argued that the ice margins were unsubstantiated by the field evidence and required the 'eye of faith'. Some support, however, for Dixon's ice marginal trend is derived from an abandoned channel system extending from immediately south-east of Penkridge to Hatherton (956112) near Cannock, incised into an essentially flat terrain. There seems little doubt that this is a remnant of a subglacial meltwater channel system, as originally proposed by Dixon, draining towards the ice margin which was probably perpendicular to the former ice front.

When the fluvioglacial depositionary patterns are examined in relation to the hydrological conditions pertaining during the deglaciation it becomes apparent that the watershed position in relation to the ice margin is critical. When the meltwaters were able to drain freely away from the glacier, the fluvial activity consisted of a varying combination of incision and the deposition of valley outwash trains or sandurs. Thus in the Worfe area Wills (1924) was able convincingly to demonstrate successively abandoned outwash sandurs which today occur as a complex of terrace remnants, including the Severn Main Terrace, in the Bridgnorth area. By contrast, when the ice was receding from the watershed, the meltwaters were often impounded between the ice margin and the bounding perimeter, producing glacier lakes. Meltwaters draining into the glacial lake would experience a reduction in velocity invoking deposition of their load. Within the ice margin bed load aggradation would build up eskers, whilst at the margin small deltas would be produced. Hence the best developed eskers and associated deltas are found north of the Iron-bridge–Gnosall watershed, particularly south-east of Newport in the Back Brook valley. The allied areas of solely suspended load deposition with laminated or varved sediments are not known, save for occasional limited outcrops, but this could be due to poor exposure.

The initiation of the permanent diversion of the Severn valley between Buildwas and Stourport via the Ironbridge gorge is a classic example of glacial derangement of a former drainage pattern, the full details of which have been elaborated by Wills (1924, 1948). Before derangement the upper Severn flowed north, presumably into the Dee, whilst the Ironbridge area formed the watershed between the north-flowing Dee system and the streams that flowed south into the Bristol Channel. This watershed was overridden by southward moving ice at the maximum of the Last Glaciation. As the ice receded north of the overridden watershed meltwaters became impounded. The initial pattern was probably a series of small ice-dammed lakes which were progressively enlarged as the retreat continued. The excess meltwater would have overflowed the watershed at its lowest point, which may have been a pre-existing col or a glacially produced feature, the exact situation being unknown. South of the Wrekin an ice-dammed lake was formed in the area now occupied by the Severn, Wills called this Lake Buildwas and its overflow produced the embryonic Ironbridge gorge. Concurrently another lake was formed in the Newport area, the height of which was determined by the watershed col at Gnosall. From this point an outwash train originates and follows the Church Eaton Brook into the Penk–Trent system.

9 The exposed Weichselian succession in the east face of Farm Wood Quarry, Chelford. Just above the water the organic bed with arrowed tree stump may be seen, within the Chelford Sands. The dashed line indicates the unconformity between the Chelford Sands and the overlying glacigenic sediments.

10 An in situ stump of *Pinus sylvestris* overlain by a thin peat at the locality marked in Plate 9.

11 Undisturbed fluvial gravels overlying coarser cryoturbated gravels in sea cliff near Llanon. Forming the top of the lower series is a fine, discontinuous bed (at the hammer), which may be a buried weathered horizon.

12 Involutions south of the above. The weathered loamy bed appears thrust down into the lower gravels which are bedded concentrically. Above is the undisturbed fluvial gravel.

Eventually with further ice recession the Newport lake amalgamated with Lake Buildwas, the new combined feature being called Lake Lapworth by Wills. At this stage the Ironbridge overflow became dominant and succeeded in capturing most of the excess meltwaters from the wasting glacier such that the Gnosall Spillway was abandoned.

Along the Severn valley westwards from Buildwas to Shrewsbury and southwards towards Church Stretton occurs an extensive area composed largely of fluvioglacial sediments resulting from dead ice. Within this zone well-developed kames, kettle holes, esker and crevasse infill ridges may be found between ridges of bedrock. These would seem to indicate the detachment of masses of ice from the active glacier as it retreated, followed by slow wastage *in situ* and the aggradation of fluvial materials around them. The infills of several of these kettle holes have been investigated in an attempt to discover floristic and faunal evidence which would give an upper age for the deposits and thereby obtain data to aid the dating of the Whitchurch moraine. None have so far yielded any material other than of a late- or post-Weichselian age. The existence of this south Shropshire dead ice zone and the fresh-looking if somewhat restricted development of the fluvioglacial erosion and depositional phenomena in the Newport–Penkridge area is not very consistent with an age other than Main Weichselian. Yet the leaching data are suggestive of an age difference and original relief forms *have* been attributed to Saale deposits in Britain (Sparks and West, 1964). In addition an area covered by a late Saale ice advance retaining depositional features has been reported from Poland and this prompted the suggestion by Boulton and Worsley (1965) that the North German Plain situation might be reflected in Britain.

The discovery of organic beds within fluvial sediments beneath a till at Four Ashes (915082) 8 km (5 miles) north of Wolverhampton by A. E. Wright (Shotton, 1967b) has enabled a reappraisal of the regional evidence. This site reveals a basic stratigraphy of Upper Bunter sandstone overlain by 1·2–4·5 m (4–15 ft) of gravels with subordinate sands largely derived from the Bunter pebble beds but including rare northern derived pebble material. Over this sequence a sandy red brown till 0·3–1·4 m (1–8 ft) thick with typical Irish Sea material extends across the site. The organic horizons lie within the gravels and the insect fauna contained within them closely matches that from the sites correlated within the Upton Warren Interstadial,

4

that is Mid-Weichselian. This conclusion was substantiated by two radiocarbon age determinations yielding dates of 36,340, and 30,650 years B.P. respectively (Birm 24, 25). From this lithostratigraphic, faunal and radiometric evidence it would appear that the overlying till is most probably of Main Weichselian age and tentatively this can be related with the advance to the Wolverhampton–Bridgnorth line.

An acceptance of this interpretation clearly involves several other conclusions. Perhaps the most important is that the 'Newer drift' limit in the Cheshire–Shropshire area may well be Main Weichselian in age and the areas of fresh surface conformation which were seemingly anomalous are not necessarily so. Once this is appreciated the magnitude of post-Weichselian maximum erosion and landscape modification becomes apparent, for the entire Severn valley from Welshpool to Bridgnorth must have been cut in a period of approximately 20,000 years. Seen in this revised perspective the Whitchurch moraine appears as a probable retreat feature despite its massive development in comparison with the almost feather edge nature of the maximum ice limit. The variation in the leaching data remains somewhat anomalous.

After mapping the Shrewsbury Geological Survey sheet, Whitehead (in Pocock *et al.*, 1938) identified an appreciable area of Welsh till and associated outwash which overlay glacial sediments of Irish Sea derivation. This area of Welsh glacigenic materials was interpreted as probably indicating a distinct advance of Welsh ice as far east as Shrewsbury. The location of the Irish Sea ice during the period of Welsh ice advance cannot as yet be confidently determined, but the eastern limit of Welsh materials has been traced from Shrewsbury north-westwards towards Ellesmere. There in the Wood Lane quarry sections (422327) a Welsh till can be seen to overlie and be the source of complicated tectonic structures in a characteristically Irish Sea type glacial sequence. The general impression gained is that the two ice-sheets were probably contemporaneous at Ellesmere and from this it follows that the event which produced the Whitchurch moraine is probably synchronous with the advance of Welsh ice to Shrewsbury. The latter has been termed the Little Welsh Glaciation by Wills (1950), and Whitehead tentatively correlated it with the Uffington terrace, the highest identified west of the Ironbridge gorge. In the Shrewsbury area the Welsh tills and outwash gravels are well exposed in the Mousecroft Lane quarry at Meole Brace (474110) and other exposures are found in the bluffs overlooking the River Severn (e.g.

455157). The height of Lake Lapworth at the time of the Little Welsh advance is unknown but some structures in the till are suggestive of deposition in an aqueous environment (Shotton, 1962). Correlations have been proposed for the three major terraces between Shrewsbury and Ironbridge with features along the lower Severn but these remain speculative.

In the light of the previous discussion it is desirable to examine the Whitchurch moraine and its relationship to the possible presence of Lake Lapworth. Again the style of morphological and depositionary response would have been largely determined by the absence or presence of ponded water in the immediately proglacial area. In the former case an extensive outwash sandur linked to part of the Severn terrace system might reasonably be expected and Yates and Moseley (1967) support this concept. They have proposed that the spillway breach in the Whitchurch moraine north of Market Drayton grades with the Worcester (Uffington) terrace of the Severn. When the distal side of the moraine is examined closely it is found that extensive fringing sandurs are conspicuous only by their rarity. The only good example of a sandur occurs at Prees Heath immediately south of Whitchurch, but even this is restricted in its extension southwards and appears unrelated to the Severn valley terrace system. Work currently in progress by the writer in the Tern valley suggests that the outwash from the Whitchurch moraine in the Market Drayton area can be traced southwards but it terminates in the Hodnet area. When the reconstructed sandur profiles are extrapolated south of this point they hang above the present day river and its low terrace. This is provisionally interpreted as indicating that the outwash aggradation and downcutting was related to a higher base level, in all probability to a gradually lowering Lake Lapworth system. Thus only when the ice had retreated well north of the Whitchurch moraine did the lake level fall appreciably.

Throughout the entire length of the moraine the depositionary landforms are consistent with the existence of a proglacial lake rather than an unrestricted outwash sandur. For example, the magnificent belt of hummocky kame and kettle terrain south-east of Ellesmere is likely to be the product of ice stagnation and deposition of abundant fluvioglacial sediment in, above and around wasting ice blocks. This type of environment is less easy to comprehend if the meltwaters were streaming away unimpeded from the ice margin southwards into the Severn valley.

The sole extensive exposure in the Whitchurch moraine occurs in the previously mentioned Wood Lane quarry sections south-east of Ellesmere. There masses of Irish Sea type till have been thrust up into overlying gravels by a pressure acting from the NNW and this feature seen in conjunction with Welsh type till overlying the western end of the section would appear to confirm the presence of Welsh ice. Most information regarding the structure and sequences within the moraine has, in the absence of any other major exposures, to be based upon borehole and geophysical information. Utilising this kind of data McQuillan (1964) was able to construct realistic sections across the moraine. As might be expected from such a massive and extensive feature the moraine is structurally complex. The widespread occurrence of hummocky relief, of steep slopes on the northern side and fresh kettle features all corroborate the impression of a relatively recent construction.

As the ice retreated farther north from the Whitchurch morainic stage, it produced a series of hummocky moraine belts trending parallel with the ice margin. These are best developed in a zone extending along the westward facing marginal slopes of the Pennines (especially between Congleton and Macclesfield) where many exposures show a predominance of fluvioglacial sands with occasional till beds without any apparent systematic stratigraphy. On the western margins, especially in the Wrexham area, an excellently developed ice marginal lateral terrace accumulated. This feature is usually known as the 'Wrexham delta terrace' in the literature, morphologically forming a generally flat upper surface at approximately 90 m (300 ft), with zones of kettle holes, abutting westwards against the foothills leading up to Ruabon mountain. To the east the terrace terminates in a minor escarpment up to 45 m (150 ft) high overlooking the modern alluvial Dee valley. This major break of slope exhibits many of the attributes associated with an ice contact feature and is a close analogue to Flint's (1929) ice marginal terrace model. An exposure in the ice contact slope immediately east of Wrexham (362517) reveals a thick sand sequence with several till sheets interbedded towards the top. This type of exposure convincingly demonstrates the absurdity of attempting to fit the sequence into a tripartite nomenclature. The assemblage of sedimentary structures is indicative of deposition into a lake and the thick sedimentary sequence indicates a significant equilibrium phase during the ice wastage. Within the Wrexham area fine examples of ice marginal drainage may be seen near Brynteg

(312523) together with depositional forms, including an excellent esker south of Gwersyllt Park (323536).

The ice recession in the northern part of the lowland was likely to have retained a reasonably well-defined ice margin. This does not deny that mass stagnation took part in the mode of ice wastage but the evidence indicates that periodically the ice margin became re-activated. It is likely that as the ice surface wasted, the mid-Cheshire ridge would initially emerge above the ice as a small nunatak-like feature. An examination of the ridge slopes reveals several dry valley systems which appear to be abandoned meltwater channels draining down the local slopes in the manner of subglacial chutes. The most impressive multiple channel is found north of the Bickerton gap, just below Raw Head, the highest point of the ridge (515544). Three shallow intake channels run parallel down the hillside draining east-wards and soon change into steep sided ravines which eventually amalgamate into a master channel before debouching onto the plain. Farther north, east of Kelsall (543680) is an unusual meltwater channel locally known as the Urchin's Kitchen. This channel, rather obscured by woodland, has for part of its course overhanging under-cut walls of Keuper sandstone while today it carries no drainage.

In the north-east lee of the main ridge lies Delamere Forest, an extensive area composed of fluvioglacial sands. The key to the inter-pretation of the overall morphology of this area seems again to lie with an appreciation of the terrain south of the retreating ice margins. In this instance the ice was retreating northwards away from the Whitchurch moraine which dammed the meltwaters to a level deter-mined by the spillway breaches in the moraine, particularly the Adderley gap south of Audlem (645390). As a consequence, in a similar fashion to the situation noted previously, outwash sandurs were not developed.

The Delamere Forest feature is defined to the north-east by a rela-tively steep slope, again interpreted as broadly an ice contact feature. This trends north-west to south-east from Kingsley towards Sandi-way. From the slope crest the land falls steadily in height from some 75 m (250 ft) to about 30 m (100 ft) where the feature is terminated by a well-defined slope facet. The detailed relief of the forest surface is diverse especially in the north where deeply incised streams, e.g. Stonyford brook (575710), drain appreciable valleys. When the valleys are examined in detail they are found to incorporate hum-mocky areas which seem to suggest that the valley development was initially favoured by the occurrence of dead-ice blocks. The other major surface characteristics are numerous enclosed hollows with intervening knolls. At first sight the explanation is seen within the stagnation concept of deposition, but the situation is complicated by the existence of active solution subsidence, for immediately below the drift sequence in the southern part lie extensive saliferous beds. Despite this additional solution factor the salt does not affect the overall geometry of the sand body, which is a function of the depositional environment.

The association of an ice-contact slope to the north, a gently falling upper surface with kettle holes and kames together with a well-defined southerly termination, invoke an analogue with delta 'sand plains' as first defined by Davis (1898). Indeed there are strong parallels between the 'sand plain' model as proposed by Davis and the Delamere Forest morphology. The environmental conditions during its deposition are envisaged as being related to a substantial period when the ice margin maintained its position over the same area. Outwash materials would then build out a delta into the meltwater lake and, as the feature developed, braided streams may have extended across the newly produced surface which rose above lake level. Along the northern margins the sequence of ice oscillations was undoubtedly complex and the sections within this zone often reveal multiple till sequences.

A particularly interesting section occurs at Sandiway (615712) in a working sand quarry. Thompson and Worsley (1966) interpreted the interbedded sand and tills sequence as being indicative of a sub-marginal depositionary environment. Invariably an 'Upper boulder clay' occurs just below or at the top of the section as a series of laterally impersistent till lenses. Close examination shows that the till was likely to have been in a fluid or semi-mobile state similar to a mud flow at the time of deposition. It is envisaged that periodic slumps of till from the melting ice would be suddenly introduced onto the aggrading fluvioglacial sediment surface, thus multiple till sheets could be deposited within a single sedimentation episode.

## CHRONOLOGY

Ever since Joshua Trimmer first recorded whole and fragmentary shell material from these lowland drifts various localities have yielded prolific faunas. Discoveries at Sandiway are typical of most shell-bearing localities in the area and two large specimens of *Nucella lapillus* from it have yielded a radiocarbon date of 27,000 years B.P. (I 1667). This date was significant in that it substantiated the suggestion of Movius (1960) that the Aurignacian culture was of a period immediately before the Weichselian maximum. This is so because the surface lowland glacigenic sediments appeared to extend round the north Welsh coastlands and be coextensive with the till which sealed the Vale of Clwyd caves containing Aurignacian material.

Previously correlation and dating attempts had been made within the strictures of the tripartite scheme of interpretation. Thus two glaciations, the Lower boulder clay and Upper boulder clay advances were separated by 'Middle sands' which, as we have seen, were thought to contain the Chelford interstadial deposit. Hence the Last Glaciation was seen to be of post-Chelford age, that is after 60,000 years B.P. In the Midlands of England organic remains have been found at or towards the base of terraces in valleys tributary to the River Severn. On morphological criteria these latter terraces appear to grade into the Severn Main Terrace which is the outwash train from the ice when it stood just north of the Bridgnorth–Wolverhampton limit. The organic material yielded ages of 38,000 and 41,900 radiocarbon years B.P. and fell within what is now called the Upton Warren Interstadial Complex. From this date the Newer Drift limit had been assigned an age of 45,000 years B.P. on the basis of the terrace correlation. This Newer Drift glacial advance has often been called the Main Irish Glaciation in the literature. However it now appears likely that a significant hiatus occurs within the terrace aggradation, highlighting the pitfalls to be encountered in terrace correlation utilising surface height criteria alone. Before the discovery of datable material at Four Ashes the Irish Sea glacier deposits from the Whitchurch moraine southwards to Wolverhampton were generally regarded as being of about 45,000 years B.P., which in comparison with known chronologies from Europe was highly anomalous. This interpretation had previously invoked two glacial advances after the Chelford interstadial in Cheshire, one of 45,000 years in age and another of Main Weichselian, say 20,000 years (Shotton, 1967a). The major difficulty arising consequent to this was that wherever the Chelford Sands were identified, the deposits of only one succeeding glacier advance could be distinguished, and this apparently was always related to the surface glacigenic deposits of the lowland incorporating the 27,000-years-old shells.

Inevitably with the passage of time, the collection of more field data and the widespread application of new techniques, a more comprehensible picture is likely to emerge. Within the last decade the conclusions relating to the Cheshire–Shropshire lowland have been subject to a period of rapid flux as attempts have been made to fit data into inadequate interpretative models. Current thought results from an amalgam of old and new ideas and it is particularly gratifying that the basic concepts of the pioneer workers, L. J. Wills, E. E. L. Dixon and T. H. Whitehead have survived a period of rethinking.

## CONCLUSION

The picture which is now emerging is of an obscure pre-Weichselian Pleistocene period from which no known and dated deposits in the lowland have survived, although the Burland borehole data potentially promises to prove otherwise. The lower parts of thick drift sequences probably do contain 'older drift' materials but these cannot as yet be positively identified, and this must remain a task for the future. The Early and Mid-Weichselian is represented by thick alluvial fan deposits which accumulated within a general permafrost environment adjacent to the Pennine foothills. A temporary amelioration within this depositionary episode saw the establishment of interstadial conditions, the sediments of which are represented by the Chelford organic bed. Until 1967 the Mid-Weichselian Upton Warren 'Interstadial Complex' had no known representatives but the provisional results of the Four Ashes findings would seem to indicate an environment of gravel transport by periodically-flooding westward draining streams from Cannock Chase. The accumulation of organic remains probably occurred during brief interstadial intervals.

The long period of alternating permafrost and interstadial conditions was terminated by an abrupt change in the climate which resulted in the extensive development of ice-caps over northern Britain. From several ice accumulation centres glaciers moved southwards into the northern Irish Sea basin incorporating bedrock and sea-bottom material and then a lobe pushed into the lowland between Wales and the Pennines to terminate along the Wolverhampton Line. The ice soon started to withdraw, the Severn Main Terrace was formed, and extensive ice-dammed lakes were impounded. The Ironbridge glacial overflow was initiated and developed to such an extent that it was permanently able to divert the upper Severn drainage. A re-advance or equilibrium stage witnessed the construction of a massive bi-lobate terminal moraine converging on Whitchurch. An advance of Welsh ice moved eastwards to terminate at Shrewsbury. Further retreat of the Irish Sea ice to the north again produced meltwater lakes which profoundly influenced the morphological response of the glacier deposition. The progress to the warmer conditions of the Flandrian postglacial was not continuous and periods of climatic deterioration induced permafrost régimes and associated solifluction and nivational activity. Many of the kettle holes produced by the melting of buried ice blocks and solution subsidences were progressively filled by late and postglacial organic matter (Birks, 1965; Shotton and Strachan, 1959).

REFERENCES

BINNEY, E. W. (1848) 'Sketch of the drift deposits of Manchester and its neighbourhood', *Mem. Manchr. lit. phil. Soc.*, Ser. 2, 8, 195–234.

BIRKS, H. J. B. (1965) 'Late-glacial deposits at Bagmere, Cheshire and Chat Moss, Lancashire', *New Phytol.* 64, 270–85.

BOULTON, G. S. and WORSLEY, P. (1965) 'Late Weichselian glaciation of the Cheshire–Shropshire Basin', *Nature, Lond.* 207, 704–6.

COOPE, G. R. (1959) 'A Late Pleistocene insect fauna from Chelford, Cheshire', *Proc. Roy. Soc.* B, 151, 70–86.

DAVIS, W. M. (1890) 'Structure and origin of glacial sand plains', *Bull. geol. Soc. Amer.* 1, 195–202.

FLINT, R. F. (1929) 'Stagnation and dissipation of the last ice sheet', *Geogr. Rev.* 19, 256–89.

GRESSWELL, R. K. (1964) 'The origin of the Mersey and Dee Estuaries', *Geol. Jl.* 4, 77–86.

HARMER, F. W. (1928) 'The distribution of erratics and drift; with a contoured map of erratics in England and Wales', *Proc. Yorks. geol. Soc.* 21, 79–150.

HOWELL, F. T. (1965) 'Some aspects of the sub-drift surface of some parts of north-west England', unpublished Ph.D. thesis, Univ. of Manchester.

HULL, E. (1864) *Geology of the country around Oldham, including Manchester and its suburbs.* Mem. geol. Surv. U.K. H.M.S.O.

LEGGET, R. F., BROWN, R. J. E. and JOHNSTON, G. H. (1966) 'Alluvial fan formation near Aklavik, Northwest Territories, Canada', *Bull. geol. Soc. Amer.* 77, 15–30.

MCQUILLAN, R. (1964) 'Geophysical investigations in a line of seismic shot holes in the Cheshire Basin', *Bull. geol. Surv. G.B.* 21, 197–203.

MOVIUS, H. L. (1960) 'Radiocarbon dates and Upper Palaeolithic archaeology in Central and Western Europe', *Curr. Anthrop.* 1, 355–91.

OWEN, D. E. (1950) 'The Lower Mersey', *Proc. Lpool geol. Soc.* 20, 137–48.

POCOCK, R. W., WHITEHEAD, T. H., WEBB, C. B. and ROBERTSON, T. (1938) *Shrewsbury District.* Mem. geol. Surv. U.K. H.M.S.O.

4*

POCOCK, T. I. (1938) 'Glacial deposits between North Wales and the Pennine range', *Zeit. Gletscherk. Eiszeitforsch. Gesch. Klimas.* **26**, 52–69.

POOLE, E. G. (1968) 'Age of the Upper Boulder Clay glaciation in the Midlands', *Nature, Lond.* **217**, 1137–8.

POOLE, E. G. and WHITEMAN, A. J. (1966) *Geology of the country around Nantwich and Whitchurch.* Mem. geol. Surv. U.K. H.M.S.O.

SHOTTON, F. W. (1962) 'A borehole at Conduit Head, Shrewsbury', *Trans. Caradoc Severn Vall. Fld Club*, **15**, 1–4.

SHOTTON, F. W. (1967a) 'The Problems and contributions of methods of absolute dating within the Pleistocene Period', *Q. Jl geol. Soc. Lond.* **122**, 356–83.

SHOTTON, F. W. (1967b) 'Age of the Irish Sea glaciation of the Midlands', *Nature, Lond.* **215**, 1366.

SHOTTON, F. W. and STRACHAN, I. (1959) 'The investigation of a peat moor at Rodbaston, Penkridge, Staffordshire', *Q. Jl geol. Soc. Lond.* **155**, 1–16.

SIMPSON, I. M. and WEST, R. G. (1958) 'On the stratigraphy and palaeo-botany of a Late-Pleistocene organic deposit at Chelford, Cheshire', *New Phytol.* **57**, 239–50.

SPARKS, B. W. and WEST, R. G. (1964) 'The drift landforms around Holt, Norfolk', *Trans. Inst. Br. Geogr.* **35**, 27–35.

THOMPSON, D. B. and WORSLEY, P. (1966) 'A Late Pleistocene molluscan fauna from the drifts of the Cheshire Plain', *Geol. J.* **5**, 197–207.

WHITEHEAD, T. H., ROBERTSON, T., POCOCK, R. W. and DIXON, E. E. L. (1928) *The country between Wolverhampton and Oakengates.* Mem. geol. Surv. U.K. H.M.S.O.

WILLS, L. J. (1912) 'Late glacial and post-glacial changes in the lower Dee Valley', *Q. Jl geol. Soc. Lond.* **68**, 180–98.

WILLS, L. J. (1924) 'The development of the Severn Valley in the neigh-bourhood of Iron-Bridge and Bridgnorth', *Q. Jl geol. Soc. Lond.* **80**, 274–314.

WILLS, L. J. (1950) *The palaeography of the Midlands*, 2nd edn. University Press of Liverpool.

WORSLEY, P. (1966) 'Some Weichselian fossil frost wedges from East Cheshire', *Mercian Geol.* **1**, 357–65.

YATES, E. M. and MOSELEY, F. (1967) 'A contribution to the glacial geo-morphology of the Cheshire Plain', *Trans. Inst. Br. Geogr.* **42**, 107–25.

# CHAPTER FIVE

# The Lower Severn Valley

Nicholas Stephens, M.Sc., Ph.D.

The Severn river system provides a means, as yet imperfect, of cor-relating Pleistocene events in north-east Wales, the Cheshire–Shropshire lowland, the West Midlands and the Bristol Channel area. It is proposed here only to discuss some of the problems of elucidat-ing the Pleistocene history of the lower Severn valley between the Ironbridge gorge and the mouth of the Bristol Avon, and to com-ment upon a possible chronology for events in the Axe valley (Figs. 5.1 and 5.3).

Two major Pleistocene events are known to have radically altered the river terrace systems in the lower Severn valley and given rise to many complex chronological problems. Wills (1924, 1937, 1938, 1950) showed that the Ironbridge gorge was initiated as a glacial overflow channel during the Weichselian glaciation from a series of ice-dammed lakes (including Lake Lapworth) impounded between an ice-front in the Cheshire plain and the watershed. As a result the upper Severn drainage was diverted permanently to flow southwards to the Bristol Channel instead of to the Dee estuary. The diversion is recorded in the Main (lower-M1) terrace of the Severn which traverses the gorge as a river terrace about half-way up its sides, and has been mapped in the lower Severn valley. Because the Main terrace of the Severn contains large quantities of gravels of Irish Sea, Lake District and Southern Uplands provenance, Wills inferred that the lake waters were ponded in front of a glacier which had pushed inland from the Irish Sea (Fig. 5.2).

Similarly, Shotton (1953) has indicated that the Warwickshire Avon has been added to the Severn trunk stream as a result of drain-age diversion by ice during the earlier Saalian glaciation. 'Lake Harrison' is believed to have occupied most of the drainage basin of a former north-east flowing tributary of the Trent (Fig. 5.2). When recession of the ice-front occurred water was released first in a south-westerly direction to the proto-Severn, and the Warwickshire Avon has maintained this general direction of flow ever since, at the expense of the former north-easterly drainage to the Trent.

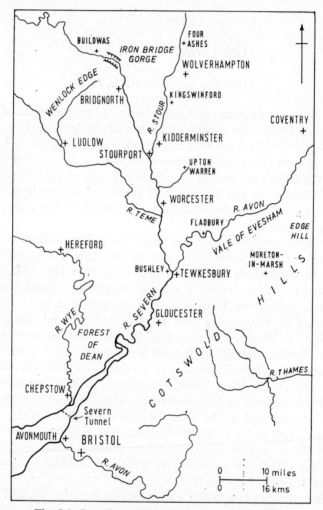

Fig. 5.1. Location map for the lower Severn valley

There are remnants of other river terraces in the lower Severn valley and its tributary valleys, especially the Warwickshire Avon, which are regarded as pre-dating the Weichselian glaciation. These higher terraces are generally more fragmentary and difficult to interpret than those at lower levels. Some of these older terraces have undoubtedly been produced by meltwater draining from the Welsh and Midland ice-sheets during the Saalian and possibly even earlier glacial periods, and by the rivers meandering across this broad lowland during the intervening interglacial periods (Wills, 1938; Tomlinson, 1925, 1935; Arkell, 1943; Shotton, 1953, 1967a, 1967b; Coope *et al.*, 1961).

Extensive terrace systems of diverse origin are among the most difficult of geomorphic features to investigate and it is necessary at the outset of any discussion of a sequence of river terraces and the buried channels found in the estuaries of our large rivers to emphasise the many problems involved, more especially when attempts are made to correlate with glacial stages, or to deduce the height of former sea-levels from river terrace data (Cotton, 1958; Dury, 1958; Stephens and Synge, 1966; Zeuner, 1959). Wills (1938) referred to this problem for the Severn valley in the following paragraph (p. 242): 'Each (terrace) represented deposits on a valley floor more or less graded to a particular sea-level, and that, so long as the sea-level and the volume of the river kept reasonably constant, this particular graded system persisted. For this reason, deposits of very different ages might, in some cases, lie at the same level, but in the absence of fossils they were extremely difficult to distinguish.' It is also important to realise that a terrace surface can be formed much later than the bulk of the gravels which lie beneath it. Shotton (in Wills, 1938) indicated a section in the Kidderminster terrace where a thin layer of horizontally-bedded gravels, forming the terrace surface, overlaid gravels deeply disturbed by solifluction—in such a case a difference of age, and of the climatic conditions, existed between the two gravel deposits and correlation with other sections was made more difficult. Assessment of the relative ages of terraces based upon the weathering characteristics of the gravel deposits can also present difficulties for the same rock type may behave differently under cold as opposed to temperate climatic conditions; but such evidence can be used sometimes to distinguish between periglacial as compared with temperate terraces (Goede, 1965). Withdrawals of the sea to low levels during glacial periods, and marine transgressions during interglacial periods have long been assumed, but such major changes of base-level take time, and the river may well alter its regime, perhaps several times, and bring about considerable alterations of any sequence of floodplain deposits and terraces.

It must be stressed that ice has advanced into the lower Severn valley from several different sources, including the Irish Sea (Northern drift), Midlands (Eastern drift), and from Wales, during the Middle and Upper Pleistocene, and each ice advance and retreat has been involved in the history of the Severn–Avon drainage. Each major source area of ice produced a characteristic suite of erratic materials, including Lake District and Southern Uplands rocks together with marine molluscs from the Irish Sea, Mesozoic rocks from the east and Uriconian volcanic rocks from Wales.

At times ice advanced southwards, obliterating parts of former river terrace systems and depositing till, while at others melting ice produced the waters responsible for carrying outwash gravels into the main channels and headwater tributary valleys. The ice-streams from the three main source areas covered different parts of the drainage basin, and were not always in close and continuous contact with one another. Thus it is possible only to provide a tentative chronology of events (Table 5.1), but one which is based largely upon the work of Wills (1938) and Shotton (1953, 1967a).

It is the dating of the southernmost extent of the Irish Sea ice, and the possibility of correlating that limit either to the Weichselian or to the Saalian glacial period which still constitutes a major problem (West, 1968). Most authorities are agreed that 'Older' and 'Newer' drifts (to use the traditional and more general terms), and a sequence of river terraces spanning at least two full glacial periods and an interglacial are present in the lower Severn valley, but there remain strong differences of opinion concerning the age of the various deposits and the river terraces and buried channels. Archaeological, geological and palynological evidence, supported by carbon 14 dating of certain organic deposits and shells, has been used in attempts to produce an acceptable chronology, notably by Coope *et al.* (1961), Shotton (1953, 1967a and b) and West (1968).

The Pleistocene history of the lower Severn valley has been divided into a number of stages, and it is convenient to begin our considera-tion of the evidence with the oldest known deposits in the area (Fig. 5.2 and Table 5.1).

*Stage* 1

The First Welsh ice and the Early Northern ice (= Bubbenhall Clay deposit, which may or may not be a true till) from the West Midlands advanced into the drainage basin. An Elster (Lowestoftian) age has

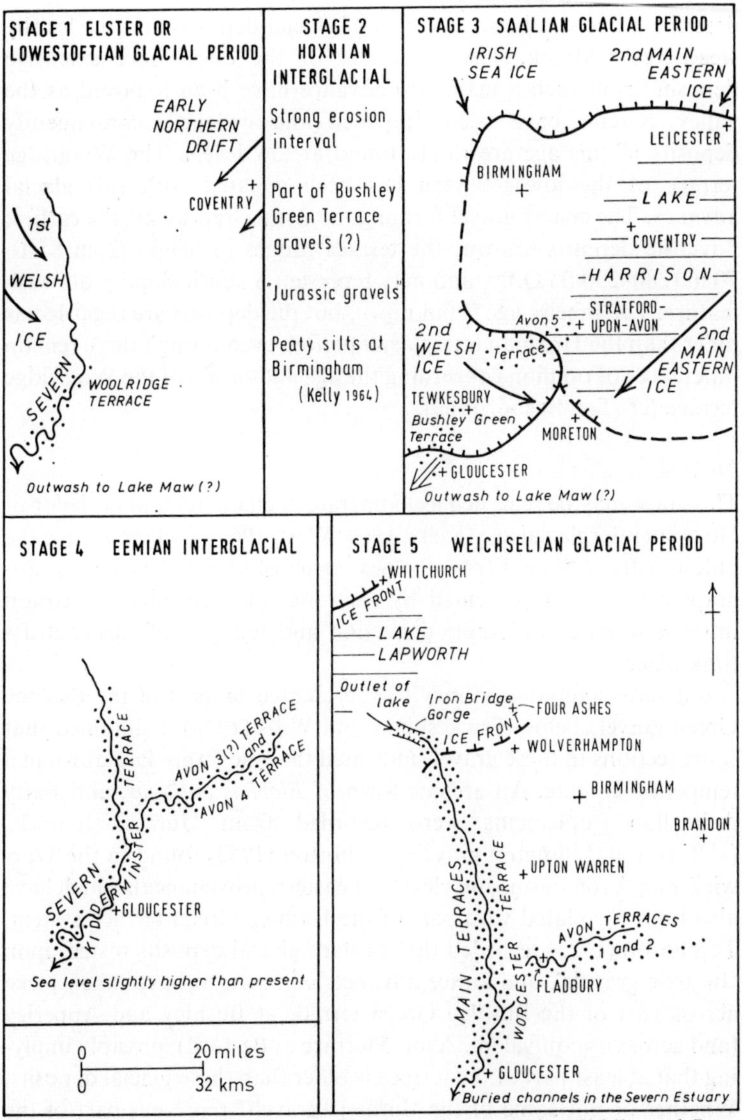

Fig. 5.2. Schematic diagram to illustrate the possible Middle and Late Pleistocene history of the lower Severn valley and adjacent areas

been assigned to these old glacial deposits, of which only remnants remain at high levels capping some interfluves; the deposits have often been referred to in the literature as the 'Plateau drifts', a term used to describe gravels, till and mixtures of glacial deposits and the locally weathered bedrock. The remainder of the once more extensive deposits from such a major ice advance have been removed as the valley systems have been deepened and widened, consequently deposits of this age are rarely found at low levels. The Woolridge terrace of the lower Severn has been equated with this glacial advance. The coarse gravel forming the terrace represents the earliest fluviatile deposits known; the terrace ranges in height from 85 to 70 m (280–230 ft) O.D., and may represent a south sloping outwash feature. The precise age is unknown, but the deposits are regarded as older than the Hoxnian interglacial period, even though there remain differences of opinion concerning the age and origin of the Woolridge terrace (cf. Tomlinson, 1963).

## Stage 2

This stage may be regarded as temperate in character and *probably* of Hoxnian interglacial age (Kelly, 1964; West, 1968), and separating the oldest drifts of Stage 1 from the next series of glacial deposits. Stratigraphically it is represented by a strong unconformity or erosion interval when considerable dissection and removal of earlier drifts took place.

But some aggradation may be represented by part of the Bushley Green gravels below Tewkesbury, and Wills (1938) maintained that some sections in these gravels indicated that they were laid down in a temperate climate. An archaic form of *Elephas antiquus*, and Early Acheulian implements were recorded from 'Jurassic gravels' (= Baginton–Lillington gravels of Shotton (1953) found in the Warwickshire Avon basin, and clearly of eastern provenance), which have also been correlated with part of the Bushley Green terrace system. Tomlinson (1925) recorded that Eastern glacial deposits rested upon 'Jurassic gravels', and an ice advance is known to have taken place across part of the Bushley Green terrace at Bushley and Apperley (and across its equivalent, Avon 5 terrace at Besford), possibly implying that at least part of the terrace is older than these glacial deposits. Whether or not some of the 'Jurassic gravels' represent part of the Hoxnian interglacial remains to be decided by future investigations. For both Stages 1 and 2 the available evidence is very limited and capable of different interpretations; and for this reason it is Stage 3 which constitutes the most acceptable beginning to the sequence of events summarised below.

*Stage* 3

The glacial period (Saalian or Gippingian) is indicated clearly by the incursion of the 2nd Welsh, and 2nd (Main) Eastern ice lobes into the lower Severn drainage basin (Fig. 5.2). This ice advance is associated with the Moreton drift at Moreton-in-the-Marsh, where a moraine has been recorded (Bishop, 1958). The Main Eastern ice (depositing the Great Chalky Boulder Clay) advanced from the Midlands and is said to be represented by the Camden Tunnel drift. Shotton (1953) emphasises that the extensive deposits left by this glaciation, which he distinguishes from an earlier 'First Welsh and Northern ice advance', are well preserved because no subsequent ice advance reached so far south. Stage 3 is also separated from Stage 1 by a strong erosion period, implying a full (Hoxnian) interglacial, although the 'Second' ice advance would have been able to obliterate much of the effects of the 'First', covering as it did the greater part of the same area.

This glacial period has been associated with a stage of the Acheulian culture (which spans the Hoxnian interglacial and Saalian glacial periods), and the ice was responsible for damming proglacial 'Lake Harrison' against the north-west facing scarp of the Cotswolds with a shoreline at about 132 m (400 ft) O.D. Several stages of this lake are known for Main Eastern ice advanced to the south-west across former lake sediments (the lower part of the Wolston Series of Shotton, 1953), and various overflow channels developed to carry lake waters across certain low cols in the scarp to the headwaters of the River Thames.

As the Welsh (Severn valley) ice retreated before the total withdrawal of the Main Eastern ice the last of a series of 'Lake Harrisons' must have found a new, lower outlet towards the south-west and the lower Severn. Thus, eventually the present Warwickshire Avon began to flow south-westwards and extended its regime over much of the former lake floor, dissecting and removing some of the lake deposits (Wolston Series, consisting largely of pebbly clay and interpreted by Shotton (1953) as a sediment of still water). Similar lake clays are known elsewhere, for example at Moreton-in-the-Marsh. The former north-east flowing drainage of the area occupied by ice and by the lake waters was prevented from re-establishing itself and the 'new' Avon succeeded in capturing a terrain formerly drained north-eastwards to the River Trent.

In the lower Severn valley an ice limit is placed at about Tewkesbury (Fig. 5.2) and the Bushley Green terrace is seen only to the south of this town. For this reason the greater part of the terrace appears to represent outwash from the ice-front and Avon 5 terrace is regarded as being the corresponding terrace in the Avon valley, post-dating the disappearance of the last 'Lake Harrison'.

## Stage 4

The following Eemian (Ipswichian) interglacial appears to be repre-sented by further aggradation along the lower Severn and the de-velopment of the Kidderminster terrace. This terrace slopes seawards less steeply than the succeeding Main terrace (Weichsel phase) and is said to occur at 30 m (100 ft) O.D. near Gloucester. Wills (1938) esti-mated from the gradient of the surface of the terrace that mean sea level was at about + 6 m (+ 20 ft), and he correlated the terrace with the marine sands of the Burtle beds in Somerset (Bullied and Jackson, 1937). Most authorities appear to accept that Avon 4 terrace, with an associated Late-Acheulian or Levalloisian industry, can be cor-related with the Kidderminster terrace. There is less agreement con-cerning Avon 3 terrace, where the correlation is doubtful because of the paucity of field evidence. The Kidderminster terrace contains a distinctly temperate fauna, including *Elephas antiquus*, and because of its low gradient and high level at Tewkesbury and Gloucester it may grade to a sea level higher than present. However, Wills (1938) seems to have expressed some doubts concerning the age of the Kidder-minster terrace, recording that 'Salwarpe outwash' grades to it, indi-cating that it too may be a composite feature comprising 'temperate' and 'cold' elements. Once again, the problem of using river terraces to construct a chronological table is illustrated.

## Stage 5

The onset of the last glacial period (Weichselian) must have brought about a lowering of base-level, probably allowing downcutting to take place as withdrawal of the sea occurred in the Bristol Channel. In the headwaters of the main streams considerable aggradation took place as periglacial processes produced vast quantities of debris to the south of the main ice front. Fluvioglacial meltwaters also laid down quantities of sands and gravels, some of which were carried southwards as 'Lake Lapworth' discharged through the Ironbridge gorge (Wills, 1938).

Shotton's (1967b) most recent account suggests that the Smestow or Wolverhampton line (Mitchell, 1960) marks the southern limit of

Weichselian Irish Sea ice in the Severn valley (Fig. 5.2, stage 5), even though such a line depends for its position upon an erratic limit (Lake District, south-west Scotland and Irish Sea material), and there is no recognisable fresh end-moraine landscape. In fact, the Ellesmere–Whitchurch moraine complex is the most southerly occurrence of an extensive fresh glacial landscape, to the south of which the drifts are much more deeply weathered (Worsley, Chapter 4, Boulton and Worsley, 1965). However, organic silt and peat recovered from gravels below Irish Sea till at Four Ashes, 8 km (5 miles) north of the Smestow–Wolverhampton line, has provided carbon 14 dates ranging from about 30,000 to 42,000 years B.P. Shotton has concluded therefore that, 'the gravels (containing the organic material) belong to the Mid-Weichsel or Upton Warren interstadial complex' (Shotton, 1967a; Coope and Sands, 1966), and consequently the overlying Irish Sea till (said to be in primary position and not a soliflucted till) is regarded as Late-Weichsel in age. Those who consider the Main terrace of the Severn to begin just within the Wolverhampton line as outwash gravels must then accept this terrace to be of Late-Weichsel age.

Unfortunately, we now appear to have a distinct breakdown in the chronology of the Weichselian river terraces, with the Main terrace of the Severn, and its geomorphological equivalents in the tributary valleys, spanning not one but two glacial advances. The terraces have a different age range depending upon whether the Upton Warren or the Four Ashes carbon 14 dates are accepted. The organic deposits incorporated in terrace gravels of valleys tributary to the Severn at Upton Warren (terrace grading to the Main Severn terrace), Brandon and Fladbury (Avon 2 terrace, grading to the Main Severn terrace), have yielded dates ranging from 41,000 to 28,000 years B.P. (Coope and Sands, 1966). Consequently the Weichselian glacial limit (Smestow–Wolverhampton line) has been dated at about 45,000 years B.P., on the basis that the Main Severn terrace represents outwash from this ice-front, and the tributary valley terrace gravels grading to the Main terrace can be expected to contain organic material slightly younger in age. Not only was such a date (45,000 years B.P.) out-of-phase with mainland European dates for the maximum extension of the Weichselian ice, but now it is contradicted by the evidence from Four Ashes. These latest dates (30,000–42,000 years B.P.) from gravels underlying the till regarded by Shotton as indicating the most southerly extension of Irish Sea ice indicate an even younger age (of about 20,000 years B.P.) for the Smestow–

Wolverhampton line and the associated Main Severn terrace, and its geomorphological equivalents in the tributary valleys, including the Warwickshire Avon. In the absence of *fresh* morainic land-forms along the Smestow–Wolverhampton line (which might best be regarded as simply a 'limit' of Irish Sea erratics), and in the presence of deeply weathered drift (as compared with the drifts along the Ellesmere–Whitchurch moraine), one may perhaps express some slight doubt as to the reliability of the Four Ashes evidence. Could it be that the till overlying the gravels containing the organic deposits is not in primary position; and, furthermore, could not the Ellesmere–Whitchurch line represent the maximum southerly extent of Weichselian ice in the Cheshire plain? Alternatively, the use of the Upton Warren and Fladbury deposits to provide an approximate age for the Main Severn terrace, and consequently a date about 45,000 years B.P. for the Weichselian glacial limit, must be abandoned. If the latter is the correct interpretation it is certain that the existing river terrace chronology will be changed radically as research proceeds.

Indeed, Wills (1938) recorded two stages of the Main Severn terrace, M1 (lower) representing a cold period, and M2 (higher) a warm period, and he commented upon the difficulties of interpretation of the sequence. Although the Lower Main terrace (M1) is said to traverse part of the Ironbridge gorge as an outwash terrace from 'Lake Lapworth', the Higher Main terrace (M2) remains an enigma. In the Avon valley, where Avon 2 terrace is regarded as equivalent to the Main Severn terrace it contains material dated at 40,000 B.P., at Fladbury (Coope, 1962), and a cold-climate fauna, which agrees with the presence of *Elephas primigenius* and *Rhinoceros tichorhinus* in the gravels of the Main terrace, presumably the lower M1 stage. The gradient of the Main terrace is steep, and it reaches − 11 m (− 35 ft) O.D. at Severn tunnel, clearly indicating that the sea had withdrawn to a lower level.

The recession of the Weichselian ice in the Cheshire plain may be indicated by the cutting of a valley within the older gravels and the development of the Worcester terrace (− 7·5 m : − 25 ft O.D. at Severn mouth). It seems likely that all the terraces below the Main terrace between Tewkesbury and Gloucester belong to this period, and many of the remnants are probably part of the Worcester terrace.

There are other implications for such correlations which cannot be discussed here (see chapters by F. M. Synge and P. Worsley), but it seems likely that the 'Older' ice limit (Fig. 5.2) is at least as old as the

Saalian glacial period. It follows therefore that the Fremington clay of Devon must also be of 'Older' drift (Saale) age, and so too are the remanié till deposits of Croyde, Trebetherick and the Scilly Isles (Mitchell and Orme, 1967). In southern Ireland the limit of 'Older' Irish Sea ice is associated with the Ballycroneen (Eastern General) till, seen at Garryvoe, and lying well outside the line of the Curraghcloe (Screen Hills)–Tipperary or southern Ireland end-moraine (Stephens, Chapter 11). The latter must now be regarded as the southern limit of the Irish Weichselian ice, presumably corresponding either to the Smestow–Wolverhampton line or to the Ellesmere–Whitchurch line. The large outwash terraces from the southern Ireland end-moraine, in the valleys of the Suir, Barrow and Lee (the latter from the equivalent Weichsel limit of the Lesser Cork–Kerry Glaciation), are probably equivalent in time to the Lower Main (M1) and Worcester terraces of the Severn.

### THE PROBLEM OF THE 'CHARD GAP' AND THE LOWER AXE VALLEY TERRACE GRAVELS

The sequence of events outlined for the lower Severn valley in the Middle and Late Pleistocene may help to explain certain anomalous features near Chard and in the lower Axe valley. There is a striking 'dry gap' at Chard with a floor at 84–88 m (275–290 ft) O.D., which provides a low level routeway between the Somerset levels to the north and the lower Axe valley to the south (Fig. 5.3). A combination of Irish Sea and Welsh ice may have succeeded in blocking the Bristol Channel in 'Older drift' times (probably during the second Welsh and second Main Eastern Glaciation [= Fremington till = Irish Sea Ballycroneen till]) as ice pressed southwards against part of the high cliffed coast of north Devon (Stephens, 1966). If this happened then outwash from ice in the Bristol Channel and the Severn valley, and runoff from the surrounding hills, may have formed 'Lake Maw' (Maw, 1864), and as Mitchell (1960) has also suggested (Fig. 5.3; see also Fig. 11.1). Assuming that such a lake once existed, its limits controlled by an ice-front in the Bristol Channel and surrounding high ground, much in the same way as 'Lake Harrison' was formed (Shotton, 1953), it would have overflowed at the lowest point of outlet to the south, in this case across a low col at Chard. The trench-like form of the Chard gap, and its low level, differs considerably from any nearby watershed col except near Crewkerne, where a low 'gap' is present also. About 6 km (4 miles) south of the intake of the channel, a magnificent terrace begins near Chard Junction railway station and extends southwards along the lower Axe valley. Several pits reveal deep sections in the gravels, which consist mainly of flint, greensand, chert, and chalk, together with a small percentage (5–10 per cent) of

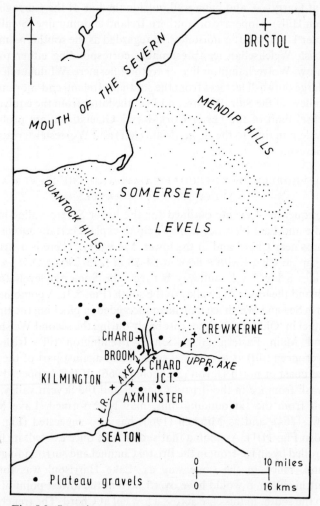

Fig. 5.3. Location map to show the relationship between the Somerset Levels (dotted area below 61 m (200 ft) O.D., approximately), the Chard 'Gap' and the Axe valley gravel pits at Chard Junction, Broom and Kilmington

'foreign' pebbles. The latter include rock types derived from Carboniferous and Devonian strata in north Devon and Somerset (Salter, 1899; Green, 1947).

The stratigraphy of the large pit near Chard Junction showed:

0·6–0·9 m (2–3 ft) silt (brickearth).

1·5 m (5 ft) gravels disturbed by frost action.

9·1 m (30 ft) (visible) undisturbed gravels.

Unknown thickness of gravels below water level in the pit.

The level of the terrace at this point is 61–65 m (200–215 ft) O.D.; at Broom, between Chard Junction and Axminster, the level is about 46 m (150 ft) O.D., and near Seaton it is about 16 m (54 ft) O.D.

At Broom numerous Palaeolithic implements of early Acheulian type have been found (made of dark Upper Greensand chert, some sharp-edged and some water-worn), with an ovate type predominating (Evans, 1897; D'Urban, 1878). The various authorities seem to agree on an Early-Middle Acheulian age for the implements, which allows them to be regarded as ranging in age from the Hoxnian interglacial to the Saalian glacial period. Their incorporation into copious water-transported gravel deposits (comprising up to 90 per cent of flint and chert pebbles) in the lower Axe valley (and found east of Chard Junction in the upper Axe valley only as far east as the Crewkerne 'gap') can be accounted for by water pouring through the Chard gap and carrying 'foreign' pebbles from north Somerset and the Bristol Channel area. It is possible that some of the 'erratics' could have been derived from yet older 'plateau gravels (drift)' capping the interfluves in the immediate neighbourhood of Chard (Waters, 1960), but not, it is believed, in a manner to allow such quantities to be incorporated in the enormous gravel terrace deposits.

For the present the existence of 'Lake Maw' and the utilisation of the 'Chard gap' by water from the lake, which was receiving Severn–Avon outwash, can be regarded as merely a tentative working hypothesis. The most serious problem arises when attempting to connect events in the lower Severn valley with the deposition of an enormous quantity of gravels in the Axe valley. Which of several Severn terraces could be correlated with the Chard Junction–Broom–Kilmington gravels and terrace? Because of its great height it is convenient to suggest that the Woolridge terrace 70–85 m (230–280 ft) may represent aggradation of outwash gravels in to 'Lake Maw', which acted as a local base-level. But the Woolridge terrace has been regarded as pre-dating the Hoxnian interglacial, and if this is correct then it certainly could not account for outwash gravels near Chard containing Acheulian hand axes. Similarly, it is possible that the Bushley Green terrace, at 46–52 m (150–170 ft) below Tewkesbury, may also represent aggradation into 'Lake Maw', partly because of the great height of the terrace and its low gradient. Furthermore, if this terrace is largely of Saalian age, as outwash from an ice-front in the lower Severn valley, and is correlated with 'Lake Maw', then a more correct time sequence is achieved for the Axe valley gravels. Outwash into and out of 'Lake Maw' during the Saalian cold period would have flowed southwards through the Chard gap and picked up the Early Acheulian hand axes (representing the artifacts made by Palaeolithic man during the Hoxnian interglacial) and incorporated them in the great terrace.

The archaeological dating of the Axe valley gravels places the Axe valley terrace as no earlier than the Hoxnian and no later than the Saalian glacial period. But if 'Lake Maw' existed and received outwash from the Severn valley the first lake level must have been higher than 84–88 m (275–290 ft) O.D. (i.e. higher than the highest Severn terrace), so as to spill through the col and to enable the water to erode the floor of the Chard gap to about its present level. It is possible that the 'Gap' was eroded with the ice all but filling the Bristol Channel and Severn estuary and thus the precise heights of the higher Severn terraces (Woolridge and Bushley Green) may indicate only later stages of the deposition of outwash gravels.

If this was the case, and in order to achieve the necessary water level to spill across the original col, it is tempting to correlate Lake Harrison and Lake Maw in time, and to suggest that as Welsh ice gave way the two water bodies became one. The highest shoreline for Lake Harrison is *c.* 132 m (400 ft) O.D. and the lake is firmly correlated with the Stage 3 Saalian ice advance (Fig. 5.2). Even if Lake Maw did not achieve this great height it may still have been high enough to breach the Chard col. The odd patches of gravel known at Bleadon (85 m: 260 ft), Brean Down (80 m: 242 ft) and at Portishead (90 m: 273 ft) may represent old shoreline or lake deposits. Such a correlation would allow the water escaping through the Chard gap to post-date the Hoxnian interglacial, when Palaeolithic man may have been living in the Axe valley and making Early Acheulian hand axes. The escaping water, carrying vast quantities of material, some ice-rafted, but most probably eroded from the sides of the gap, poured through the gap and picked up the discarded hand axes as the great terrace was built in the lower Axe valley.

Whatever may have been the timescale for these events it is suggested that the existence of a combined Lake Harrison and Lake Maw and the presence of an ice-barrier in the Bristol Channel would help to explain:

1. The origin and form of the Chard gap, with its floor at about 84 m (275 ft) O.D. at its northern end, and constituting a major break in the continuity of the east–west trending watershed which exceeds 130 m (400 ft) O.D.
2. The sudden beginning of a huge gravel terrace at the southern end of the gap, and the absence of a similar feature throughout the upper Axe valley except south of Crewkerne.
3. The presence of 'foreign' pebbles in the gravels, probably derived from the Carboniferous and Devonian rocks in north Devon and Somerset, as well as from the pre-existing plateau drifts.
4. The presence of ovate hand axes of Early-Middle Acheulian type in the gravels at Broom and elsewhere.
5. The known height (> 61 m: 200 ft O.D.) of the Fremington till and outwash gravels on Lundy Island (> 107 m: 350 ft O.D.) which indicate the presence of an ice-front of sufficient dimensions to completely block the Bristol Channel (Mitchell, 1968); this implies correlation of the Fremington till with the second Welsh Glaciation and the Saalian glacial period.

It may be noted that Zeuner (1959) has reported the presence of Mid-Acheulian (and Early Clactonian) hand axes in beach gravels at the base of the '30 m (100 ft) beach' at various localities in Sussex. Zeuner also asserts that there is 'corroborative evidence for the Mid-Acheulian being the industry of the Hoxnian Interglacial'. In the lower Axe valley the terrace appears to grade to about 16 m (54 ft) O.D. at the coast, and cannot therefore be correlated with the '30 m (100 ft) beach' of Hoxnian interglacial age. Consequently, it may be suggested that the included Early Acheulian hand axes (which are not generally recovered from Eemian interglacial deposits) were picked up as the gravels were deposited during the Saalian glacial period as waters poured through the Chard Gap from Lake Maw. Cryoturbation of the gravels near the surface and the deposition of brickearth can be attributed to the latter part of the Saalian and the Weichselian glacial periods.

The implications of the possible existence of a combined Lake Harrison and Lake Maw are far reaching, and research is in progress in an attempt to solve the problem of the origin of the Chard gap and in the search for possible lake sediments and shorelines. The Chard gap (and possibly the Crewkerne gap) may have a composite origin (and other possible gaps may yet be found), perhaps originating as an abandoned low col after river capture. But undoubtedly the final cutting of the gap, as we see it today, was accomplished by the running water responsible for the deposition of the vast gravel terrace in the lower Axe valley. Because the terrace gravels can be dated approximately by the included Early Acheulian implements this enormous terrace is regarded as being no older than the Hoxnian interglacial and no younger than the end of the Saalian glacial period. There is no indication that the Chard gap has ever been used again by meltwaters from the north. It may be suggested therefore that at no time since the Saalian glacial period has ice extended sufficiently far south and in such volume as to block the Bristol Channel and produce another 'Lake Maw'. Furthermore, the more complete sequence of terraces in the lower Severn valley from the Kidderminster terrace onwards also suggests no major blockage by Irish Sea and Welsh ice of the Bristol Channel and Severn estuary during the Weichselian glacial period.

### REFERENCES

ARKELL, W. J. (1943) 'The Pleistocene rocks at Trebetherick Point, north Cornwall: their interpretation and correlation', *Proc. Geol. Ass.* **54**, 141–70.

BISHOP, W. W. (1958) 'The Pleistocene geology and geomorphology of three gaps in the Midland Jurassic escarpment', *Phil. Trans. Roy. Soc.* B, **241**, 255–305.

BOULTON, G. S. and WORSLEY, P. (1965) 'Late Weichselian glaciation in the Cheshire–Shropshire basin', *Nature, Lond.* **207**, 704.

BULLIED, A. and JACKSON, J. W. (1937) 'The Burtle Sand Beds of Somerset', *Proc. Somerset Arch. Nat. Hist. Soc.* **83**, 171–95.

COOPE, G. R. (1962) 'A Pleistocene Coleopterous fauna with arctic affinities from Fladbury, Worcestershire', *Q. Jl geol. Soc. Lond.* **118**, 103–23.

COOPE, G. R. and SANDS, C. H. S. (1966) 'Insect faunas of the last glaciation from the Tame Valley, Warwickshire', *Proc. Roy. Soc.* B, **165**, 389–412.

COOPE, G. R., SHOTTON, F. W. and STRACHAN, I. (1961) 'A Late Pleistocene fauna and flora from Upton Warren, Worcestershire', *Phil. Trans. Roy. Soc.* B, **244**, 379–417.

COTTON, C. A. (1958) 'Eustatic river terracing complicated by seaward downflexure', *Trans. Edinb. geol. Soc.* **17**, 165–78.

DURY, G. H. (1958) 'Tests of a general theory of misfit streams', *Trans. Inst. Br. Geogr.* **25**, 105–18.

EVANS, J. (1897) *The Ancient Stone Implements, Weapons and Ornaments of Great Britain*, 2nd edn, London.

GOEDE, A. (1965) 'Geomorphology of the Buckland Basin, Tasmania', *Proc. Roy. Soc. Tasmania*, **99**, 133–55.

GREEN, J. F. N. (1947) 'Some gravels and gravel-pits in Hampshire and Dorset', *Proc. Geol. Ass.* **58**, 128–43.

KELLY, M. R. (1964) 'The Middle Pleistocene of north Birmingham', *Phil. Trans. Roy. Soc.* B, **247**, 533–92.

MAW, G. (1864) 'On a supposed deposit of boulder clay in North Devon', *Q. Jl geol. Soc. Lond.* **20**, 445–51.

MITCHELL, G. F. (1960) 'The Pleistocene history of the Irish Sea', *Adv. of Science*, **17**, 313–25.

MITCHELL, G. F. (1968) 'Glacial gravel on Lundy Island', *Trans. Roy. geol. Soc. Cornwall*, **20**, 65–8.

MITCHELL, G. F. and ORME, A. R. (1967) 'The Pleistocene deposits of the Isles of Scilly', *Q. Jl geol. Soc. Lond.* **123**, 59–92.

SALTER, A. E. (1899) 'Pebbly and other gravels in southern England', *Proc. Geol. Ass.* **15**, 264–86.

SHOTTON, F. W. (1953) 'The Pleistocene deposits of the area between Coventry, Rugby and Leamington, and their bearing upon the topographic development of the Midlands', *Phil. Trans. Roy. Soc.* **237**, 209–60.

SHOTTON, F. W. (1967a) 'The problems and contributions of methods of absolute dating within the Pleistocene period', *Q. Jl geol. Soc. Lond.* **122**, 357–83.

SHOTTON, F. W. (1967b) 'Age of the Irish Sea glaciation in the Midlands', *Nature, Lond.* **215**, 1366.

STEPHENS, N. (1966) 'Some Pleistocene deposits in north Devon', *Biuletyn Peryglacjalny*, **15**, 103–14.

STEPHENS, N. and SYNGE, F. M. (1966) 'Pleistocene shorelines', in *Essays in Geomorphology*, ed. G. H. Dury. Heinemann. pp. 1–51.

TOMLINSON, M. E. (1925) 'River terraces of the lower valley of the Warwickshire Avon', *Q. Jl geol. Soc. Lond.* **81**, 137–63.

TOMLINSON, M. E. (1935) 'The superficial deposits of the country north of Stratford-on-Avon', *Q. Jl geol. Soc. Lond.* **91**, 423–60.

TOMLINSON, M. E. (1963) 'The Pleistocene chronology of the Midlands', *Proc. Geol. Ass.* **74**, 187–202.

D'URBAN, W. S. M. (1878) 'Palaeolithic implements from the valley of the Axe', *Geol. Mag.* **5**, 37–8.

WATERS, R. S. (1960) 'The bearing of superficial deposits on the age and origin of the upland plain of east Devon', *Trans. Roy. geol. Soc. Cornwall*, pp. 26–8.

WEST, R. G. (1968) *Pleistocene Geology and Biology*. Longmans.

WILLS, L. J. (1924) 'The development of the Severn valley in the neighbourhood of Ironbridge and Bridgnorth', *Q. Jl geol. Soc. Lond.* **80**, 274–311.

WILLS, L. J. (1937) 'The Pleistocene history of the West Midlands', *Brit. Ass. Advmt. Sci.*, pp. 71–94.

WILLS, L. J. (1938) 'The Pleistocene development of the Severn', *Q. Jl geol. Soc. Lond.* **94**, 161–242.

WILLS, L. J. (1950) *The Palaeogeography of the Midlands*, 2nd edn, Liverpool University Press.

ZEUNER, F. E. (1959) *The Pleistocene Period*, 2nd edn, Hutchinson.

# CHAPTER SIX

# The Cardigan Bay Area

Edward Watson, M.A., Ph.D.

For a considerable time the picture of mid-Wales under a large ice sheet which streamed outwards from centres in the Cader Idris to Aran and Plynlimon areas has been generally accepted as typical of the Pleistocene. It is still true that, as Jones and Pugh (1935) stated, 'the effects of the glaciation have not been worked out in detail'. It is therefore proposed to examine the field evidence in some detail, taking the area in two parts; firstly, the Dovey–Dysynni basins together with the high ground stretching southwards from Plynlimon and eastward of the Teifi, where the centres of accumulation have been postulated and secondly, the coastal plateau with the Teifi basin (Fig. 6.1).

## THE EASTERN AND NORTHERN UPLANDS

In addition to the cliffed amphitheatres at valley heads, such as those of the Dovey tributaries, which probably marked the heads of valley glaciers, cirque basins with lakes enclosed by moraines are found where a considerable area exceeds 600 m (2000 ft). Both features face north and east, while the south- and west-facing valley heads at the same level appear to be little modified by ice action. One of the best examples of the effects of orientation is provided by the Cader Idris group of cirques (Watson, 1960). At elevations of about 150 m (500 ft) below that of the glacial cirques and with the same northerly and easterly orientation, there are cirque-like depressions which are generally considerably smaller in size (Fig. 6.2). There is no evidence of erosion on their floors, which are thickly covered with superficial deposits. It has been suggested that they developed around perennial masses of inert snow and ice, not thick enough to undergo plastic deformation and flow, and therefore incapable of evacuating debris from their basins (Watson, 1966). They consist of two types. The first has a rather straight moraine-like ridge, the 'protalus rampart', close to the back wall, so that the 'basin' is long and narrow and parallel to

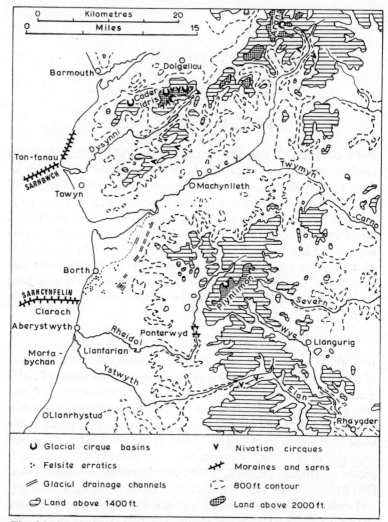

Fig. 6.1. South Merioneth and north Cardiganshire. Contours at 800, 1400 and 2000 feet are at approximately 244, 427 and 610 metres

the back wall whereas in a glacial cirque the basin is normally elongated perpendicular to the back wall. The second has no moraine-like enclosure but a smooth concave floor of drift which may be accordant with the surface of solifluction deposits outside the cirque or have a scarped or terraced front at the cirque mouth. The former

occur on the north side of the Cader Idris escarpment and the head of
the Tal-y-llyn valley (Watson, 1968), the latter in the Dovey–
Dysynni watershed and the Dovey headwaters area. The most impressive example in the area is Cwm Du in the upper Ystwyth valley
(Fig. 6.2), one of the second type, which has built up a terraced fan on
the Ystwyth floor, blocking the valley which is some 210 m (700 ft)

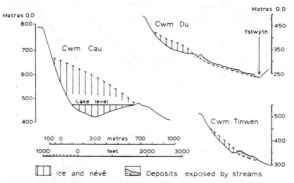

Fig. 6.2. Glacial and nivation cirques. The reconstruction
in Cwm Cau is by W. V. Lewis (1949), and in Cwm Du and
Cwm Tinwen by the author. The vertical scale is twice the
horizontal. From *Geography at Aberystwyth* by permission
of the University of Wales Press Board

wide, and rising 90 m (300 ft) from the Ystwyth to the cirque floor.
Figure 6.2 shows the size of a glacial cirque, Cwm Cau (below Cader
Idris), in relation to the two nivation cirques in Cwm Ystwyth; Lewis
(1949) estimated that when the ice in it reached the lip it was 190 m
(620 ft) thick. Cwm Tinwen illustrates the effects of a steeply sloping
bank of névé; much of the debris from the frost-shattering of the
headwall slides over the steep snow surface and builds up the protalus at its foot. Cwm Du results from a gently sloping névé mass;
debris from the headwall largely falls into the chasm which opens in
summer between the snow and the headwall, is broken up by thaw-
freeze and migrates as a thin layer between the base of the névé and
the frozen ground below towards the foot of the névé mass (Fig. 6.3).

The trough valley, one of the hall-marks of glaciation, is cut in
rock. Many of the valleys in central Wales have concave sides and
some are largely of rock; e.g. Cwm Gwarin, north of Plynlimon, Cwm
Berwyn, east of Tregaron, or the Irfon valley above Abergwesyn. In
very many cases, however, the existing concave sides reflect the

presence of thick screes. The Tal-y-llyn valley is such; it may have been
a true trough as is suggested by well-marked spur facets and the basin
form of its floor (Watson, 1962), but everywhere exposures show thick
screes on the lower slopes, as much as 18 m (60 ft) at the head of the
valley and 12 m (40 ft) some 7·2 km (4½ miles) south-west of Tal-y-

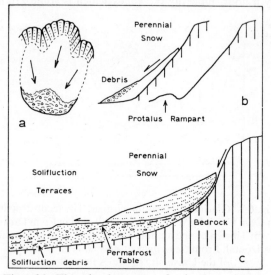

Fig. 6.3. The development of nivation cirques.
(*a*) Plan of steeply sloping snow bank (based on Cwm
Tinwen in the upper Ystwyth valley); (*b*) the same in
section, together with its post-glacial profile; (*c*)
section of a gently sloping snow bank in the Urals,
after S. G. Botch, 1946

llyn. Exposures in the Dovey basin, in Cwm Rheidol, Cwm Ystwyth
and similar valleys suggest that these conditions are very widespread.
The concave slopes are not so much evidence of recent glaciation as
of scree accumulation under permafrost conditions as shown by the
presence of frost wedge-structures in the screes (Watson, 1965).
More than simple gravity fall is involved; with an impervious frozen
substratum, slope wash especially from snow melt, mud flows and
solifluction may occur so that the foot of the scree may lie at a very
low angle.

Valley lakes, usually indicating severe glacial erosion, are represented by Tal-y-llyn lake. This has often been described as moraine-
dammed from the hummocky nature of the deposit impounding it.

13 Solifluction terrace on the right faces a south-facing rocky slope in the Nant-y-moch valley. The terrace scarp rises rapidly to 18-22 m (60-70 ft), at X as the rock slope at its rear rises sharply.

14 The same valley above X. Though rock outcrops on left, solifluction terraces occur on both banks in the upper basin.

15 Cwm Gwarin, a trough 2½ km (1½ miles), NNE of Plynlimon summit. The trough form is cut in rock, which outcrops down the valley side and occurs at intervals on the floor.

16 Stratified scree, on the valley side 800 m (½ mile), below Tal-y-llyn lake. Beds of coarser, loose openwork scree alternate with finer, more silty, closely packed beds. Marks on hammer shaft 10 cm apart.

This deposit consists of angular rock debris and large masses of shattered rock; it lacks the matrix of a till and the worn and bedded character of a kame. The material is typical landslip debris; its position directly below a large hillside scar of apparently larger dimensions than the lake dam, seconds this interpretation (Watson, 1962). Hence this lake can no longer be cited as evidence for glacial action. Where rock basins might be expected, in the deeper stretches of the main valleys, the Dovey, Rheidol and Ystwyth, there is an infill of terraced water-laid gravels rising up to 30 m (100 ft) above the river. These terraces, unpaired except for the highest, are due to the erosion of a gravel fill. The aggradation this represents might be the result of a periglacial environment; Arctic rivers are aggrading their beds at the present day. It might also result from the summer flooding accompanying the presence of névé and glaciers on the uplands. In any case the absence of such terraces in Snowdonia, in conjunction with the presence of lakes there, indicates a different history from that of the Cardigan Bay area.

Though deposits described as till and boulder clay occur at many points in the upland valleys, those with an indisputable glacial morphology, such as end moraines, are rare. In the upper Rheidol basin at Ponterwyd, two ridges of stony clay have diverted the river into rock gorges. The upper of these gorges now contains Dinas Dam, the lower turns the Rheidol into the narrow rock gorge just below the village. These probably mark the limit of upper Rheidol ice at some stage. From this to the coast there are no moraines in the Rheidol valley. The 'lateral moraines' of Keeping (1882) on the south side of the upper Ystwyth are terraces of head (Watson, 1968); no landforms suggesting end moraines occur above Llanfarian. At the latter, Llanrhystud Road Station, the highest parts of Keeping's 'esker' have proved to be rock; the gravels in the railway cutting dip downstream and consist of local debris. The gravels here owe their survival to protection from river erosion by rock outcrops at several points at water level, so that it is difficult to say what was the original accumulation form. The Dovey valley contains no moraines while in the Dysynni the only morainic deposits which occur are at the coast and contain north Wales erratics showing that they belong to ice moving down the coast.

In the great majority of cases the drift of the uplands takes the form of smooth terraces sloping towards the river and usually occurs on one side of the valley only, with a rocky slope opposite. This terrace

5

development depends firstly on lithology. On rocks which disintegrate into joint blocks producing little fine material, such as the volcanic rocks north of the Dovey, they do not form at all, while on mudstones which break down to produce much silt, thick deposits occur. On the latter type of rock, the higher the slope at the rear the thicker the accumulation; the drift terrace height rises and falls as the height of the interfluve behind changes. Where the valley sides are of the same rock and rise to the same height, the deposits are thicker at the foot of a slope facing NW, N or E than at the foot of one facing SE, S or W. This close relationship with lithology, height and orientation of the valley slope can only mean that they are slope deposits produced by weathering and transport on the slope itself. The character of the deposits, mainly stony clays, suggests a soliflual origin; 'solifluction sheets are composed of an unsorted till-like mass of stones and fines of all sizes' (Washburn, 1947). Interbedded with these are discontinuous but widespread gravel horizons, bedded parallel to the slope, formed in intervals when slope wash predominated over solifluction (Watson, 1969).

The long axes of elongated stones in these deposits lie predominantly parallel to the slope. If the drift were laid down by ice moving down the valley, the statistical preference of the long axes, or the 'preferred stone orientation' would be parallel to the valley; if laid down by an ice-cap with a regional flow, it should be broadly parallel to this flow. As Fig. 6.4 shows, in the upper Rheidol it fits neither of these patterns; it is parallel to the slope, and changes with changes in the valley direction. The terraces undoubtedly contain glacial material, e.g. glacially-shaped and striated stones, and some *in situ* glacial deposits may be buried below the slope deposits. Nevertheless the close relationship between the thickness of the terraces and the character of the slope at the rear suggests that the greater part of the material is of slope origin.

In addition to the upper Rheidol basin above Ponterwyd, these terraces also occur along the upper Wye above Llangurig, the upper Elan, and some of the tributaries of the Dysynni, where a slope origin is confirmed by the preferred stone orientation; they are equally well developed along the upper Clywedog and some of the Ystwyth and Dovey tributaries.

Throughout this upland area there are few end moraines or other glacial deposits, and a limited number of fresh erosional glacial landforms. Since the ice last covered it, a lengthy period of periglacial conditions has been experienced; an interpretation of the landforms based on periglacial models may be consistently applied (Watson, 1967b).

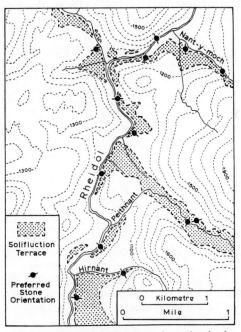

Fig. 6.4. The preferred stone orientation in the upper Rheidol basin. The sites are all in the upper bed of a series forming smooth terraces of 'drift' above Ponterwyd, between N832 and N876. The preferred stone orientation is subparallel to the slope behind each site, though the slope orientation varies considerably. Movement of material during deposition was consistently down-slope. Contours at 1100, 1300, 1500 and 1900 ft are at approximately 335, 396, 457 and 579 m

### THE COASTAL PLATEAU AND THE TEIFI BASIN

The rest of the Cardigan Bay area consists of the coastal plateau south of the Dovey estuary and the Teifi basin (Fig. 6.5). Between the latter and the coast, elevations of 300 m (1000 ft) are reached at points south-west of the Aeron and 335 m (1100 ft) in the Mynydd Bach to the north of it. Both Charlesworth (1929) and Jones (1965) stressed the relative scarcity of glacial deposits outside the narrow coastal belt and the Teifi floor. Considering the low elevation it is not surprising that erosion forms are almost unrepresented.

Over much of the area and especially to the south-west of the Aeron up to the highest elevations, convex interfluves pass down into a con-

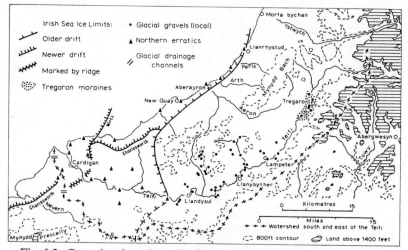

Fig. 6.5. Central and southern Cardiganshire. Contours at 800 and 1400 ft are at approximately 244 and 427 m

stant slope, which becomes gently concave and merges with the valley floor, a form produced by long continued solifluction, according to Tricart (1963). Though they probably owe their perfection in part to a dominant lithology of well-cleaved mudstones, their development supports the generally held view that this area lay beyond the ice limits during the Last Glaciation—Charlesworth (1929), Wirtz (1953), Mitchell (1960) and Synge (1961). The Mynydd Bach, which all of these authors also put south of the ice, has a surface of ridges and swales parallel to the cleavage. The bedrock consists of the Aberystwyth Grits and the ridging seems due to the arches and troughs of the folds being more shattered by the tight folding than the limbs, which are consequently more resistant to weathering and form the ridges. Between Aberystwyth and Llanrhystud the same relief forms are developed on the Aberystwyth Grits, while to the east there are smooth interfluves on mudstones. There is no justification in terms of landscape for the ice front drawn by Charlesworth from Tregaron round the north end of the Mynydd Bach to the lower Aeron, for which he admitted he could find no evidence in the form of glacial deposits. This applies equally to Sissons's line (1964, p. 137).

These deductions from slope form are borne out by the superficial deposits. On the interfluves the rock is often at the surface and quarries show a thin cover of angular rock debris quite close to the summits. As in the higher areas to the east, the steeper slopes are mantled with screes while more clayey types of head occur in the valley bottoms. Available relief is generally less, so that slope deposits are thinner. This may be compared with the higher, more 'senile' parts of the High Plateau, where valleys are often shallow and the deposits on their floors thin, in spite of altitude.

The deposits most frequently exposed are the screes, which are often of small calibre and sometimes stratified, showing not only size sorting but an alternation of openwork and compact muddy beds. These have been most closely examined in the Aberystwyth area (Watson, 1967b). They have also been observed at widely scattered points throughout the area, as far south-west as Newcastle Emlyn. In the more open valleys a stony clay whose angular debris lies parallel to the slope may be seen, e.g. in the lower Ystwyth valley near Llani-lar, where it overlies alluvial gravels, and in the Llanfarian area, in the Carrog valley north-east of Llanrhystud (Fig. 6.6), and in the Teifi valley south-west of Lampeter. Similar terrace deposits are known elsewhere in the lower Rheidol and Teifi basins, but the pre-ferred stone orientation has not yet been determined.

Although the landscape suggests a periglacial rather than a glacial origin, the Teifi basin has many exposures of washed sorted gravels, which, from their distribution and dead ice forms, must be of glacial origin. Their main concentration is between Lampeter and Llandysul, where they occur not only in the main valley but also far up the tribu-taries on the north-west bank. Above Llangeler no erratics have been found though Irish Sea ice material occurs west of this (Fig. 6.5). These gravels may represent the furthest limits of the central Wales ice, which in the retreat stage, passed into a stagnant state in its forward zone.

In the bottom of the main valley these gravels form gently undulat-ing masses, trimmed by the streams, with numerous closed depres-sions which may contain water or marsh. They may fill the valley floor as at Pencarreg, above Llanybyther; elsewhere an isolated mound may occur. It is very doubtful that they can be interpreted as

halt-stage moraines as Griffiths (1940) suggested. Some of the deposits
in the tributary valleys on the north-west bank have a ridge form
transverse to the valley and may represent marginal accumulations
against downwasting ice. More often they form a terrace whose

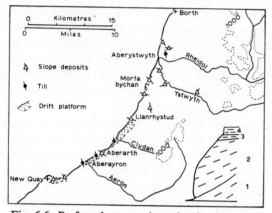

Fig. 6.6. Preferred stone orientation in the coastal
area between Aberystwyth and New Quay. The
slope deposits inland have been formed on valley
slopes so that the preferred stone orientation is
variable. The orientation of the coastal slope
between Morfa-Bychan and New Quay changes
only gradually, therefore there is more constancy
in preferred stone orientation. In the Morfa-Bychan
area the stone orientation might be used as an
argument for deposition by west-flowing ice,
nevertheless, it does change with changes in slope
orientation in the Bay as a whole. The decreasing
dip of the beds, (inset: (1) a fossil scree; (2) a solifluction deposit; (3) slope wash gravels; (4) a solifluction deposit), is typical of slope deposits. The
contour shown is at 1000 ft or 305 m

upper surface merges into the hill slope without a physical break. In
some cases the gravels occur on the interfluves where they form a
very subdued, gently undulating surface with closed depressions.
Where the gravels are well exposed, the bedding usually dips towards
the west and north, against the general direction of stream flow.

The merging of the gravel surface into the hill slope may be due to
head from the latter deposited on top of the gravels. This is difficult to
demonstrate as the gravel pits are nearly always in the outer part of a

terrace, but in one pit close to the valley side, a well-sorted gravel is overlain by head which forms a smooth concave surface above the pit.

Above Lampeter these washed gravels are restricted to the valley bottom. Charlesworth (1929) regarded the arcuate ridge enclosing Tregaron Bog and the moundy mass of gravels extending 5½ km (3¼ miles) downstream, as the only moraines marking a limit of Welsh ice in the Teifi valley. This limit he dated to the Last Glacial, but as yet there is no independent evidence for this. Pollen analysis has shown that there was open water in the basin now filled by the bog, as late as the beginning of the postglacial period.

In the area stretching some 20 km (12 miles) west of Llandysul, the 'gravel pits' on Ordnance Survey maps have been found to be in scree and rock rubble. Erratics of the Irish Sea ice have been found, usually in secondary positions (for example, in scree, and in a delta-like deposit) and this, in addition to the complete lack of morainic forms has led to general acceptance of the view that this erratic material is older than the Last Glacial period. The limit of its distribution on Fig. 6.5 crosses the Teifi at Llangeler where Jones (1965) reported the easternmost erratic. Three km (2 miles) further east, he found only local debris in a till on the valley bottom at Gilfach-wen. South of the Teifi the position of this line is very doubtful owing to the lack of evidence, but on the basis of Griffiths's map (1940, Plate II), showing a fan of erratics stretching south and south-east of Prescelly (305–518 m: 1000–1700 ft) the ice may have gone over the low watershed area to the east which is in part below 225 m (800 ft). The glacial deposits of this period are probably largely confined to the deeper sections in the valley floor. Figure 6.7 shows the plan of an infilled meandering valley (recalling that of the Dee in the Llangollen area), with a floor at 32½ m (107 ft) below sea-level at Cardigan (Jones, 1965).

The feature which stands out against the very subdued relief of the lower Teifi valley is the moundy ridge of shelly sand and gravel (with Irish Sea erratics and clay balls of calcareous till), at Bancywarren to the north-east of Cardigan (Fig. 6.7). After a break where the surface falls quickly to the Teifi bottom, these deposits appear to be continued south-westwards by a low ridge, which south of the river, consists of the brown calcareous till of the Irish Sea ice. This ridge ends against a steep rise to the Teifi–Nevern watershed; on the watershed, to the north-west, are sand and gravel deposits. Charlesworth (1929) regarded these and the Bancywarren deposits as marking the limit of the Irish Sea ice in the Last Glacial. The site of a later smaller ice lobe is defined by a curved ridge of calcareous till (overlain by blown sand), which causes the contraction of the estuary 2 miles below Cardigan. Charlesworth believed that this lobe was responsible for the later stage of proglacial 'Lake Teifi' which discharged by the Cippin channel intaking at 107 m (350 ft); at the earlier stage it drained by the Pontgareg channel intaking at 'about 450 feet' or 137 m (see

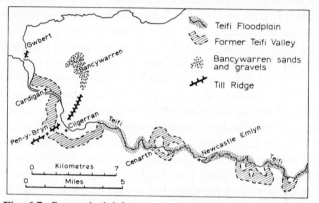

Fig. 6.7. Some glacial features in the lower Teifi valley. The former Teifi valley is filled with superficial deposits, its rock floor being at −12 m (−40 ft) O.D. at Cenarth; −32 m (−100 ft) O.D. at Cardigan

Fig. 10.7, below). No detailed study has been made of 'Lake Teifi', but it may be noted that in addition to deltaic gravels (quoted by Charlesworth), at Llanwnen, 3 km (2 miles) NNW of Llanybyther, an extensive delta with well-developed foreset bedding underlies Lampeter; Llanybyther is on another, while there is a striking gravel terrace at the mouth of the valley opposite; a fourth, with the gravels well exposed, lies 3 km (2 miles) further downstream at the mouth of the Cledlyn. The surface of all these is at about 122 m (400 ft), but before any correlation with drainage channels can be discussed, the height of the rock floors of the latter needs to be determined.

The best exposures of Irish Sea drift are in the coastal cliffs; the northernmost, a small area of calcareous till, occurs at Llanrhystud. Here and further south, the erratics found in this till, flint, jasper, serpentine, quartzites and various igneous rocks, are scattered through gravels. At the Arth and further south, these erratics and the shelly calcareous till are found in considerable quantities. The inland limit of this Irish Sea ice material is not fixed by any morainic feature; together with considerable thicknesses of gravels and slope deposits, it forms a sloping terrace, the 'boulder clay platform', which is backed by a rock slope, the seaward edge of the coastal plateau.

The calcareous till is always overlain by later deposits and in most places does not occur more than about 1 m (3–4 ft) above the modern beach. It passes up into gravelly deposits, largely fluvioglacial, laid down in the withdrawal stage by the ice which deposited the till underneath. The gravels may form moraine-like features, as just north of Llanon, but these are always very subdued and when seen in cliff sections the included sands and silts are deeply weathered and cryoturbated. The latter features, consisting of well-developed involutions and vertical stones, are marked where surface slopes are gentle. Where the higher ground approaches the coast, thick slope deposits laid down by solifluction and slope wash are banked against the rock. Just north of Aberarth the lower part of the slope deposits has calcareous lenses and erratics showing it to be soliflucted Irish Sea till. This passes up into solifluction deposits formed by frost weathering of the local rock. The Irish Sea till had been weathered yellow grey and largely decalcified before it was soliflucted. No fresh till overlies it so it seems probable that the Irish Sea till was deposited by ice before the Last Glaciation, the solifluction deposits, like the involutions and vertical stones of the *in situ* Irish Sea drift, being of Weichsel or Würm age. Near the river mouths the glacial deposits have been removed and in the case of the more steeply graded streams such as the Arth, the Peris and the Wyre, fluvial gravel fans have replaced them. The local fluvial origin is shown by the marked imbrication of these boulder gravels into the present stream flow. The fans are probably contemporary with the fluvial terraces of the more gently graded rivers, the Clarach, Rheidol and Ystwyth. They show at least three phases of deposition, the middle series displaying well-developed vertical stones and involutions, which may be compared with those in the nearby glacial deposits, in the upper terrace of the Rheidol, or in the delta gravels at Lampeter (Watson, 1965). These indicate an active layer in permafrost conditions, usually of 2 m (6–7 ft).

In the Morfa-bychan area north of Llanrhystud where the coastal slope deposits have been examined in some detail, the material is all locally derived (Watson, 1967a). The great proportion of it was formed by solifluction and sheet wash from the rock slope behind but the

5*

lower part of the solifluction beds contain yellowish weathered material which may, like the base of the Aberarth deposits, be soli-flucted till. The lack of Irish Sea erratics suggests that only local ice affected the area when it was last glaciated. A local till covered by solifluction deposits occurs at Aberystwyth Hospital (592818). This has a preferred stone orientation approximately NNW–SSE, similar

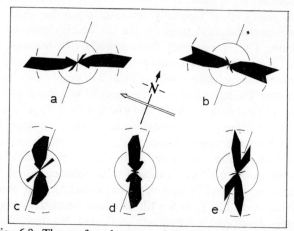

Fig. 6.8. The preferred stone orientation of the superficial deposits at Bronglais Hospital, Aberystwyth (SN/592818). The circles represent 10% and the arcs 20% of the sample, and the arrow shows the surface slope. The stone orientation in *c*, *d*, and *e* is oblique to the slope but parallel to that of the calcerous till of the Irish Sea ice to the south; it is probably a local till. The stone orientation in *a* and *b*, from the overlying stonier bed, is almost parallel to the slope and represents a later solifluction deposit. Reproduced from the *Geological Journal* by permission of the editor

to that of the Irish Sea till further south (Figs. 6.6 and 6.8), suggesting that local ice was here flowing alongside the main Irish Sea ice stream. No beds with this stone orientation were found in the Morfa-bychan deposits, though counts were taken at thirty-six sites, includ-ing ten on the shore platform. All showed a stone orientation more or less perpendicular to the hill-slope behind. This reflects the bedding of the deposits which is regularly off-slope at a decreasing angle from the bottom to the top of the deposits (Fig. 6.6, inset). Similar deposits of local material occur 3–5 km (2–3 miles) north of Aberystwyth at Clarach and Wallog.

The drifts of the coastal fringe of Cardiganshire are complex. Williams's (1927) subdivision into Lower boulder clay, sands and gravels, and Upper boulder clay is no longer tenable. His Upper boulder clay includes such diverse elements as an angular rubble resembling the Upper and Main Head of north Devon and the boulder alluvium of the Arth fan.

At the northern limit of the coastal plateau, a concentration of erratic felsite blocks and marginal drainage channels on the south side of the Dovey estuary have been interpreted by Jones and Pugh (1935) as marking a limit of North Wales ice (Fig. 6.1). The felsite might have come from the Arans at the head of the Dovey; it might also have come from further north and west, for example the Cader Idris range. As these felsite erratics are not associated with any morainic forms, it is probable that, although the drainage channels may mark the limit of a North Wales ice advance, it was not during the Last Glacial.

This conclusion is borne out by the occurrence of large involutions and vertical stones similar to those between Llanrhystud and Aber-aeron on the coast between Towyn and Tonfanau. They are absent however in the cliff exposures north of this, suggesting that the deposits there are of younger age and may represent the limit of North Wales ice, during the Last Glacial (Watson, 1967b). The morainic ridge at Tonfanau appears to turn seawards into Sarn Bwch, a large ridge of gravels on the sea floor. A similar ridge, Sarn Cynfelyn occurs some 3 km (2 miles) north of Aberystwyth. These two may represent former limits of ice advance on the present floor of Cardigan Bay but there is insufficient evidence to prove this (*vide* Chapter 2, p. 28).

## THE GLACIAL SUCCESSION

Any attempt to trace the movement of Welsh ice in the area is hindered by the lack of distinct erratics except in the northern part, owing to the relatively uniform nature of the bedrock, which consists of cleaved mudstones and grits. Few striae have been recorded. This is in part due to the nature of these rocks which weather rapidly, especially under periglacial conditions. The most significant striae records are from the area south of the lower Dovey where Jones and Pugh (1935) noted two sets, one east–west, the other approximately north–south. The latter they interpreted as showing that the ice from central Wales at first flowing westward was turned southward by the pressure of Irish Sea ice in the Bay. At Aberystwyth the preferred stone orientation counts in local till indicate a flow from NNW, suggesting that this pressure was forcing the local ice inland again. It seems possible that south of the Mynydd Bach this local ice pressed as far inland as the lower Teifi basin, while above Lampeter the valley was filled with ice from the upper Teifi area. This would explain the distribution of sands and gravels in the middle Teifi, as due to ice which had come in from the Bay being cut off and becoming stagnant as the coastal watershed emerged during downwasting. The ice from the upper Teifi became stagnant only after it had thinned considerably so that kames are confined to the valley bottom.

The extent of Welsh ice just described probably represents conditions during the Saale or Riss Glaciation. At this period Irish Sea ice which flowed southwards across the Bay invaded the lower Teifi valley below Llandysul, where its erratics are found inland of the Cardigan moraine. The calcareous till which occurs in the narrow coastal strip south of Llanrhystud is probably of the same age. In the first place, the morainic features there are very subdued and consist of deeply weathered and strongly cryoturbated material, unlike deposits of Main Würm (Weichsel) age elsewhere; secondly the volume of slope deposits (in part incorporating the till) and multistage alluvial fans overlying the glacial deposits seems too great for the Late Glacial period. The widely used limit of the 'Newer drift' for this area drawn by Charlesworth (1929) from the Bancywarren deposits round the north end of the Mynydd Bach to Tregaron was described by him as 'conjectural'—'moraines and indubitable evidence of marginal drainage are alike singularly scanty'. The evidence for 'Lake Aeron' on which this limit depends, is very flimsy, with no lacustrine deposits in it being known, and no convincing drainage channels leading from it. The Bancywarren deposits may nevertheless be of Würm age as Wirtz (1953) suggested and the possibility of a proglacial 'Lake Teifi' may be regarded as still open. Indeed recent radiocarbon dates obtained from organic material in the Bancywarren sands, 31,800 B.P. (Brown *et al.*, 1967) and 33,750 B.P. (John, 1967), would imply an age younger than the middle of the Last Glaciation.

During the Last Glaciation Wirtz envisaged ice from the Irish Sea basin crossing Lleyn and flowing south to reach the coast again in the area of the Teifi estuary. The coast to the north of this was free of Irish Sea ice and local ice, though there were local ice-caps over north and east Wales. The morainic ridge at Tonfanau may mark the limit of the former, but there is no palynological or radiocarbon evidence of its age.

With regard to the extent of ice in the central Wales uplands at this stage very little can be said with certainty. The evidence points to a lack of vigorous ice-sheets, which Charlesworth (1929), in discussing the high ground east of Tregaron, suggested was due to its being near the climatic snow line of 425–455 m (1400–1500 ft). Manley's (1951) calculations put this somewhat higher, probably 485–520 m (1600–1700 ft). The only consistent evidence for ice erosion is the distribution of small lakes which are scattered across the higher parts of the plateau from the Dovey–Rheidol watershed to a point east of Lampeter. These are very largely above 455 m (1500 ft). Figures 6.1 and 6.5 show the distribution of land above 425 m (1400 ft); because of the western location of the main watershed, ice forming in this area would flow mainly eastwards as Wirtz implied. West of the watershed it seems to have been largely confined to the upper Rheidol basin; the Tregaron moraine (152 m: 500 ft O.D.), may mark its maximum and the Ponterwyd moraines (225 m: 800 ft O.D.), a later readvance stage. In such conditions the upper Dovey probably also carried ice though no moraines to show its limits are known.

The strong remodelling of the upland surface by periglacial processes and the thickness of the slope deposits suggests that these represent a considerable part of the Last Glacial. Just how much is not known, in the absence of material for dating by pollen analysis or radiocarbon determinations. It may be that they formed in the latter part. Certainly a large part is later than the Ponterwyd moraines as they are well developed in the valley bottom further upstream.

The main features relating to the glaciation of the Cardigan Bay area are as follows. The Irish Sea ice is traceable by its erratics, but Welsh ice south of the Dovey basin is not. Furthermore, there is no evidence for depicting a continuous ice front to Welsh ice at any stage, and no evidence of a system of valley moraines. The only stage that is clearly defined by moraines is that of the cirque glaciation, and this affects only the very highest parts. Alongside this is evidence for prolonged periglacial conditions. These have greatly reduced any chance of tracing the direction of movement of Welsh ice by preferred stone orientation studies. 'Drifts' of local material almost always prove to be slope deposits; till, where it occurs, is invariably at considerable depth. The main hope of establishing a chronology in this situation would seem to be by pollen analysis and radiocarbon determinations, but at the moment none exist earlier than the Late Glacial, north of the Cardigan district.

## *Appendix*

### PREFERRED STONE ORIENTATION

In a deposit the particles have a characteristic arrangement known as its fabric. This arrangement may be studied in a horizontal plane, when the bearings or azimuths of the axes of the pebbles (or, under a microscope, sand grains) are recorded. When a representative number of azimuths of the long axes are plotted they may be found to group, sometimes closely, sometimes with a considerable spread, around a mean. This grouping is the preferred orientation of the stones in the deposit, the orientations which they have adopted in preference to all others.

The preferred stone orientation in ground moraine has been shown by Richter and Hoppe to be parallel to the direction of glacier flow in the case of existing ice and to the direction of striae, drumlin axes, etc. in the case of Pleistocene ice, by Holmes, Lundqvist, Wright and others. In solifluction deposits, the preferred stone orientation is parallel to the movement of the material, that is to the slope (Lundqvist). In both cases, where the stones are elongated they should give a clear indication of the direction of movement, which may be used to distinguish between solifluction deposits and ground moraine. In Fig. 6.4 the preferred stone orientation is consistently sub-parallel to the hill slope at the rear. Deposition by ice would not produce this pattern. Valley glaciers would cause it to be parallel to the valley axes, that is to the present streams. A regional ice-flow, say from the east, would cause a preference for an orientation broadly parallel to this flow.

Where azimuth readings only are used, these are usually shown diagrammatically on a rose diagram (Lundqvist, 1949; cf. Fig. 6.8). The number of readings in each 10° interval (351–360°, 1–10°, 11–20°, etc.), expressed as a percentage, is plotted on the centre ray for each 10° arc, (i.e. 355°, 5°, 15°, etc.), on a scale reading outwards from 0 per cent at the centre of the rose. Figure 6.8 demonstrates how effective this data may be in differentiating between two deposits. But these data from one site do not prove the deposit with the stone orientation diagrams *a* and *b* to be a head; it might be a second till laid down by ice from the east. Only data from sites over an area of changing slope orientations such as Figs. 6.4 and 6.6 can demonstrate a slope origin for a deposit.

In addition to the azimuth of the long axis of the stones, the dip angle of this axis may also be recorded and plotted on an equiareal polar graph (Krumbein, 1939) as shown on Fig. 6.9. The resulting pattern may throw light on the origin of a deposit. For instance, screes and solifluction deposits show a marked constancy of dip angle as well as direction so that the points on the polar graph show a close grouping near the dip angle of the bed (Fig. 6.9a). In a till the dip

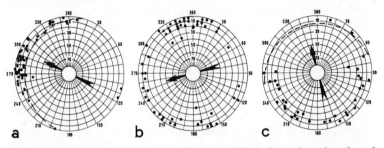

Fig. 6.9. Three-dimensional fabric diagrams. These show the azimuth and dip of the long axes in a sample of 50 stones in: (*a*) a solifluction deposit, and (*b*) a till; (*c*) shows the azimuth and dip of the maximum projection plane of 35 stones in a slopewash gravel. The arrow shows the slope direction and the curved dashed line the dip of the bedding.

directions and amounts are much more varied (Fig. 6.9*b*); the pattern (Fig. 6.9*a*), shown by the Morfa-bychan deposits is a further reason for regarding them as slope deposits. Where slope orientation and expected ice-flow direction coincide the dip data may help to differentiate between them.

In waterlaid gravels the azimuth and dip of the long axis are very variable so that a measure of the direction of imbrication is the best method of showing the direction of flow during deposition. In some places such as the Nant Iago valley near Cader Idris (Watson, 1969), or the upper Rheidol valley (Watson, 1967b), gravels interbedded with solifluction deposits dip downslope with the latter, suggesting deposition by slope wash. Figure 6.9*c* shows the direction and amount of the maximum dip of their maximum projection planes (Krumbein, 1939); this is upslope, confirming deposition by water flowing downslope.

REFERENCES

BROWN, M. J. F. (1967) 'A new radio-carbon date for Wales', *Nature, Lond.* **213**, 1220–1.

CHARLESWORTH, J. K. (1929) 'The South Wales end-moraine', *J. geol. Soc. Lond.* **85**, 335–8.

GRIFFITHS, J. C. (1940) 'The glacial deposits west of the Taff, South Wales', unpublished Ph.D. thesis, Univ. London.

JOHN, B. S. (1967) 'Further evidence for a Middle Würm Interstadial and a Main Würm Glaciation of southwest Wales', *Geol. Mag.* **104**, 630–3.

JONES, O. T. (1965) 'The glacial and post-glacial history of the lower Teifi valley', *J. geol. Soc. Lond.* **121**, 247–81.

JONES, O. T. and PUGH, W. J. (1935) 'The geology of the district around Machynlleth and Aberystwyth', *Proc. Geol. Ass.* **46**, 247–300.

KEEPING, W. (1882) 'The glacial geology of central Wales', *Geol. Mag.* **19**, 251.

LEWIS, W. V. (1949) 'The function of meltwater in cirque formations: a reply', *Geogrl Rev.* **39**, 110–28.

MANLEY, G. (1951) 'The range of variation of the British climate', *Geogrl J.* **117**, 43–68.

MITCHELL, G. F. (1960) 'The Pleistocene history of the Irish Sea', *Advmt Sci.* **17**, 313–25.

MITCHELL, G. F. (1962) 'Summer Field Meeting in Wales and Ireland', *Proc. Geol. Ass.* **73**, 197–214.

SISSONS, J. B. (1964) 'The glacial period', in the *British Isles: A systematic geography*, ed. J. W. Watson and J. B. Sissons, pp. 131–51.

SYNGE, F. M. (1961) 'Drifts in west Wales', Rep. 2, *Welsh Soils Discussion Group*, pp. 15–18.

TRICART, J. (1963) *Géomorphologie des régions froides*, Paris.

WASHBURN, A. L. (1947) 'Reconnaissance geology of portions of Victoria Island and adjacent regions', *Geol. Soc. Am.* Mem. 22.

WATSON, E. (1960) 'Glacial landforms in the Cader Idris area', *Geogr.* **45**, 27–38.

WATSON, E. (1962) 'Glacial morphology of the Tal-y-llyn valley', *Trans. Inst. Br. Geogr.* **30**, 15–31.

WATSON, E. (1965) 'Periglacial structures in the Aberystwyth region of central Wales', *Proc. Geol. Ass.* **76**, 443–62.

WATSON, E. (1966) 'Two nivation cirques near Aberystwyth, Wales', *Biul. peryglac.* **15**, 79–101.

WATSON, E. (1967b) 'The periglacial element in the landscape of the Aberystwyth region', unpublished Ph.D. thesis, Univ. Wales.

WATSON, E. (1968) 'The periglacial landscape of the Aberystwyth region', in *Geography at Aberystwyth*, ed. E. G. Bowen, *et al.*, University of Wales Press.

WATSON, E. (1969) 'The slope deposits in the Nant Iago valley, near Cader Idris, Wales', *Biul. peryglac.*—in the press.

WATSON, E. and WATSON, S. (1967a) 'The periglacial origin of the drifts at Morfa-bychan, near Aberystwyth', *Geol. J.* **5**, 419–40.

WILLIAMS, K. E. (1927) 'The glacial drifts of western Cardiganshire', *Geol. Mag.* **64**, 205–27.

WIRTZ, D. (1953) 'Zur Stratigraphie des Pleistocäns im Westen der Britischen Inseln', *Neues Jahrb. für Geol. und Paläont.* **96**, 267–303.

## Appendix

HOLMES, C. D. (1941) 'Till fabric', *Bull. geol. Soc. Am.* **52**, 1301–54.

HOPPE, G. (1953) 'Nagra iakttagelser vid Isländska jöklar, Sommaren 1952', *Ymer Stockh.* **4**, 241–65.

JOHANSSON, C. E. (1965) 'Structural studies of sedimentary deposits', *Geol. Forens. Forhand.* **87**, 4–61.

KRUMBEIN, W. C. (1939) 'Preferred orientation of pebbles in sedimentary deposits', *J. Geol.* **47**, 673–706.

LUNDQVIST, G. (1949) 'The orientation of the block material in certain species of flow earth', *Geogr. Annlr*, **31**, 335–47.

RICHTER, K. (1936) 'Gefüge studien im Engebrae, Fondalsbrae und ihren Vorlandssedimenten', *Zeitschr. für Gletsch.* **24**, 22–30.

SED. PET. SEMINAR BLOOMINGTON (1965) 'Gravel fabric in Wolf Run', *Sedimentology*, **4**, 273–83.

WRIGHT, H. E. (1957) 'Stone orientation in Wadena Drumlin Field, Minnesota', *Geogr. Annlr*, **39**, 19–31.

# CHAPTER SEVEN

# The Upper Wye and Usk Regions

Colin A. Lewis, B.A., Ph.D.

During the Pleistocene the mountainous interior of Wales was affected only by local ice. This developed over a number of different mountain centres, whence it moved towards the coasts and into the eastern borderlands.

The glaciation of Brecknock was described by Professor Edgeworth David as long ago as 1883. He showed that ice from the Brecon Beacons/Fforest Fawr uplands once passed down the coalfield valleys to debouch on to the Vale of Glamorgan around Cardiff. This important finding ended any doubts about the origin of the drift which covers much of the region: Ramsay (1846) had noted that the drift around Kington 'is not less than 84 feet deep' and that it was composed of 'the debris of the neighbouring rocks' (p. 325), but he believed that this drift had been formed by the action of an erosive sea; 'the loose drifty deposit...being...but the dregs of the matter removed from the rising land [by the sea]' (p. 327).

In 1894 Mellard Reade wrote about the cirque-moraines that exist in the upper portion of Cwm Llwch in the Brecon Beacons and within ten years the study of glacial features in Breconshire had become fashionable. Many who pursued this taste confined their attentions to the river valleys, which were, after all, easier of access than the area Reade had described. Whilst much of the work which appeared in the early 1900s is of little interest, there was one most important publication. This was entitled 'Notes on glacial action in Brecknockshire and adjoining districts', and it was the work of F. J. Howard (1903–4). Howard noted the distribution of erratic material and of striae and, on this evidence, was able to show that ice from northerly sources (which were the plateaux of Plynlimon and its southern extensions), moved down the valleys of the Wye and Usk, although 'the range of mountains, from Hay to the source of the Tawe on the west, formed an impassable barrier...except at one point—where the escarpment is broken by the Usk gorge and the adjoining passes of the Bwlch and

147

Dinas Castle'. In addition Howard stated, correctly, that ice from the Wye valley moved up the Llynfi valley and thereby invaded the Usk.

Howard and his predecessors wrote in terms of one period of glaciation, but ere long the number of glaciations multiplied. Pocock (1925) thought there were two major glaciations of the area, as well as 'two limited and recent glaciations'. 'It seems probable that these last correspond with different stages of the Würm period, the more general glaciation with the Riss period and the maximum with the Mindel period' (p. 38). Charlesworth (1929), in his classic paper on the 'South Wales end-moraine', attempted to define the limits of the most recent glaciation of the area. Unlike Pocock, he believed in only two periods of glaciation, whilst Dwerryhouse and Miller (1930), writing in the following year, discussed only one glaciation. The Rev. B. B. Clarke (1936–7), who studied the Black Mountains and their Herefordshire foothills, delimited two glaciations, Older and Newer, without ascribing them to any particular stage of the Alpine glaciations (which Pocock had attempted to do). But the last definition of glaciations had yet to come. In 1940 Pocock described the drift and river terraces of the Wye, defining three terraces at Glasbury. Of these he believed the lowest two to be Würm and the highest one Mindel.

Although all these divisions and classifications of the drifts make interesting reading, they contribute relatively little to the true determination either of the number of glaciations or of their age. Pocock was obviously influenced by what had been found in the Alps and sought, perhaps too readily, to find parallels in Wales. The important contributions, partly because they were based on fieldwork in suitable areas, were those of Charlesworth and Clarke. Although they may appear cautious in only defining two glaciations, in comparison with Pocock's four, they were probably correct, and we can today find evidence for both glaciations.

### Regional relief

Brecknock and the adjoining areas are composed of a series of south and east flowing valleys which are entrenched in the middle and high plateaux of Wales (Brown, 1960). The most important valleys, by virtue of their size, are those of the Wye and Usk. In addition, the upland of Mynydd Eppynt, which reaches 470 m (1 560 ft), is divided from the central mountainous core of Wales by the Irfon–Ithon depression. Both the erosion surfaces and the major portions of all the valleys date from the Tertiary period and this pre-existing relief greatly influenced the later formation and movements of ice in the Pleistocene.

## THE GLACIATIONS

Almost the whole of our area bears evidence of glaciation in some form or another and the lowlands are generally covered with glacial deposits. These divide, on a morphologic basis, into the discontinuous rounded spreads of drift that exist in eastern Monmouthshire and in many areas of south-west Herefordshire, and the fresh landforms of the middle and upper Usk and Wye valleys. The divide between the two areas is very distinct, coinciding with the Penpergwm moraine, i.e. the area around Llangattock nigh Usk (Strahan and Gibson, 1927, pp. 103–6), in the Usk valley* and the Stretton Sugwas moraine, near Hereford (see Chapter 8), in the Wye. Both moraines, and the material up-valley from them, appear fresh; outside them the deposits are generally rounded and fragmentary. It therefore appears that there has been a twofold division in the glaciation of this part of Wales, with Older and Newer stages.

## THE OLDER STAGE

The course of the Older stage of glaciation is a matter for speculation except where Older deposits are seen outside the limits of the Newer glaciation. Occasional patches of weathered drift in the Vale of Gwent and south-west Herefordshire, which lie outside the fresh deposits of the more recent glaciation, indicate that the Older stage was characterised by more extensive ice-cover than existed in the Newer. The evidence so far discovered of the Older stage is so fragmentary that it is impossible to describe its course or extent in any detail.

## THE DIVISION BETWEEN STAGES

There is, as yet, no evidence to show the nature of the period that divided the two stages. Although it is likely that they each comprised separate glaciations, it cannot at present be proved that such was the case.

* The Geological Survey memoir on Monmouth and Chepstow (Welch and Trotter, 1961) states that 'the Usk glacier penetrated as far east as Raglan' (p. 130). Williams (1968) has shown that there is an end-moraine at Kemeys Commander in the Usk valley and this is probably the southern extension of the Raglan deposits. If this moraine proves to be of Newer stage age it must, almost certainly, be the outer limit of that stage, in which case the Penpergwm moraine is a retreat phase of that glaciation.

THE NEWER STAGE

The Newer stage of glaciation (Fig. 7.1) was on a smaller scale than that of its predecessor and it has left extensive deposits and well-marked landforms to testify to its work of deposition in the middle Usk and Wye valleys. By comparison, in the uplands north and west of the Irfon–Ithon depression, it has scoured the landscape, so that the land is often poor and rocky, although sometimes mantled with periglacial deposits.

The main ice-centre during the Newer period lay on the high plateau that forms the core of mid-Wales. This plateau is at an altitude of 450 to 510 m (1500 to 1700 ft) and is open to the westerly winds which, as Manley (1961) has shown, were the main precipitation bearing winds of Newer drift times. An ice-cap apparently developed over the most isolated section of the plateau, located between Llangurig, Tregaron, Rhandirmwyn and Abergwesyn. Names of physical features such as Drum yr Eira (Hill of the Snow), testify, even today, to the isolated and inhospitable nature of this upland. Unlike the Older stage, ice-caps did not develop over the Black Mountains, although some ice fed off the Brecon Beacons/Fforest Fawr range into the Usk. Although the pre-existing relief was important in guiding the development of the early glaciation, it was even more important during the Newer stage. Even at its maximum the ice-flow followed, with few exceptions, the line of the major valleys.

*The Wye valley (Stretton Sugwas and Bredwardine Stages)*

As the glaciation developed, ice from the mid-Wales plateau poured down the valleys of the Irfon, Cammarch, Dulas, Chwefru and upper Wye to gather in the Irfon–Ithon depression. The main outlet from this depression was the Wye valley itself, but below Builth Wells the Wye runs through a deep and narrow gorge for a distance of over 13 km (8 miles). This formed a bottleneck which prevented rapid dispersal of ice from the depression, so that it was forced to seek alternative routes of escape. The critical altitude which the ice had to reach before it could overtop the Wye gorge and flow along the open ground east of the gorge, was 330–360 m (1100–1200 ft). This was 45–75 m (150–250 ft) more than the watershed at the Sugar Loaf, which divides the Wye drainage from that of the Towy. It was only natural, therefore, that as ice pressure developed in the depression some ice sought, and found, an outlet into the Towy. This is evidenced by the presence of erratics from the Builth Wells area, which

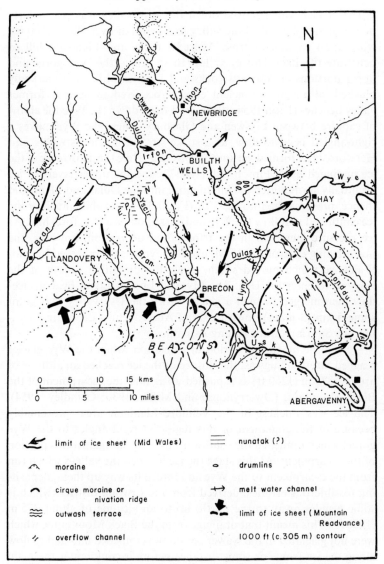

Fig. 7.1. The Newer stage glaciation of the upper Wye and Usk area

have been found in the Towy valley. At the same time the eastern section of the ice sheet moved east to escape through some of the valleys of Radnor Forest and through valleys further north (see Chapter 8).

The glacier which moved down the Wye valley was particularly powerful and soon filled the valley to a depth of over 180 m (600 ft), whereupon it spilled across Aberedw rocks and Llandeilo hill to penetrate the Bach Howey valley. In so doing, the Wye gorge and upper portions of Aberedw rocks and Llandeilo hill were markedly plucked, although insufficient erosion occurred to destroy former valley benches (Rice, 1954, 1957a, b). Lower down the Wye the elbow of capture between Llyswen and Glasbury, where the Wye makes a right-angled bend to flow, as a strike stream, north-eastwards into Herefordshire, formed yet another obstacle. Here the main force of the ice impinged upon Pipton Hill (150370) and the upper layers of ice eventually surmounted the hill to move into the Dulas and Llynfi and thence, presumably, into the Usk (Fig. 7.3). Meanwhile much of the ice continued to follow the Wye and passed on into Herefordshire. As ice pressure increased in the Irfon–Ithon depression, however, the north-facing escarpment of Mynydd Eppynt was submerged by mid-Wales ice which moved south, down the dip slope valleys of the Eppynt, to enter the Usk. The Mynydd Eppynt/Llynfi-Beacons ice-flow then merged to send a glacier down the Usk valley to a point some 5 km (3 miles) downstream of Abergavenny, depositing fossi-liferous Silurian erratics along its path.

During the maximum period of glaciation, when the Wye glacier reached Stretton Sugwas, the surface of the ice reached an altitude of around 375 m (1250 ft) as it pushed up against the escarpment of the Black Mountains (Dwerryhouse and Miller, 1930; Grindley, 1954). The ice was thickest in the Velindre–Llanigon area, presumably because of the alignment of this region at right-angles to the Wye gorge, which funnelled the southward moving ice against this portion of the escarpment. At this stage the ice invaded the valleys which run from the escarpment to the Wye and forced its way up them, deposit-ing fossiliferous erratics, derived from the Erwood district (which is some 6 miles away to the north) up to an altitude of about 375 m (1250 ft). This meant that drainage from the Black Mountains, which were not glaciated in the Newer period, was impeded and had to flow along the ice-front for some distance before it could enter the main glacial drainage channels. The east-flowing channel on the common near Twyn y Beddau (240387) and the even finer meltwater channel which cuts across the spur between the Digedi and Cilonw valleys (at 230375) were utilized, although they might have been formed origi-nally in the Older stage of glaciation. East and west of this area the level of the ice-sheet declined so that ice was unable to spill down the Monnow or Escley valleys, although it did pass over the divide at Castell Dinas (180300) to pour down the Rhiangoll.

Throughout the glacial maximum the Radnorshire valleys, between Llanbedr and Glascwm hills and the Wye, were submerged by the ice-sheet. The direction of movement of the ice was largely parallel to the trend of the valleys, whilst the higher parts of the Begwns (410 m: 1361 ft) were only just covered. This is reflected in the thick drift mantle and multiple moraines of the valleys (such as that of the Arrow and its easterly continuation, the Bach Howey), and the thin and stony drifts found on the upper Begwns. Many of the deep valleys which exist between the Begwns and the Wye may have been formed during this and subsequent stages of the glaciation as subglacial drainage poured down-slope towards the main valley floor. On the other hand, pre-existing valleys, some of which were at least 60 m (200 ft) deep and completely gorge-like, were infilled with glacial detritus so that they now appear mature.

As the ice-sheet retreated from Stretton Sugwas to Bredwardine (see Chapter 8) it is reasonable to assume that the level of the ice further upstream fell, although there is little evidence for this. As the retreat continued for an unknown distance, so the ice-margin moved upstream from Bredwardine only to readvance again to form a moraine at Hay-on-Wye. At this stage (the Hay stage), the ice-sheet of earlier times was reduced to little more than a swollen valley glacier. It still covered the flanks of the Begwns and attempted to re-enter the Bach Howey, forming drumlins around Llan Bwch-llyn (120465). The ice also covered the eastern edge of the Eppynt and spread, as a lobe, through the Llynfi Basin and into the Usk. As the decline continued a number of complicated advances and retreats occurred in the Llynfi basin before the glacier finally retreated to the confines of the Wye. Before we can discuss these complications, however, we need to know more about the history of the Eppynt and the Usk.

## Mynydd Eppynt and the Usk valley

During the Newer stage some ice entered the Usk from the main cirques of the Fforest Fawr and Brecon Beacons, as is shown by the morainic deposits in such valleys as the Tarell, where a glacier emerged from the spectacular cirque of Cwm Cerrig-gleisiad. Nevertheless, the main ice-streams that entered the Usk valley apparently came from northerly sources and entered the valley after flanking Mynydd Eppynt.

The Eppynt escarpment varies in height from 470 m (1550 ft) to as little as 300 m (1000 ft) at the watershed between the south flowing Cilieni and the north flowing Dulas (885385). Except during the maximum of the Newer stage it appears, from the paucity of glacial deposits and from the prevalence of solifluction spreads, that the highest part of the escarpment, from Tafarn-y-mynydd to the Drover's Arms, may have remained above the ice. This section (the 'nunatak' area shown on Fig. 7.1) lies entirely at an altitude of above 420 m (1400 ft), while individual eminences rise to over 450 m (1500 ft). When the ice in the Irfon–Ithon depression rose to an altitude of as little as 330 m (1100 ft) it was able to overflow the escarpment at the head of the Cilieni, and by the time it reached 360 m (1200 ft) it was able to escape from the depression via Cwm Owen (020440) and down the Nant yr Offeiriad valley to the Wye. As the glaciation developed, ice pressure in the depression increased until the ice front attained an altitude of almost 420 m (1400 ft). This enabled ice to cross Mynydd Bwlch-y-Groes in the west, to spill down Cwm-y-Glynn, Nant Gwennol and the tributaries of the Bran into the Towy system, and to surmount the watershed between the north-flowing Duhonw and the south-flowing Honddu in the east.

As the glaciation proceeded towards its maximum the glaciers in the Eppynt valleys coalesced with ice from the Wye area and from the Brecon Beacons/Fforest Fawr uplands to form a continuous ice-sheet stretching from Llanbedr hill in Radnorshire to Mynydd Illtyd south of the Usk. This *mer-de-glace* extended up the Usk and into the Towy area leaving widespread deposits over the whole region, forming in its easterly extension, the Usk glacier.

## The Usk glacier

Although the main ice-sheet ended against the Black Mountains/ Mynydd Llangors/Brecon Beacons escarpment, it sent powerful glaciers down both the Usk and Wye valleys. At its maximum the Usk glacier covered the whole valley west of Crickhowell to an altitude of about 300 m (1000 ft), although the depth of ice decreased eastwards as evidenced by deposits on the valley sides. The glacier was sufficiently large for it to bifurcate at Crickhowell. The lesser branch moved across the lower Grwyne Fechan valley to enter the Grwyne Fawr/Cwm Coed-y-cerrig trough at Llanbedr (Fig. 7.2). It then forced its way through the trough to emerge in the lower portion of the Vale of Ewyas, where it formed the magnificent Llanfihangel–Crucorney moraine. The main Usk glacier, during this stage, ended at Penpergwm.

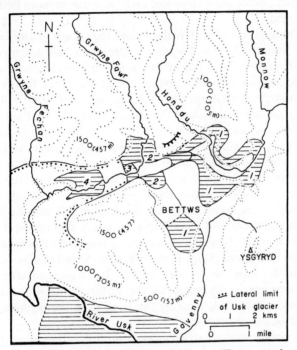

Fig. 7.2. The Cwm Coed-y-Cerrig/Grwyne Fawr trough. The shaded areas are end or readvance moraines and the numbers, 1–4, refer to the stages of glaciation of the trough as described in the text

We can detect little of the minor waxings and wanings of the main Usk glacier until it reached a stage of marked retreat, but deposits in the Grwyne Fawr/Cwm Coed-y-cerrig trough indicate that the maximum stage of glaciation was followed by considerable variations in the size of the glacier.

As the glacier moved through the trough it deposited a lateral moraine across the mouth of the Grwyne Fawr in the neighbourhood of Patrishow. This feature, plastered on bedrock, extends to an altitude of around 300 m (1000 ft), thereby indicating the size of the glacier. At the same time that this lateral moraine was being formed the glacier extended into the Vale of Ewyas, where it deposited its terminal moraine around Llanfihangel–Crucorney (Stage One). At this stage the glacier also sent an offshoot through the Bettws valley. This offshoot just reached the right bank of the Gavenny and, in its stagnation, deposited a kame and kettle moraine between Bettws and the Gavenny.

As the ice-sheet of mid-Wales waned the trough glacier began to retreat from the Vale of Ewyas. Its retreat was halted, temporarily, when it reached the shade of the trough, and it deposited a stony jumbled moraine below Coalpit hill (Stage Two). The Bettws glacier was isolated at this stage although, during the Coalpit hill stand, some ice managed to push up the valley for almost 1 km (half a mile). However, this advance soon proved abortive, and the ice was left to stagnate *in situ* giving rise to another small area of kames and kettles.

Following the Coalpit hill stand, the glacier retreated to a point immediately west of the junction of the Bettws valley and the trough, where it deposited a well-marked terminal moraine (Stage Three). The marshy area between this and the Coalpit hill moraine is probably the last remains of a small lake formed at that period. Like the previous moraine, Stage Three was only a temporary halt, and the ice retreated to Dyffryn (Stage Four), where it formed another terminal moraine. By now the ice was well on the wane and after a few minor resurgences (indicated by the surface form of the Dyffryn moraine), it retreated, vacating the trough for the confines of the Usk valley.

While its distributary trough glacier pushed as far east as the Vale of Ewyas, the main Usk glacier stretched down valley to the neighbourhood of Penpergwm, where its terminal moraine spreads across the valley floor from Penpergwm to Gilwern. The outermost part of the moraine is composed of marly material which has the typical shape of an end-moraine formed at a glacier snout. Up-valley, above Abergavenny, the debris assumes the form of kame and kettle moraine, and much of the deposit is gravelly. These differences of material indicate that the glacier changed in character from an advancing feature to one in which the snout, at least, was subject to considerable stagnation. It may well be that, in this difference of material, we see reflected Stages One and Four of the trough glacier, with the intermediate stages hidden in the general mass of debris. The most obvious area of marly terminal moraine is best seen near the roundabout which links the Heads of the Valleys road with the A40. Kames and kettles are well displayed in the grounds of the new Abergavenny Hospital.

Although the initial waning of the glacier was on a minor scale, the accelerated melting, evidenced by the kame and kettle moraine, increased to such an extent that the ice-front retreated up-valley from Gilwern to Crickhowell before halting. The gravelly moraine at Crickhowell, stretching in arcuate fashion across the valley to Llangattock, is small in relation to the Abergavenny–Penpergwm moraine. It probably represents a temporary halt in the recession of the glacier, rather than any advance. As the Usk glacier retreated towards Crickhowell it decreased in size so that it was no longer able to send its former distributary glacier down the trough. From Crickhowell the glacier retreated again, but a later readvance to the gorge section of the Usk between Llangynidr and Buckland House (near Talybont) formed the Llandetty moraine (Lewis, 1966b) which is partly a kame and kettle deposit. Kettle holes, now converted into ornamental ponds, are well displayed in the grounds of Buckland House.

During this Llandetty stage, the Usk glacier penetrated the Caerfanell valley, south of Talybont, as far as the site of the present reservoir dam. The great thickness of sandy material on the valley floor beyond the glacial limit (exceeding 50 m; 160 ft; North, 1955) was probably deposited in a proglacial lake.

During the early stages of glaciation it is likely that ice from the Wye passed up the Llynfi valley and into the Usk. As the glaciation intensified ice from westerly sources forced its way down the Usk, thereby obstructing the passage of Wye ice, some of which spilled over the pass at Castell Dinas (180300) to enter the Usk via the Rhiangoll. When the level of ice in the Irfon–Ithon depression decreased, the flow of ice over the western part of Mynydd Eppynt ceased. This ended the main flow of ice from westerly sources down the Usk and left the Usk valley open to invasion by Wye–Llynfi ice.

The portion of the Usk that was most vulnerable to penetration by Wye–Llynfi ice lay between Brecon and Talybont, and had no high ground to protect its left bank. In addition, this was the section that was most isolated from the Eppynt/Bwlch y Groes glaciers, and from those of the Fforest Fawr range and the Tarell on which the Usk glacier relied for its nourishment. As the flow of ice across the Eppynt declined Wye ice was able to invade this section of the Usk. The Wye ice, at this period, spread across the Llynfi basin in a great lobe which apparently ended at the Hay moraine to the north and the Llandetty moraine to the south. The Hay and Llandetty moraines are therefore contemporaneous.

As well as reaching Hay and Llandetty, Wye ice moved westwards across the lower slopes of the eastern Eppynt to mingle with Beacons ice in the area around Bishops Meadow (055300), blocking the lower Honddu valley. This caused a lobe of Honddu ice to diverge westwards, via the gap at Penoyre between Pen y Crug and the southern edge of the Eppynt, to enter the Usk at Aberyscir (Fig. 7.1). Here it came into contact with the retreating Usk glacier, and a large moraine, cut through by the Bran, formed between the two glaciers. This was the furthest point up the Usk valley that was reached by 'eastern' ice. As the lobe retreated it did so unevenly, depositing spectacular terminal and lateral moraines in the Penoyre/Cradoc area. The golf course at Penoyre (020310), for instance, has been constructed almost entirely on the gravelly material of the glacier's right-bank lateral moraines.

The history of both the Usk and Wye glaciers from the Hay/ Llandetty stage until their final disappearance is one of uneven retreat, as is instanced by the formation of the Llanfrynach/Groesffordd moraine. The retreat of the Usk glacier above Aberyscir was marked by a number of temporary halts, during which a confused series of end-moraines was formed across the valley.

The retreat of the main ice sheet was followed by a later readvance of local glaciers in the Senni, Trewern, Crai, Hydfer and Tarell valleys which led to the formation of multiple moraines in those valleys, although the readvance was insufficient for ice to descend once more the main Usk valley. Fine terminal moraines cross that valley, however, at the confluences of the major tributaries and exceedingly well-developed terminal moraines exist, for example, between Defynnog and Sennybridge. During the retreat of this late mountain glaciation moraines were formed around Blaenau-isaf in the Hydfer valley, Onen-fawr ford in the Crai and below Y Gaer in Cwm Trewern, as well as around Heol Senni in the Senni. At all these points morainic debris is deposited across the valley floors and the streams run through channels incised into the debris. This is particularly well seen downstream from Heol Senni. The Tarell glacier advanced almost to Brecon and its western edge spread onto Mynydd Illtyd, where it coalesced with ice from the Senni valley to form a piedmont glacier. During the retreat of this ice a complicated series of morainic ridges was formed on the plateau, one of which enclosed a peat bog in which there is evidence of organic deposition starting in pollen Zone Ia. This provides us with a reasonable idea of the age of this mountain glaciation, for it must have ended by Zone Ia to allow organic deposition to take place at that time.

During this mountain readvance the main ice-sheet began to increase in size once more, and the last abortive effort of mid-Wales ice to reach the Usk resulted in the deposition of spectacular moraines around Pentre'r-felin (920304) in the Cilieni valley. Further to the east a moraine was formed immediately above the Pont-faen gorge in the Yscir-fawr valley (010340), and, as the ice later retreated from that valley, lobate drift ridges developed in the col between the Yscir-fawr and the Honddu (995395). In the Honddu itself the readvance led to the formation of a large moraine near Llandefaelog Fach church before the ice began its final retreat from the Eppynt. During this retreat the Nant yr Offeiriad valley function-ed as a meltwater channel (Dwerryhouse and Miller, 1930; Thomas, 1959) and fluvioglacial material was flushed through it to cover extensive tracts of the valley floor around Gwenddwr. The incised nature of the stream, particularly from Crickadarn to Erwood, is probably partly due to erosion caused by this meltwater.

*The Wye valley and Llynfi basin (Hay stage to the disappearance of the ice)*

Although, at the maximum of the Hay stage, Wye ice covered the whole of the Llynfi basin and invaded the Usk, this was only a tem-porary event before the ice retreated to the confines of the Wye valley. Four stages of this retreat, each evidenced by morainic deposits, can be seen in the Llynfi basin (Fig. 7.3).

Stage One is evidenced by an inconspicuous mass of drift which is banked against the hill behind Llanfihangel Tal-y-Llyn and against the eastern flanks of Yr Allt. There is no distinguishable continuation of this feature across the floor of the Llynfi basin which, as the ice retreated eastwards, was occupied by proglacial Lake Llangors. The shorelines of this lake, which are followed in part by both the Pennorth–Bwlch and Llangors–Bwlch roads are at an altitude of about 190 m (630 ft). The lake overflowed both at Pennorth, where the railway line follows the floor of the outlet, and at Bwlch. The narrow, twisting and wooded channel, across which the A40 road is carried by an embankment 180 m (200 yd) west of Bwlch, was the main outlet through which lake waters entered the Usk. In the

Pennorth outlet there are clay deposits, that may date from that period.

During retreat from the Usk to Stage One quantities of dead ice became isolated on the plateau between Felin-fach and Brecon. This

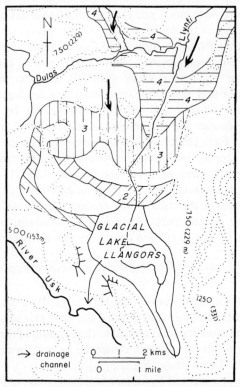

Fig. 7.3. The Llynfi basin. The shaded areas approximately delimit the glacial deposits referred to in the text as stages 1–4 in the glaciation of the basin

had been an important means of entry to the Usk, the Wye ice forcing its way through the col now followed by the A438 road. Evidence of dead ice here is provided by gravelly lobes which point towards Brecon, one of which (formed, however, by Brecon Beacons ice), is clearly seen near the secondary schools playing fields above the town (055297). In addition, some of the melt-water from the ice during Stage One poured down the northern flanks of Yr Allt, to enter the

17 Llyn Cwm Llwch, Brecon Beacons. Notice the moraine which encloses the lake, and the fact that the lake lies in the most sheltered portion of the valley head, away from the sun's rays. Older and much less prominent moraines may be seen on the bottom right, almost buried by more recent solifluction debris.

18 Altiplanation terraces, Torpantau, Brecon Beacons, looking south.

19 Residual hillock and block field, Torpantau, Brecon Beacons.

20 Solifluction terrace, Slaettaratindur, The Faeroes.

Dulas valley, forming what were possibly subglacial valleys which may be seen on either side of A438 near Felin-fach.

Stage Two is shown by the feature which crosses the Llynfi north of Llangors, whilst the Stage Three moraine exists near Tre-Walter (129297). It is probable that glacial Lake Llangors ceased to overflow into the Usk during Stage Two, and it probably drained englacially through the ice to the Wye.

Whilst all the moraines in the Llynfi basin were formed by re-advances of the Wye glacier following retreats, the readvance of Stage Four was particularly prominent. This was responsible for the deposition of the material that blocks the Llynfi valley south of Talgarth. Prior to Stage Four Wye ice had surmounted Pipton hill, crossed the Dulas valley, moved over Llanfilo hill below the village and entered the Llynfi basin. In Stage Four the invading ice was split in three. One section passed into the Dulas valley by following the low ground from Llyswen to Trephilip (127346). This glacier moved down the Dulas to meet the southward moving ice that had entered the lower Llynfi as two glaciers, one crossing Pipton hill at Coldbrook, the other entering the Llynfi from Pipton itself. Between the Dulas and Llynfi glaciers there developed a thick moraine on which Bronllys and its sanatorium are now located. The main Llynfi glacier pushed up-valley almost to Trefeinon, whilst its lateral moraines spread down the east side of the valley and alongside the Wye to Hay. The spectacular drainage channels between Glasbury and Tregoyd (Lewis, 1966b) were probably active at this period, although they may have been formed much earlier.

Following Stage Four the ice retreated, for the last time, from the Llynfi basin and became confined to the Wye valley. Initially the Wye glacier retreated to the Llyswen moraine, downstream of which, as far as Pipton, there are great thicknesses of outwash. The Llyswen moraine is largely a kame and kettle deposit, as is best seen in the grounds of the Dderw, where some of the kettle holes are flooded to form small fish ponds. Later the ice retreated to the Llanelwedd/Alltmawr moraine* from which outwash terraces run downstream. No further moraines exist on the valley floor below Builth Wells, although lateral moraine and kame material occurs across the mouth of the Edw and in the valley between Llanelwedd and Hundred House, where it has caused local diversion of drainage.

The Irfon–Ithon depression, during all these stages of retreat, was

---

* This moraine is located between Llanelwedd and Allt Mawr (*c.* 065515).

6

occupied by a mass of largely stagnant ice, the meltwater from which was responsible for the formation of the 'in and out' channels which exist east of the Builth Wells to Llandrindod Wells road north of Llanelwedd. The Wye glacier deposited drumlin-like drift between Builth Wells and Newbridge before retreating to the Newbridge–Rhayader gorge, where it formed the Newbridge moraine. A further moraine exists immediately upstream from Llanwrthwl, whilst a complex series of moraines is located around Rhayader. These formed when the retreating Wye glacier split in three and the previously united Wye, Elan and 'Glan-Llyn' glaciers, each issuing from the uplands around the head of the Elan valley, became separate entities. The Elan glacier deposited its moraine across the valley floor immediately west of Rhayader, and this has led to the southerly diversion of the Elan river.

When at its maximum, the Wye glacier poured off the plateau of central Wales into the valley below Llangurig, and some ice forced a way up the Marteg valley into the St Harmon basin. A series of crescentic moraines, formed around the edge of this invading ice, exists between St Harmon village and the Wye. Far larger moraines exist further south in the other east–west aligned valleys east of the Wye. The great outwash or kame moraine of the Dulas valley, less than 1 mile east of Nantmel, is particularly spectacular and is at present being exploited for its sand and gravel.

### THE CIRQUE GLACIATION

During the maximum development of the Newer stage glaciation invading mid-Wales ice intermingled with glaciers from the Fforest Fawr/Brecon Beacons uplands. Nevertheless this invading ice-sheet did not enter the Brecon Beacons valleys, presumably because it was repelled by local ice. Pocock (1925) believed that the terraces flanking either side of the Beacons valleys were lateral moraines of local glaciers that fed into the main ice-sheet at the Newer stage. Lewis (1966b) queried Pocock's statement, believing these terraces to be of periglacial (solifluction) origin, and he suggested that only local cirque glaciers existed on the Beacons at that time. Probably both authors are partially correct, in that we now know that there was a long cold period (Zone Ia and later) following both the retreat of the mid-Wales ice-sheet from the flanks of the Beacons and the later re-advance into the Usk valley at Sennybridge and Brecon (and possibly elsewhere) of local Fforest Fawr/Brecon Beacons glaciers. This period, and the other cold periods that followed it, was probably long enough to enable much of the morainic material of the mountain re-advance to be reworked by solifluction and for new material to be deposited on top of the remnants of the lateral moraines, giving them the appearance today of true solifluction terraces.

Within all the large cirques of the Beacons (Fig. 7.1) there are two sets of moraines (Lewis, 1966a). The earlier group, which extend further down the valleys than their successors, are often difficult to appreciate, forming gentle swells that arc across the valleys. The inner, and therefore more recent moraines, are quite different, rising as arcuate or linear ridges up to 18–21 m (60–70 ft) above their sur-roundings. Those of Cwm Llwch, first described by Symonds in 1872, encompass a small lake in the depression between the moraines and backwall, whilst smaller lakes once existed behind moraines in many other valleys, as indicated by the lacustrine deposits (see also Howard, 1901). In some instances the most recent 'moraines' developed as scree material from the backwall slid down the surface of a snowbed to form a *nivation ridge* or *protalus* (Behre, 1933; Russell, 1933). Particu-larly well-developed nivation ridges exist under Fan Fawr at the head of Nant Penig, and under Fan Hir (Thomas, 1959, p 111) in the Carmarthenshire Fans.

The Black Mountains, unlike the Brecon Beacons, do not appear to have supplied ice to the Newer stage ice-sheet, which enveloped the mountains on three sides. Only one small glacier, located at the head of the Olchon valley, existed on the mountains in Newer stage times, and even that glacier was no more than a few hundred metres (yards) long, as its moraines testify. There were, however, three large snow-beds (or possibly diminutive glaciers) on the west side of the Honddu valley, near Capel-y-ffin. Each of these engendered the formation of a small moraine, or nivation ridge. The remainder of the Black Mountains was apparently free of ice, with the ground bared to solifluction, which gave rise to the smooth slopes and valley terraces of the area.

Outside the limits of the cirque-glaciers periglacial landforms developed. The main periglacial landforms in our area are alti-planation terraces, solifluction terraces and screes.

When snowbeds rest against a hillside, erosion often occurs between the snowbed and its backwall. This leads to the gradual retreat of the backwall. At the same time meltwater from the snowbed saturates the ground under and in front of it. This enables the waste mantle to flow downhill. The outcome of such combined erosion, removal, and

eventual deposition of material, is the development of an altiplanation terrace (Fig. 7.4). At Torpantau (055170) the bedding was so
inclined that it formed minor platforms on which snowbeds developed. These snowbeds eroded into the hillside, depositing the eroded
material further downslope, to form a series of altiplanation terraces. If the process of altiplanation is sufficiently prolonged it is
possible that snowbeds will erode their way through a mountain, and
at one point in the erosion cycle only a small tor, or eminence, will
remain to indicate the original summit. Even this remnant, as the

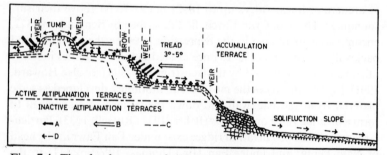

Fig. 7.4. The development of altiplanation terraces (after Botch and
Krasnov, 1951). (A) frost erosion. (B) direction of slope retreat. (C) retreat
stages. (D) evacuation of debris

process continues, is doomed to disappear (Peltier, 1950). At Torpantau the upper snowbeds almost succeeded in eroding their way
through the mountain, and only a few isolated hillocks, rising above
the general surface, remain to indicate the former summit.

Solifluction was defined by Andersson in 1906 in the following
terms: 'The slow flowing from higher to lower ground of masses of
waste saturated with water (this may come from snow-melting or
rain), I propose to name *solifluction* (derived from solum—"soil",
and fluere—"to flow")' (pp. 95–6). Climatic conditions in those parts
of Wales not covered by the Newer stage ice-caps were ideally suited
to this process and solifluction deposits are widespread in the Black
Mountains as well as on the glaciated Brecon Beacons and Fforest
Fawr. Unlike rapid forms of transport, such as sheet-flow, material
moved by solifluction seldom travels more than a few metres (yards) a
year. In addition solifluction is characterised by the formation of
terraces as layers of material, possibly 1–1·3 m (3–4 ft) deep, roll and
flow downslope.

The solifluction terraces of our area, which are fossil landforms, are usually lobate, 1–1·3 m (3–4 ft) high and little more than gentle breaks of slope, such as those of the Blaen Caerfanell valley, shown on Fig. 7.5. In other instances they are 3–6 m (10–20 ft) high, with linear fronts, as in the case of Blaen Taf Fechan.

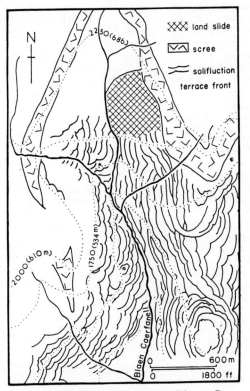

Fig. 7.5. Solifluction terraces, Blaen Caerfanell, Brecon Beacons

In addition, screes are widespread in the more mountainous parts of our region. Some of them, like those around Llyn Cwm Llwch, are active under present climatic conditions.

### GLACIAL CHRONOLOGY

The glacial chronology for the Usk and Wye regions, as far as is known at present, is shown in Table 7.1.

TABLE 7-1.  *The glacial chronology of the Usk and Wye regions*

| Deposits | Dating | |
|---|---|---|
| Postglacial deposits | Zones IV–VIII | Postglacial |
| Inner cirque moraines, Brecon Beacons. Solifluction and scree | Zone III | LATE GLACIAL |
| Organic deposits, e.g. Mynydd Illtyd, Bryniau Gleision, Rhos Goch, Waen Ddu | Zone II | LATE GLACIAL |
| Outer cirque moraines, Brecon Beacons, solifluction and scree. Organic deposits Mynydd Illtyd, Bryniau Gleision, Waen Ddu | Zone Ic | LATE GLACIAL |
| Organic deposits, e.g. Mynydd Illtyd, Bryniau Gleision, Waen Ddu, Lacustrine (?) Rhos Goch | Zone Ib Bølling | LATE GLACIAL |
| Onset of solifluction and some pollen deposition, Mynydd Illtyd, Waen Ddu | Zone Ia | NEWER STAGE |
| Re-advance of Brecon Beacons/ Fforest Fawr glaciers (?). Possibly correlated with Wye valley retreat moraines, e.g. Llyswen Allt Mawr, and major valley moraines of Eppynt, e.g. Llandefaelog Fach in Honddu valley | Possibly early Ia | GLACIAL |
| Hay-Llandetty moraines | | GLACIAL |
| Bredwardine-(?) Crickhowell moraines | | GLACIAL |
| Stretton Sugwas-Penpergwm moraines | | GLACIAL |
| Zone of weathering | | Inter stadial/ interglacial |
| Smooth glacial deposits of the Vale of Gwent and south-west Herefordshire | | OLDER STAGE GLACIAL |

In the absence of organic deposits between drifts of the Older and Newer stages it is impossible to be dogmatic about the age of either stage. In fact, at no point has drift of the Newer stage been seen to overlie that of the Older stage and this makes even the dual differentiation of the drifts difficult. The morphologic difference between areas of smooth and discontinuous drift and those of fresh and basically continuous drift, leads one to believe that two different periods of glaciation are represented. Whether these two stages were divided by an interstadial or an interglacial is still unknown.

The threefold morainic sequence in both the Wye and Usk; Stretton Sugwas, Bredwardine and Hay; and Penpergwm, Crickhowell and Llandetty leads one to believe that these moraines may be more or less contemporaneous. During these stages of glaciation it is likely that some ice fed off the Brecon Beacons into the main ice-sheet, although after the ice-sheet had retreated from the Usk valley there was apparently an advance of local Fforest Fawr/Brecon Beacons glaciers into the Usk, but this ended by Zone Ib (the Bølling), which is clearly represented in the Mynydd Illtyd and Bryniau Gleision deposits (Fig. 7.6). At Waen Ddu, above Llangattock, Anderson (1964) has shown that Zones Ia, Ib, Ic, II, III and IV are represented, which suggests that the cirque moraine there predates Zone Ia. Indeed, the glaciation of the Usk valley below Waen Ddu must have preceded that date. Separate cirque glaciers existed on the Brecon Beacons in Zones Ic and III, as is shown by the fact that the pollen record for cirques such as Cwm Cerrig Gleisiad does not start until Zone IV (Anderson, 1964; Moore, in press). Lewis (1966a) has shown that these cirque glaciers melted during Zone II, and that the Zone III glaciers were entirely new ice-bodies, and not revivified remnants of older glaciers.

The evidence for the retreat of the mid-Wales ice-sheet following the Hay–Llandetty stage is not clear. Bartley (1960) has shown that Zone I deposits rest on top of lacustrine deposits at Rhos Goch, but he does not subdivide Zone I, and one can only speculate that the lacustrine deposits date from Zone Ib. If this is so, the retreat from the Arrow valley took place before Zone Ib. The ages of the retreat moraines at Llyswen, Allt Mawr and Newbridge are unknown. They may be Ia or Ic, although it is likely that the ice-cap of mid-Wales had melted before the end of Ic. Whether this was, in fact, the case remains to be proved.

Because of the absence of earlier organic deposits than those of

Zone Ia it is inferred that Newer stage ice covered much of mid-Wales until at least that time. It is therefore apparent that that stage was equivalent to the Weichsel (Würm) of north Europe. Whether or not the Older stage was also a Weichsel (Würm) phenomenon, or a Saale (Riss) glaciation, is unknown. Certainly there remain many gaps in our knowledge of the Pleistocene history of this part of Wales and it is quite obvious that many opportunities remain for research.

## *Appendix\**

### THE POLLEN DIAGRAM FOR MYNYDD ILLTYD

The sample for analysis was taken from the centre of a small mire about 1 km in diameter marked Traeth Mawr on O.S. 1 inch sheet 147 (968256). The centre of the mire has a birch copse and the boring was made 100 m north of its edge. It is situated on a gently undulating plateau of about 2 sq km between the 1050 and 1100 ft contours. To the south lies the ridge of Fan Frynych (2047 ft) whose drainage waters do not at present affect the region of the mire.

The spacing of the samples is indicated on the diagram and the curves are based on an average of 367 grains per sample (never less than 211). Percentages have been calculated on the basis of all pollen exclusive of aquatics (*Nymphaea, Myriophyllum, Lemna*) and fern spores. *Ranunculus* pollen was included in the total pollen sum, although most of it probably belonged to the sub-genus *Batrachium*.

The zonation of the upper part of the diagram presents no difficulties: the upper four samples may be placed in Zone V because of the appearance of coryloid and tree pollen (not drawn in the diagram). Zone IV is marked by the dramatic rise in *Betula* pollen and by a Juniper maximum. The lower part of the diagram (310 cm down to the barren silt clay at 627·5 cm) is open to two interpretations:

1. If one is to be guided mainly by the stratigraphy, the uniform layer of fine red clay without sand or gravel (from 310 to 495 cm) should be placed in Zone III and the clay-mud beneath it (495–627·5 cm) in Zone II. The coarse silt-clay with sand and pebbles below this should be considered as a solifluction deposit from Zone I.

This interpretation is rejected mainly because it demands that 205 cm of sediments was deposited in Zone III. When compared with

*\*Contributed by John J. Moore, S.J., B.Sc., College Lecturer in Botany, University College, Dublin*

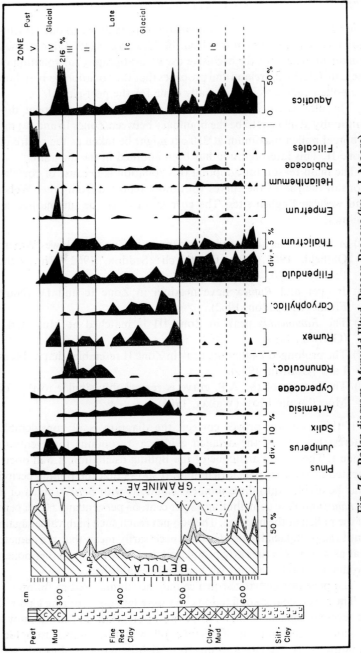

Fig. 7.6. Pollen diagram, Mynydd Illtyd, Brecon Beacons (by J. J. Moore).

6*

other British sites, this seems excessive, especially when one remembers the small catchment involved. Besides, Watts (1963) has argued convincingly against the exclusive use of stratigraphy for zonation of the Late Glacial in Ireland and proposes that the zonation be based on the pollen curves as is universally done for the post-glacial.

2. The alternative is to take the sharp pollen and sediment discontinuity at 495 cm to be the boundary between Zones Ib and Ic; the arboreal pollen maximum at 350 cm might be taken as the centre of Zone II. This interpretation has been adopted in the diagram.

The main reason for adopting this zonation is because of correlations of the run of the curves with those reported from other Welsh and western English sites. The correlations may be summarised as follows:

(*a*) *Artemesia* has an extended maximum in Zone Ic (Hawes Water: Oldfield, 1960; Llyn Dwythwch: Seddon, 1962; Rhosgoch: Bartley, 1960).

(*b*) Juniper and *Rumex* have maxima in Zone Ic and IV (Nant Ffrancon: Seddon, 1962).

(*c*) The *Ranunculus* peak in Zone III is matched at Moss Lake (Godwin, 1959).

(*d*) The prolonged *Thalictrum* peak in Zone II resembles that of Moss Lake (Godwin, 1959).

(*e*) The pattern of the A.P. curve is remarkably similar to that of Moorthwaite Moss (Walker, 1966).

If this zonation proves correct, the diagram is of considerable interest because of the detailed evidence it contains for events in Zone I. The British analogue of the Bølling oscillation is well documented in 1·3 m of pollen-rich mud. It shows that the Bølling period may be divided into three minor oscillations with *Betula* (much of it *B. nana*) and Pine contributing 58 per cent, 65 per cent and 60 per cent of the pollen at the maxima. They do not reach such high values again until the postglacial. This does not necessarily mean that the climate was as warm as in the early postglacial, but only that the other pollen producing plants had not yet been fully established.

The probable climatic sequence may be summarised as follows:

The lower, coarse solifluction clays were laid down during a period of severe periglacial climate. The lowest pollen-bearing strata show a shortlived amelioration of climate followed by a return to colder conditions. A second warming up, slightly more prolonged, was followed by a relatively long cold period with a sustained maximum of grass pollen. Then followed a final warmer period with a clear Pine maximum to be succeeded by a dramatic drop in temperature reflected in the changes in sediment type and in almost all of the pollen curves. It seems reasonable to correlate this stage with the earlier corrie glaciers which deposited the lower set of moraines in many of the cirques of the Brecon Beacons. It may be assigned to Zone Ic. During this relatively long period the vegetation was dominated by herbaceous plants and no dramatic changes are shown by the pollen curves. Finally the classical sequence of the Allerød oscillation is apparent. The Pine and Birch maximum of Zone II is followed by a colder period probably corresponding to the most recent corrie glaciation of Cwm Cerrig Gleisiad, 3·5 km to the south on the other side of the Fan Frynych ridge where the first organic deposits clearly belong to the following Zone IV. Zone IV in the Mynydd Illtyd diagram shows the typical abrupt changes in the curves associated with the beginning of the postglacial period.

## REFERENCES

ANDERSON, D. (1964) 'Data for Late Glacial and Post Glacial history in South Wales', unpublished Ph.D. thesis, Univ. Wales (Swansea).

ANDERSSON, J. G. (1906) 'Solifluction, a component of subaerial denudation', *J. Geol.* **14**, 91–112.

BARTLEY, D. D. (1960) 'Rhosgoch Common, Radnorshire: stratigraphy and pollen analysis', *New Phytol.* **59**, 238–62.

BEHRE, C. A. (1933) 'Talus behaviour above timber in the Rocky Mountains', *J. Geol.* **41**, 622–35.

BOTCH, C. G. and KRASNOV, I. I. (1951) 'Process of altiplanation and origin of mountain terraces' (translated title), *Priroda*, **5**, 25–35.

BROWN, E. H. (1960) *The Relief and Drainage of Wales.* University of Wales Press.

CHARLESWORTH, J. K. (1929) 'The South Wales end-moraine', *Q. Jl geol. Soc. Lond.* **85**, 335–58.

CLARKE, B. B. (1936–7) 'The Post-Cretaceous geomorphology of the Black Mountains', *Proc. Bgham nat. Hist. phil. Soc.* **16**, 155.

DWERRYHOUSE, A. R. and MILLER, A. A. (1930) 'The glaciation of Clun Forest, Radnor Forest and some adjoining districts', *Q. Jl geol. Soc. Lond.* **86**, 96–129.

EDGEWORTH DAVID, J. W. (1833) 'On the evidence of glacial action in South Brecknockshire and East Glamorganshire', *Q. Jl geol. Soc. Lond.* **39**, 39–54.

GRINDLEY, H. E. (1954) 'The Wye Glacier' in *Herefordshire*, Woolhope Nat. Fld Club Centenary Volume, chapter 3.

HOWARD, F. J. (1901) 'Observations on the lakes and tarns of south Wales', *Trans. Cardiff Nat. Soc.* **32**, 29–43.

HOWARD, F. J. (1903–4) 'Notes on glacial action in Brecknockshire and adjoining districts', *Trans. Cardiff Nat. Soc.* **5**.

LEWIS, C. A. (1966a) 'The Periglacial landforms of the Brecon Beacons, Wales', unpublished Ph.D. thesis, Nat. Univ. Ireland.

LEWIS, C. A. (1966b) 'The Breconshire end-moraine', *Nature, Lond.* **212**, 1159–61.

MANLEY, G. (1961) 'The range of variation of the British climate', *Geogr. J.* **117**, 43–68.

MOORE, J. J. In press.

NORTH, F. J. (1955) 'The geological history of Brecknock', *Brycheiniog*, **1**, 9–77.

PELTIER, L. C. (1950) 'The geographical cycle in periglacial regions', *Ann. Ass. Am. Geogr.* **40**, 214–36.

POCOCK, T. I. (1925) 'Terraces and drifts of the Welsh border and their relation to the drift of the English midlands', *Z. für Gletscherkunde* **14**, 10–38.

POCOCK, T. I. (1940) 'Glacial drift and river-terraces of the Hereford Wye', *Z. für Gletscherkunde* **27**, 98–117.

RAMSAY, A. C. (1846) *On the denudation of South Wales and the adjacent Counties of England*, Mem. Geol. Sur. of G.B. London, **1**, 297–335.

READE, MELLARD (1894–5) 'The moraine of Llyn Cwm Llwch, on the Beacons of Brecon', *Proc. Lpool geol. Soc.* **7**, 270–6.

RICE, R. J. (1954) 'A contribution to the geomorphology of the Upper Wye Basin', unpublished M.A. thesis, Univ. London.

RICE, R. J. (1957a) 'The erosional history of the Upper Wye Basin', *Geogr. J.* **123**, 357–70.

RICE, R. J. (1957b) 'The drainage pattern and upland surfaces of south-central Wales', *Scot. Geogr. Mag.* **73**, 111–22.

RUSSELL, R. J. (1933) 'Alpine landforms of western United States', *Bull. geol. Soc. Am.* **44**, 927–50.

STRAHAN, A. and GIBSON, W. (1927) *Geology of the South Wales Coalfield, Part II: Abergavenny* (2nd ed.). Mem. Geol. Surv. H.M.S.O.

STRAHAN, A. and GIBSON, W. (1932) *Geology of the South Wales Coalfield, Part V: The country around Merthyr Tydfil* (2nd ed.). Mem. Geol. Surv. H.M.S.O.

SYMONDS, W. S. (1872) *Records of the Rocks*. London.

THOMAS, T. M. (1959) 'The geomorphology of Brecknock', *Brycheiniog*. **5**, 55–156.

WELCH, F. B. A. and TROTTER, F. M. (1961) *Geology of the Country around Monmouth and Chepstow*. Mem. Geol. Surv. H.M.S.O.

WILLIAMS, G. J. (1968) 'The buried channel and superficial deposits of the lower Usk, and their correlation with similar features in the lower Severn', *Proc. Geol. Ass.* **79**, 325–48.

## Appendix

BARTLEY, D. D. (1960) 'Rhosgoch Common, Radnorshire: Stratigraphy and pollen analysis', *New Phytol.* **59**, 238–62.

GODWIN, H. (1959) 'Studies in the post-glacial history of British vegetation. XIV. Late-glacial deposits at Moss Lake, Liverpool', *Phil. Trans. Roy. Soc., London*, B, **242**, 127–49.

OLDFIELD, F. (1960) 'Studies in the post-glacial history of British vegetation: Lowland Lonsdale', *New Phytol.* **59**, 192–217.

SEDDON, B. (1962) 'Late-glacial deposits at Llyn Dwythwch and Nant Ffrancon, Caernarvonshire', *Phil. Trans. Roy. Soc., London*, B, **244**, 459–81.

WALKER, D. (1966) 'The late quaternary history of the Cumberland lowland', *Phil. Trans. Roy. Soc. Lond.* B, **251**, 1–210.

WATTS, W. A. (1963) 'Late Glacial pollen zones in western Ireland', *Irish Geogr.* **4**, 367–76.

# CHAPTER EIGHT

# The Hereford Basin

## B. B. Luckman, M.A.

Ice nourished in central Wales was able to escape eastwards into Herefordshire and south Shropshire unimpeded by other ice masses. The Clun, Teme, Lugg, Hindwell and Arrow valleys all contained small glaciers fed by ice spilling over watershed cols from the west. Further south, ice from the Wye valley spread over the Hereford basin in a large piedmont lobe, known as the Wye glacier, which is the main subject of this chapter.

Since most of the Hereford basin is underlain by Devonian rocks, the 'drifts' of the area soon became known to local geologists because of their included Silurian, Ordovician and igneous erratics from central Wales. Murchison first described these drifts in 1834 and further observations by local workers were reported in the Transactions of the Woolhope Naturalist's Field Club, Hereford and other local journals from 1854 onwards. The accounts of T. Curley, F. W. Merewether, W. S. Symonds, T. S. Aldis, H. E. Grindley and many others are important sources of information on which all later authors have drawn.

Only four major papers have been published dealing with the glaciation of this area. The most important is that of Dwerryhouse and Miller (1930) which deals with the glaciation of a large area of central Wales and the borderland, including an important section on the Wye glacier. Two papers by T. I. Pocock (1925, 1940) contribute much previously unrecorded material as well as summarising earlier observations. Pocock also established a chronology, based on moraines and related outwash terraces, which he correlates throughout the Welsh borderland. The remaining paper, Grindley's essay in the *Woolhope Centenary Volume,* is largely a summary of earlier work. The area is also referred to in Charlesworth's paper on the South Wales end-moraine (Charlesworth, 1929) but few local details are given. These articles contain several conflicting views on the extent and chronology of glaciation.

The Geological Survey Memoir, *The Wells and Springs of Hereford-shire* (Richardson, 1935), contains general information on surface deposits by parishes but the only geological survey maps of the area were made in the last century and show no drift except alluvium. More recently the Soil Survey have begun mapping in the north of the county and completed a 'Rapid Reconnaissance Soil Map of Herefordshire' (Burnham, 1964).

This chapter will concentrate on a descriptive analysis of the extent and character of the glaciation of the Wye valley below Hay. The limits of the last glaciation of this area will be outlined first and other features discussed with reference to this datum. Since little is known about the stratigraphic succession of glacial deposits because of the lack of exposure and borehole records, these divisions are based mainly on morphological criteria.

## THE HEREFORDSHIRE END-MORAINE

The limit of the last glaciation in Herefordshire is marked by a major moraine which can be traced round the Hereford lowlands. This moraine approximately defines the limit of continuous glacial deposits and corresponds broadly to the course of the South Wales end-moraine (Charlesworth, 1929) in this area. As there are numerous minor differences between this line and Charlesworth's, it is suggested that the use of the terms 'Newer drift limit' and 'South Wales end-moraine' be abandoned in this area and that the moraine be referred to as the Herefordshire end-moraine (Fig. 8.1).

This moraine varies considerably in form and development throughout its length and will be divided into several sections for descriptive convenience.

### The Kington–Orleton kettle moraine

This term, coined by Dwerryhouse and Miller (1930, p. 111), refers to the morainic belt at the foot of the Silurian dipslope which marks the northern boundary of the lowland. It extends from Downfield Farm (307574) to Orleton (490670) and in places is over 1·6 km (1 mile) wide. The morainic deposits are mainly red till with much Old Red Sandstone debris and western erratics (Silurian siltstones, gabbro and dolerite) which can be seen, for example, at Titley Court (336593). However, sands and gravels do occur, as at Flintsham (352613) and Stockley Barn (369625), especially in three marked belts of kames and kettles which are thought to mark the courses of preglacial valleys

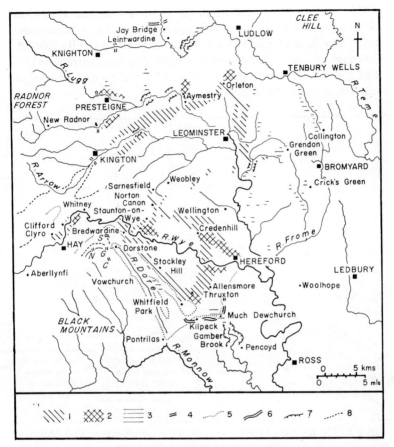

Fig. 8.1. The extent of the Wye glacier in Herefordshire. (1) Approximate extent of the Herefordshire End Moraine, Staunton Moraine and Hay Moraine. (2) Outwash deposits. (3) 'Older' glacial deposits. (4) Small valley moraines. (5) Approximate drift limit in the Dore valley area (glacial deposits and some head) from Burnham (1965). (6) Approximate drift limit in the Worm Brook area, from R. W. Pocock's maps (after Grindley, 1954). (7) Steep slopes. (8) Inferred continuation of the limits of the Herefordshire End Moraine and equivalent features. (B) Blackmoor Farm. (C) Common Bach. (N) Nant-y-Bar. (G) Gannols. (S) Sidcome Farm

masked by the moraine (Luckman, 1966, pp. 191–201). East of Mortimer's Cross the moraine is less well marked as several marl ridges break through it, as at Oaker Wood (465364). Gravel exposures are more common in this area, as, for example, at Lucton (433637).

The northern limit of drift is the Orleton moraine. This 15 m (50 ft) high arcuate ridge stretches across the low Lugg–Teme interfluve from Bircher Knoll to the flanks of Shuttocks Hill (500650). Dwerryhouse and Miller described sections of coarse, current bedded sand and gravel in this feature (1930, p. 112), and there is a spread of outwash in front of the moraine.

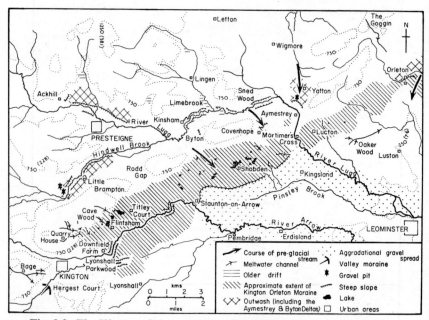

Fig. 8.2. The Kington–Orleton section of the Herefordshire End Moraine and associated drainage diversions

*Drainage diversions associated with the Kington–Orleton moraine* (see Fig. 8.2)

Dwerryhouse and Miller suggested that as the ice advanced eastwards along the dipslope, the streams which flowed through the escarpment were successively impounded and diverted. Recent work has led to a modification of some of these views. Before the last glaciation, the Lugg left the Presteigne basin by the Byton gap (370630, Dwerryhouse and Miller, 1930; Brown, 1960). This was blocked and an ice-marginal lake formed at a level of about 150 m (500 ft). Outwash terraces from contemporaneous valley glaciers in the Lugg and Hindwell valleys both end at this height and a small gravel fan occurs at Byton (372641). The lake overflowed at approximately 150–160 m (500–525 ft) at Kinsham cutting a gorge linking the Lugg to the Limebrook valley. Ice later blocked the lower part of this valley forming a moraine at Covenhope (408641). This led to the formation of the Sned Wood channel, still occupied by the Lugg. The course of this diversion was probably guided by a fault col or fault-guided tributary.

It is also believed that the Teme flowed southwards, via Aymestrey, before this glaciation (Grindley, 1915). The Aymestrey outlet, however, was blocked by the Wye glacier, creating a lake in the Wigmore lowlands supplemented by meltwaters from local glaciers in the Teme and Clun valleys. This lake reached a level of 128–131 m (420–430 ft; Cross, 1966) before overflowing and cutting the present gorge of the Teme to Downton. A large delta was built into this lake by drainage from the Wye glacier, blocking the former valley. It is well exposed in the large gravel pit at Aymestrey.

The broad valley between Ludlow and Leominster, now blocked by the Orleton moraine, was cut by the southward-flowing, preglacial Onny. From the evidence of reversal in the Teme valley between Woofferton and Knightsford Bridge, Dwerryhouse and Miller suggested that the present course of the Teme was initiated while the Wye glacier was at Orleton. However, the presence of western drift at Ankerdine Hill (735565; Grindley, 1926, 1937) suggests that this diversion may belong to an earlier glaciation or is possibly a preglacial river capture.

The other diversions described by Dwerryhouse and Miller are not considered to be glacial in origin. The gap at Knill and the abandonment of the Rodd gap (335615) are the result of river capture and the Covenhope valley is considered to be the preglacial course of the Limebrook (Luckman, 1966, pp. 40–51).

The course of the Arrow has also been considerably modified. It now flows in a gorge cut across alternating drift-infilled buried valleys and rock interfluves between Downfield Farm and Staunton-on-Arrow. At Lyonshall Park Wood (325567) a large abandoned valley over 30 m (100 ft) deep cuts through the ridge south of the Arrow. Exposures are not available in its floor but the humped profile is probably due to a drift infill. The position and size of this feature suggest that it was not cut during the last glaciation but is either a former course of the Arrow or a large meltwater channel cut at an earlier date. Minor drainage diversions occur on the dipslope at Pawpits (305585) and Cave Wood (313593).

*The moraine between Orleton and Dinmore* (Fig. 8.3)

In this area the moraine can be traced as a distinct morphological feature in the valleys east of the low escarpment forming the western edge of the plateau of north-east Herefordshire. From Orleton the moraine passes through Ashton (516644) into the Cogwell Brook valley. Glacial deposits occur in the valley around Hillside and 2·5 m (9 ft) of fine and coarse bedded gravels are exposed in a pit at 527628. Further south the moraine can be traced as a wide, debris strewn bench at about 120 m (400 ft) through Kimbolton to Grantsfield (626603) and Rowley Field (625594), filling the Yolk Brook valley (626598). The East* Stretford Brook valley is almost blocked by a mass of sands, silts and gravels at Strawberry Cottage (532579) which link up with a broad gravel flat extending southwards from Drum Farm to Stoke Prior. Several gravel pits occur in this feature, notably at Stoke Prior (527565, 527567) and a borehole at Blackwardine showed 27·5 m (90 ft) of 'running sand' (S. W. Rushgrove, personal communication). This flat continues southwards at the same height to Wisett's Farm and about 5 m (16 ft) of gravels are exposed at Risbury Bridge (540549). Similar gravels also occur in pits at Court Hill in the Holly Brook valley (553557, 555559).

These large, mainly gravel, accumulations are ice-marginal forms deposited between the ice and the upland to the east. Several stream diversions have occurred along this ice margin. The Holly, Humber and East Stretford brooks originally flowed westwards through Stoke Prior. The former two were deflected southwards cutting a gorge which joins the Lugg at Hampton Court. East Stretford Brook and, further north, the headwaters of Cogwell Brook were diverted northwards. The present course of the Lugg is probably a diversion from the strike valley now occupied by West Stretford Brook. Since glacial deposits occur in the Dinmore gorge section of the Lugg, near Newton (511540) and in a tributary valley near Hampton Court (527533) it is apparent that this diversion must have been cut during ice advance or in an earlier glaciation.

*The Hereford area*

The Burton–Westhope–Dinmore hillmass split the Wye glacier into two major lobes. Though these were probably contemporaneous, there is little morainic development on the northern flank of the hill-

---

* There are two Stretford brooks, one on either side of Leominster. For clarity they will be referred to as West and East Stretford Brook.

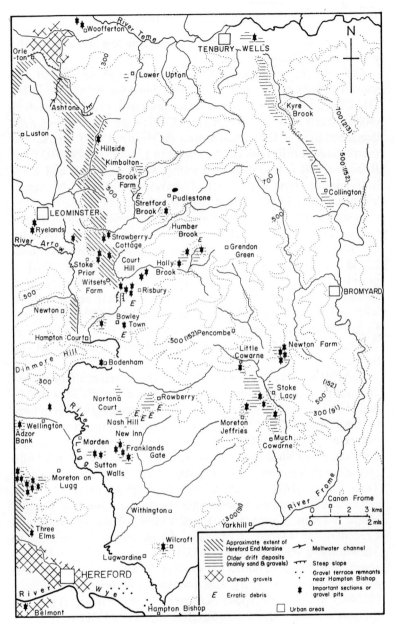

Fig. 8.3. The eastern limits of the Wye glacier. (The information for the deposits shown in the Kyre Brook valley and at Brook Farm and Lower Upton was provided by J. M. Hodgson, although the interpretation of Mr Hodgson's information is the author's)

mass. Some drift deposits do occur in this area (e.g. 473529) but the drift limit is difficult to trace.

The moraine continues further south banked up against the southern flanks of Burton Hill (395487), blocking the Yarsop valley (412473) and causing a minor drainage diversion. From about 180 m (600ft) near Yarsop and Wormsley (437477, where it occurs as a well-developed area of kames and kettles on Hereford Golf Course) the morphological drift limit can be traced to 135–150 m (450–500 ft) at Brinsop Court (446458). It then swings south-eastwards and bore-hole records show 9 m (30 ft) of gravel at Aylestone Nurseries, Tillingham and 13 m (42 ft) at Rookwood, Burghill.

East of Burghill the moraine merges into a series of low marl hills running northwards from Hereford. Between Burlton Court (486440) and Minor's Park (480457) the highest section of these hills (about 122 m, 400 ft) is covered with glacial deposits. The 6 m (20 ft) of gravels, sands and silts (Grindley, 1904a; Pocock, 1940) exposed at Bewdley Pitch (489447) were considered by Pocock to be fluvioglacial with two gravel terraces 49 m and 67 m (160 and 220 ft) above the Lugg further east. Recent exposures show these deposits to be con-tinuous and extensive, varying in character from gravels to silt and fine sand.

These deposits appear to be older than the moraine; although they are contiguous with it in the west they lie approximately 30 m (100 ft) higher and the eastern part is clearly a hill-top capping similar in form to the older deposits further east (see below). This distinctive mode of occurrence suggests that these too are older deposits which were par-tially disturbed when the moraine was deposited. The low marl hills to the south have no glacial capping except for a small area near Three Elms (491421; Grindley, 1918; Richardson, 1935).

The moraine attains its maximum development south of Burghill where it forms a wide arcuate belt running from Swainshill and Credenhill (452437) via Clehonger, Allensmore and Kingstone to Thruxton and Whitfield Park (421336). Over most of this area the pre-glacial relief has been buried but some of the hills within the moraine are solid rock, e.g. Perry Hill (469386) and Breinton Common (455400). Many small exposures formerly occurred in gravel pits, but the best sections today are at Stretton Sugwas and Old Weir.

The Stretton gravel pits (458420) have been considerably expanded since the observations of Grindley, Richardson and Moore (all 1905). Richardson and Grindley refer to a conical core of sandy clay surrounded by a mass of boulders and Moore (1904) and Aldis (1904) both mention a basal till. Their information indicates the drift is 18–30 m (60–100 ft) deep at this locality. Present quarrying exposes 12–18 m (40–60 ft) of ill-sorted, poorly bedded gravels and boulders, with some sand and silt. These deposits have been folded into gentle anticlines and synclines over most of the quarry. In places these folds are capped with 1·5–2 m (5–6 ft) of red till (see Plate 23) which suggests that the folding is due to a minor oscillation of the ice front overriding earlier outwash. Similar gravels occur in the 15–18 m (50–60 ft) high river cliffs at Old Weir (446414), but these show little sign of disturbance apart from one small fold in the upper part of the face.

The moraine is broken in three places by outwash gravel fans. The largest of these is the broad flat between Stretton and Burghill. This continues into Hereford where it forms a marked terrace on which most of the city is built. Borehole evidence shows it to be about 9 m (30 ft) thick (Richardson, 1935). However, apart from large terrace remnants near Hampton Bishop (542493, 555385), Holme Lacy (545360) and near Dinedor (540367), all about 21–24 m (70–80 ft) above the river, there is very little evidence of terraces below Hereford and the river occupies broad alluvial flats.

A second fan begins near Clehonger and is well exposed in the Belmont gravel pit (487384) where 6 m (20 ft) of outwash can be seen. This feature falls rapidly eastwards and is only 6 m (20 ft) above the river at the western end of the city. Several other small gravel deposits occur nearby but they do not appear to be connected with the outwash and may be older features, that at Broomy Hill (496395) is known to be 6–7 m (19–22 ft) deep (Curley, 1863). The other fan occupies the broad valley between Thruxton and Thruxton Vallets but is not very well exposed.

*Whitfield to Hay*

From Whitfield Park, where it blocks a small valley, the moraine can be traced northwards at the foot of the Dittonian (Old Red Sandstone) escarpment. Gravel pit exposures along the scarp foot have been described from Whitfield (419335?; Grindley, 1954), Kingstone Grange (418317; Grindley, 1918), Great Brampton (405367; Grindley, 1954) and Blakemere (359409; Richardson, 1935). However, the maximum extent of ice is shown by higher kame deposits in the scarp face valleys. Fluvioglacial material, exposed at 372389, blocks the upper part of the Stockley Hill valley between 137 and 167 m (450 and 550 ft). Similar features occur up to 180–210 m (6–700 ft) in Pentre Dingle (332432) and a large drift bench occurs above Finestreet Dingle at approximately 210 m (700 ft), just below Crafta Webb (320442). This evidence is thought to occur only in the valleys because they provided a more favourable environment for the deposition and preservation of ice marginal phenomena. The adjacent scarp face slopes were too steep to retain such deposits and deposition only occurred at the base of these slopes where conditions were more favourable.

At Bredwardine the younger Staunton moraine flanks the base of the escarpment but higher deposits can be traced round into the subsequent section of the Wye valley. Large amounts of boulder clay cover the lower step of the Black Mountains at altitudes up to 360–390 m (1200–1300 ft; Dwerryhouse and Miller, 1930; Grindley, 1954) and the northern flanks of the valley, where occasionally they form kame-like benches, as at Cwm-yr-Eithin (223480; 274 m: 900 ft). In addition, Wye ice was probably confluent with the ice in the Arrow valley in places across the watershed. The upper limit of drift has not been mapped in detail but in places it reaches considerable thickness. A borehole at 229474 (229–236 m: 750–775 ft) shows at least 39 m (127 ft) of drift, presumably filling a preglacial valley.*

There is also considerable evidence of meltwater drainage along the valley sides in the Llyswen–Whitney section of the valley. Many north bank valleys appear to have carried much greater volumes of water than at present, whilst several valley network plans suggest meltwater modification, for example at Brilley and Millhalf (280488) and Cilkenny Dingle-Moity (170415). Several channel systems occur on the south bank, especially between Aberllynfi and Hay. These channels, together with the sand and gravel deposits in the valley bottom, indicate that fluvioglacial erosion and deposition, in association with stagnant ice, were important factors during the deglaciation of this particular section of valley (Lewis, 1966, and Chapter 7).

## The Dore valley

Frequent erratics show that this valley has been glaciated by ice pushing through the low broad col (170–180 m: 550–600 ft) at its head but the broad flat-floored valley displays little morphological

---

* 0·6 m (2 ft) subsoil, 37·5 m (123 ft) dirty gravel, 0·6 m (2 ft) fine gravel, un-bottomed (S. W. Rushgrove, personal communication).

evidence of drift deposits. Small amounts of drift occur near Dorstone and Llanafon (326416) and the area between Vowchurch and Blackmoor Farm (398346) is heavily covered with boulder clay. In contrast, well-developed small moraines block tributary valleys near Nant-y-Bar (290 m: 950 ft; 283411) and Common Bach (200 m: 650 ft; 307405); glacial deposits also occur between Gannols and Pant-y-Weston (304471) and near Sidcome Farm (about 168 m: 550 ft; 297444). Since the ice in the Wye valley reached at least 210–240 m (700–800 ft) near Bredwardine, it must have passed through the valley-head col during the last glaciation, despite the lack of morphological evidence. The features at Common Bach, Gannols and Sidcombe probably relate to this stage but the Nant-y-Bar moraine may be older (see below).

THE AREA WITHIN THE HEREFORDSHIRE END-MORAINE

Other moraines lie within the limits of the Herefordshire end-moraine marking later stages in the glacial history of this area. Apart from these however, there are few distinctive glacial landforms within this area.

### The Staunton moraine

This is a well-developed, continuous, single ridge up to 30 m (100 ft) high and in places is nearly 1·6 km (1 mile) wide. Though it has frequently been recognised as a moraine its full extent has not always been realised. The Oaker Hill–Brobury Scar ridge is solid rock and divides the moraine into two parts; a smooth, flat-topped, lobate ridge running via Staunton-on-Wye, Norton Canon and Eccles Green towards Hyatt Sarnesfield (380550) and a smaller less well marked feature from Tin Hill (353452) to Bredwardine.

This moraine is poorly exposed and only superficial debris and shallow sand and silt exposures occur. T. I. Pocock described a section of 'stiff red loam full of pebbles and angular blocks' (1940, p. 104) at Moorhampton Station (390468). At Bredwardine the moraine merges into an area of well-developed kames and kettles with the kettle holes as much as 20–24 m (70–80 ft) deep (e.g. 315457). Stream sections occur at Quinta, Old Court and Turner's Boat. The Quinta section (339453) is mainly fine sand and silt but about 2 m (6 ft) of till, derived mainly from the underlying marls, occurs at the base of the section. The 15 m (50 ft) section of sand and silt at Old Court (333454; Grindley, 1923b) is now overgrown but 3·5 m (12 ft) of foreset-bedded sand and gravels occur at Turner's Boat (313459) in the lower part of a 6–7·5 m (20–25 ft high) section.

South-east of Bredwardine this kame and kettle area continues to Cross End (347431) and merges downstream into a kettled outwash terrace at Moccas. A similar terrace occurs north of the river at Monnington, heading in a gap in the moraine at Tin Hill (357453). These terraces cannot be traced downstream as they merge into undulating drift around Byford and Preston-on-Wye.

The Wye is superimposed between the Staunton moraine and Hereford. In places it has cut deeply into the underlying rock, as at Brobory Scar (354444) and the gorge between Breinton Common (453400) and Warham (486393). The preglacial course was probably further north. The depth of drift at Stretton (18–30 m: 60–100 ft), and in the nearby Corngreaves (457426) and Isolation Hospital boreholes (probably 472437; 48 and 30·7 m: 160 and 101 ft respectively; Richardson, 1935) indicate a deep buried valley passing between Breinton and Credenhill–Burghill where bedrock is at or close to the surface. The most probable course of this valley is followed by the old Hereford and Brecon Railway but further borehole evidence is needed to confirm this.

There is no apparent continuation of the Staunton moraine northwards across the broad strike valley between the Westhope–Burton Hill mass and the Silurian escarpment.

*The valley above Staunton*

Few glacial deposits have been found in the area between Whitney and the moraine. Large expanses of the valley floor are covered with the alluvial and possibly lacustrine deposits of Letton Lakes. Further upstream however, glacial deposits flank both sides of the valley between Whitney and Glasbury. Several sections have been reported in these features (Pocock, 1940 and references therein) and up to 3–4·5 m (10–15 ft) of till and gravels were exposed in road widening sections along the B4350 road in 1967.

Two major features have been described in this section of valley. Both T. I. Pocock (1925, 1940) and Dwerryhouse and Miller (1930) refer to a moraine at Hay. This is a large irregular drift mass occupying the valley floor between Clyro (214437), Bronydd (227452) and Tir-Mynach (224434). A road cutting at Clyro Castle (214435) in 1964 showed 6–7·5 m (20–25 ft) of bouldery gravels with occasional sand or silt beds. East of Wyecliff the Wye is cutting bedrock and appears to have been diverted across the former Wye–Cusop Brook interfluve. Upstream of the moraine, kame terraces and other ice marginal phenomena occur on both sides of the valley (T. I. Pocock, 1925, 1940; Lewis, 1966).

Further downstream an irregular gravel deposit occurs at Upper Cabalva (235460) opposite a much larger kettled gravel spread on the east bank south of Clifford. At Clifford this grades into a flat-topped terrace 9–12 m (30–40 ft) above the river which falls eastwards until it is truncated at Clifford Place. No trace of this has been found downstream. Pocock considered that these features marked another moraine and outwash terrace, separate from the Hay moraine. An alternative hypothesis would be to consider these features (Bronydd, Cabalva and Clifford) as the dissected remnants of a once continuous spread of kettled outwash laid down over a stagnant glacier snout in the valley bottom between Hay and Clifford. At present the evidence is inconclusive but the author favours the latter interpretation.

No evidence is available at present to indicate whether these two moraines (Staunton and Hay) belong to separate glaciations or are merely readvances of ice from central Wales. No equivalent stages occur in the valleys to the north. Dwerryhouse and Miller (1930) and Lewis (1966) have suggested that ice in the Wye valley remained active for a long period because it was the lowest outlet of the central Wales ice-cap, and therefore these moraines record stages in the history not found elsewhere.

### THE EVIDENCE OF EARLIER GLACIATIONS

No evidence has yet been found within the limits of the Herefordshire end-moraine to indicate an earlier glaciation. The only two observed sections showing the drift-solid junction are at Quinta (see above) and New Weir (435419) where 2 m (6–7 ft) of gravels underlie a badly slumped section of till and gravels. These basal gravels are probably outwash from the advancing glacier. However, outside the moraine there is both morphological and depositional evidence for an earlier glaciation.

### *North of Hereford*

Scattered erratics are found on the crest of the Silurian escarpment west of Shobden, in places over 180 m (600 ft) higher than the Kington–Orleton moraine. Also several large meltwater channels occur on the dipslope above or outside the limits of the moraine (e.g. Bage, 270560; Quarry House, 295582 and the Goggin, 468703). This suggests that the escarpment was ice-covered at an earlier time (Luckman, 1966, pp. 203–7).

East of the Lugg there are several isolated high level gravel deposits containing Silurian and igneous erratics of western origin, as Grindley (1954) and other workers have reported. These can be conveniently divided into two groups; the lower deposits mainly in Marden parish and the higher deposits further east.

Nash Hill (537485), Sutton Walls (525463), Frankland's Gate (540465), the spur above Norton Court (537495) and the Redbridge–Rowberry area (555490) are all flat topped areas between 90 and 99 m (300 and 325 ft) which have a gravel capping. The exposures at Norton Court, Frankland's Gate and possibly Wilcroft (565416; see Symonds, 1869, 1872, 1889; T. I. Pocock, 1940) show coarse, ill-sorted and at times bouldery gravels with some sand and silt. In the western face of the Frankland's pit up to 12 m (49 ft) of these deposits are exposed and in places the beds are faulted or vertical. In contrast, the nearby deposits at Sutton Walls (Clarke, 1953) are fine, well-bedded gravels and sands. Grindley (1917, 1918) also describes a gravel capping on Adzor Bank and about 3 m (10 ft) of ill-sorted gravels are exposed at 483478. It is tempting to consider that the bulk of these deposits, plus those at Bowley Town (541532) and Bodenham Moor (Dwerryhouse and Miller, 1930), represent the terminal deposits of Charlesworth's Wellington Lobe (1929, Plate XXI) and thus are part of the Hereforshire end-moraine. More detailed mapping is needed to substantiate this but their distinctive mode of occurrence together with the intervening drift-free areas suggest that these deposits are older, remanié deposits. Several higher deposits occur further east. The highest of these is at Hampton Wafer where a large area of valley side and the adjacent plateau are mantled with gravels up to 220 m (725 ft).

Further north 6–7·5 m (20–25 ft) of partially cemented* gravels, sands and silts form an isolated spur capping near Pudleston Court (564591). Other isolated deposits also occur at Brook Farm (548609; 152–168 m: 500–550 ft) and near lower Upton (545661; about 122 m: 400 ft). Extensive remnants of outwash gravels containing local and

---

* Many of the gravel deposits, e.g. Clifford, Court Hill and Bewdley Pitch are cemented or have calcareous stringers or concretions. As these occur in deposits of all ages they are not thought to be important in the relative dating of deposits. All of the gravels contain large amounts of cornstone (Devonian limestones).

Silurian stones with occasional igneous erratics are found in the Kyre Brook valley north of Collington (650600). These appear to have been deposited by a stream flowing northwards (J. M. Hodgson, personal communication). The only deposit of western origin east of this is the isolated deposit of boulder clay with Devonian erratics in a quarry on Ankerdine Hill (approximately 735565; Grindley, 1926, 1937). The most extensive of these remanié deposits is near Stoke Lacy. The pits at Moreton Jeffreys (607487) and Red Witchend (622484) have long been known and a large pit has recently been opened on Stoke Hill (613495). Field mapping indicates that these are all part of the same deposit, extending from Ash Pollards (608502) to Much Cowarne as a gently sloping interfluve capping. The sands and gravels exposed behind the Three Horseshoes Inn, Little Cowarne (604507) may be part of the same deposits. Up to 12 m (40 ft) of gravels, sands and silts occur at Moreton Jeffreys and at least 15–18 m (50–60 ft) at Stoke Hill where the deposit varies considerably from bouldery gravels to thick beds of silt showing steep dips to the east (commonly 60–70° and in places vertical). In contrast, 3 m (10 ft) of laminated clays occur in a recent exposure at 616589. In the pit a large west–east channel-fill indicates that these deposits were emplaced either before the Lodon valley was cut or when it was full of ice.

Subdued mounds of sand, gravel and silt have been exposed in pits at Newton and Tuthill farms, Crick's Green (627515, 627513, 627517; 152–175 m: 500–575 ft; Dwerryhouse and Miller, 1930; Pocock, 1940). Field debris suggests that this deposit extends at least to Roxpole (621516) and Grove Farm (626519). Clarke (1934) has also noted erratics in the Frome gravels at Yarkhill (608427) and Canon Frome (652437).

The variety of deposits and their differing modes of occurrence suggest that these features are merely the dissected remains of formerly more continuous deposits. Dwerryhouse and Miller (1930), noting the presence of western erratics in these deposits concluded that 'The furthest limit of the Wye Glacier can be traced by these gravels... probably once continuous...[which]...can be seen banked against the hillsides in a great semicircle from Orleton by Pudleston, Grendon Green, Crick's Green, Windmill Hill and Wilcroft to Hereford' (p. 112). This line could not possibly be a morainic limit as they and later authors imply (R. W. Pocock and Whitehead, 1948, Fig. 38, p. 76; Charlesworth, 1957, Fig. 244, p. 1200). The arcuate plan bears no relationship to the relief of the area; the ice is presumed to have reached its highest point, 210 m (700 ft), in the lee of the Burton–Dinmore hillmass which would form a major barrier to the effective passage of ice eastwards while, at the same time, this ice mass was unable to penetrate through broad, low level gaps into the Teme and Frome valleys at about 75–90 m (250–300 ft). The deposits in most cases are not 'banked against the hillsides' but are spur—and summit —cappings, often separated by considerable drift-free areas. They are obviously only remnants of a once greater cover which has been largely removed by erosion. The mode of preservation is clearly analogous to many of the 'older drifts' of the Midlands and lower Severn valleys (Wills, 1938; Tomlinson, 1963) and indicates a far greater age than the more continuous low level deposits further west. Further work is obviously needed to establish the limits of the glaciation and its relationship with the glacial deposits of eastern origin on the flanks of the Malvern Hills (Hey, 1959).

### South of Hereford

East of the Herefordshire end-moraine sand and gravel deposits are piled up against the broken line of sandstone hills from Much Dewchurch to Bullinghope. The relief is very subdued and this drift is known mainly from shallow pits and borehole records at Bullinghope, Haywood (at least 12 m: 40 ft of gravel), Haywood Lodge and Much Dewchurch (Richardson, 1935). A long, low ridge of sand and gravel occurs between Kivernoll and Much Dewchurch and a recent pit near the eastern end of this feature (489307) showed 4·5 m (15 ft) of horizontally bedded fine gravels. R. W. Pocock's field maps (quoted in Grindley, 1954) show this drift continuing to Kilpeck and Wormbridge. The map in Burnham (1964) extends this limit to Pontrilas but no glacial landforms have been observed in this area. This accumulation is separated from the Herefordshire end-moraine by the large open valley drained by the Worm Brook. It is differentiated from those drifts on the grounds of morphology and texture of dissection.

The sands and gravels at Much Dewchurch extend to the col at the head of the Gamber valley near Wormelow Tump (493303). The Gamber valley has large valley meanders even at its head and a grossly underfit stream, which suggest that it has been considerably modified by the discharge of meltwaters from the ice front. Clarke (1934) records erratics at Pencoyd (516266) further down this valley,

Little is known about the drifts of the Ross area. Clark (1934) refers to a few high level gravels near the river and Pocock mentions 'older drift more than 100 ft [30 m] above the river in the country around Ross' (1940, p. 107). W. S. Symonds (1872, p. 438) found erratics on the edge of the Forest of Dean but considered them of fluvial origin.

The fragmentary nature of this evidence underlines the need for further work, particularly in the Wye valley below Hereford where both head deposits (see below) and occasional terraces occur.

## PERIGLACIAL PHENOMENA

No accounts of periglacial phenomena in this area have been published though several descriptions occur in the literature from as early as 1839 (Murchison). In the Silurian uplands north of the Wye glacier the most common periglacial deposits are valleyside screes of frost shattered debris which are probably the source of the extensive aggradational gravels which fill the valley floors. These screes occur both inside and outside the equivalents of the Herefordshire end-moraine. A fine section of this scree occurs in the Forestry Commission road in Sned Wood (405678) where up to 4·5 m (15 ft) of crudely bedded, angular, tabular scree debris can be seen. Solifluction terraces also occur in Radnor Forest (Luckman, 1966, pp. 217–31).

Extensive head deposits flank the Woolhope Dome where they were originally described as 'local drifts' (Murchison, 1839; Merewether, 1870, 1877, 1878). Good sections by the Moon Inn (572373), Mordiford Bridge (568365) and near Letchmer's Ley, Fownhope (573351), show several interbedded angular heads of Silurian materials and laterally impersistent lenses of red micaceous sands and silts. Boreholes at Fownhope (13 m: 42 ft, of sand and gravel; 5·5 m: 18 ft, of clay) and Woolhope (15 m: 50 ft, of a 'mixture of clay and stone' at the Butcher's Arms) indicate that these deposits reach a considerable thickness. Similar deposits, derived from the Old Red Sandstone, are exposed near Bullingham (565327) and comparable thicknesses of head occur in the Forest of Dean (Trotter and Welch, 1960).

Several gravel exposures show small-scale disturbance due to frost action but only isolated small frost wedges have been seen. The largest of these, about 1 m (3 ft deep), is at Sutton Walls and several smaller wedges also occur here and at Moreton Jeffreys and Stoke Prior. Small-scale involutions occur in the pits near Much Dewchurch (489307) and Hillside (527628).

## DATABLE DEPOSITS

As yet no organic material has been recovered from the drift succession. Grindley (1954) described a quarry section at Woolla containing a 'buried soil', but the section was infilled in 1965 before the author could examine it.

Several mammalian remains were recovered from drift deposits by early workers, such as Curley (1866, 1880), W. S. Symonds (1869, 1872, 1889), Dawkins (1869) and Richardson (1905). Molars of *Rhinoceros tichorhinus* and *Elephas primigenius* are the most common, but the remains of the Eemian interglacial (Butzer, 1965, p. 252). *E. antiquus* have also been found. Unfortunately it is almost impossible to use these to date the deposits since, as well as possible confused identification, in many cases the horizon or exact location of the find is not given. The best documented finds are those in King Arthur's Cave (537156) about 90 m (300 ft) above the Wye below Ross (W. S. Symonds, 1872; P. B. Symonds, 1924, 1935; Clarke, 1954; Trotter and Welch, 1960 and references therein) where two highly mammaliferous cave-earths, separated by a bed of red sand with dolerite pebbles, underlie a series of Aurignacian cave hearths. Unfortunately these deposits cannot yet be correlated with the sequence of events upstream.

The most problematical deposit in this area is a 'marine clay' discovered in the banks of the Wye by Grindley (1923) at Bredwardine and Breinton. This clay contained shallow water foraminifera which were identified as marine in origin (Wright, 1905; Grindley, 1923). The author has not been able to locate this deposit but it seems probable that the foraminifera were mistakenly identified. The deposit lies within the Staunton moraine, just below the alluvium and there is no supporting evidence for a marine incursion at that time.

## CORRELATIONS AND CONCLUSIONS

Within the Hereford lowlands several distinct episodes of glaciation have been recognised. The oldest glacial deposits are isolated remnants of drifts in the northern, eastern and southern parts of the county. These represent at least one glaciation which was separated by an extensive period of erosion from the 'last' glaciation, the limit of which is marked by the Herefordshire end-moraine. Two other moraines occur within this limit at Staunton and Hay. These probably represent readvances or stages during the 'last' glaciation rather than distinct glaciations. Lewis (Chapter 7) has shown that at least

21 The Sned Wood Channel looking east from a small tributary valley near Shirley. The valley crossing the middle ground is the preglacial course of the Limebrook which used the Covenhope Valley (to the right). When this was blocked by ice the Sned Wood Channel (centre) was adopted.

22 Section in the Aymestrey Delta at Yatton, May, 1964. About 6 m (20 ft) of outwash unconformably overlies a similar thickness of deltaic fine gravels, sands and silts, dipping northwards. Boreholes indicate these gravels are at least 21 m (70 ft) thick and the Aymestrey Vicarage borehole (Grindley 1915) suggests there may be 46 m (150 ft) of drift fill in this gap.

23 Folded outwash in the north face of the pit at Stretton Sugwas, May, 1965. Till can be seen in the core of the syncline at A and as a 2 m (5-6 ft) capping to the section at B. These folds continue for another 200 metres to the east (right on the photograph).

24 Part of the periglacial scree section at Sned Wood. The closely packed, angular, tubular debris of Silurian Siltstone and shales is dipping towards the camera (i.e. downslope). The lower part of the section is obscured by silt washed from above. The hammer is 25 cm (10 in) long.

five other moraines exist upstream from Hay, emphasising the complexity of glaciation in the Wye valley.

Further north several small glaciers occupied the valleys which drain eastwards to the Severn and Wye. These glaciers were formed by ice from central Wales spilling over the watershed cols at their heads: they were not the products of small local ice-caps on Radnor Forest, Clun Forest and the Kerry Hills. Small moraines and outwash terraces indicate that ice advanced to Hergest Court (281554) in the Arrow valley, Little Brampton (305615) in the Hindwell valley and Ackhill (291658) in the Lugg valley. Similar evidence suggests that ice advanced at least to Knighton in the Teme valley and down the Clun valley almost to Jay Bridge (395794). These glaciers stagnated *in situ* producing meltwater channels and fluvioglacial deposits in the Lugg, Arrow and Hindwell valleys. The moraines are thought to be equivalent in age to the Herefordshire end-moraine (Luckman, 1966). Scattered isolated deposits outside this limit, for example at Shirley (383653), Steadvallets (465755), and Vernold's Common (473805; C. P. Burnham, personal communication) together with high level erratics, indicate an earlier more extensive glaciation involving ice from the north (Longmyndian erratics occur in the drift near Ludlow) as well as from the west.

Irish Sea ice from the Shropshire–Cheshire basin reached as far south as Church Stretton and Rushbury (R. W. Pocock, *et. al.*, 1938; Mackney and Burnham, 1966). A well-marked outwash terrace containing distinctive northern erratics can be traced down the Onny and Teme valleys as far as Ludlow and Woofferton (518687; T. I. Pocock, 1925; Dwerryhouse and Miller, 1930; Geological and Soil Survey Sheets No. 166). Specimens of Eskdale granite and Ennerdale granophyre can be seen in the pits on Ludlow racecourse and at Bromfield. Woofferton is only 1·5–3 km (1–2 miles) north of the Orleton moraine and Pocock considered that the terrace must have been laid down against the moraine and therefore postdated it. Thus the maximum advance of Irish Sea ice postdated the Welsh maximum at the Orleton moraine. The relationship of these two ice-sheets has not yet been definitely determined further north, though in places their deposits are in juxtaposition. The most common opinion expressed in the literature (e.g. Charlesworth, 1957) is that they are more or less contemporaneous.

The chronology of glaciation in the Cheshire–Shropshire basin is still being revised as new evidence and radiocarbon dates become

7

available. Boulton and Worsley (1965) have identified a major re-advance (the Wrexham–Whitchurch–Bar Hill moraine) of Late Weichsel age (see Chapter 4). More recently Shotton (1967) has suggested that the 'Newer Drift limit' (including the Church Stretton moraine) is of similar age, contrary to earlier opinions based on the dating from Upton Warren (Coope *et al.*, 1961).

There is little doubt that the Herefordshire end-moraine correlates with the Church Stretton–Bridgnorth–Wolverhampton moraine. Current views on the age of this moraine and equivalent Welsh glaciations are discussed elsewhere in this volume. This reliance on dating from other areas emphasises the need for an intensive search for new information and datable deposits in Herefordshire. Nevertheless it is apparent that the Hereford basin has been glaciated on at least two occasions and whilst the date of the earlier glaciation(s) is not known with certainty, the more recent glaciation is of Weichsel age.

## REFERENCES

ALDIS, T. S. (1904a) 'Drift in the Wye valley', *Trans. Woolhope Nat. Fld Club*, 1902–4, pp. 325–9.

ALDIS, T. S. (1904b) 'The Dry valley at Wormelow Tump', *Trans. Woolhope Nat. Fld Club*, 1902–4, p. 335.

BOULTON, G. S. and WORSLEY, P. (1965) 'A Late Weichselian glaciation in the Cheshire–Shropshire basin,' *Nature, Lond.* **207**, 704–6.

BROWN, E. H. (1960) *The Relief and Drainage of Wales*. University of Wales Press.

BURNHAM, C. P. (1964) 'The soils of Herefordshire', *Trans. Woolhope Nat. Fld Club*, **38**, 27–35.

BUTZER, K. W. (1965) *Environment and Archaeology*. Methuen.

CHARLESWORTH, J. K. (1929) 'South Wales end-moraine', *Q. Jl geol. Soc. Lond.* **85**, 335–58.

CHARLESWORTH, J. K. (1957) *The Quaternary Era*. E. Arnold. 2 vols.

CLARKE, B. B. (1934) 'Geomorphology of the lower Wye valley', unpublished M.Sc. thesis, Birmingham Univ.

CLARKE, B. B. (1953) 'A geologist looks at King Arthur's Cave', *Trans. Woolhope Nat. Fld Club*, pp. 76–82.

CLARKE, B. B. (1954) Appendix I. Report on the geology of the site at Sutton Walls and on the building stones used, p. 85 of 'Excavations at Sutton Walls, Herefordshire 1948–51' by K. M. Kenyon, Royal Archaeological Institute of Great Britain and Ireland—reprint from *Arch. Journ*, **110**.

COOPE, G. R., SHOTTON, F. W. and STRACHAN, I. (1961). 'A Late-Pleistocene fauna and flora from Upton Warren, Worcestershire', *Phil. Trans. Roy. Soc.* **B**, **244**, 379–422.

CROSS, P. (1966) 'The glacial geomorphology of the Wigmore and Presteigne basins and some adjacent areas', unpublished M.Sc. thesis, Univ. London.

CURLEY, T. (1863) 'On the gravels and other superficial deposits of Ludlow, Hereford and Skipton,' *Q. Jl geol. Soc. Lond.* **19**, 175–9.

CURLEY, T. (1866) 'Geological field address', *Trans. Woolhope Nat. Fld Club*, pp. 170–4.

CURLEY, T. (1880) 'Extinct animals and British fossil oxen discovered in Herefordshire', *Trans. Woolhope Nat. Fld Club*, pp. 248–51.

DAWKINS, W. BOYD (1869) 'On the distribution of the British postglacial mammals', *Q. Jl geol. Soc. Lond.* **25**, 192–217.

DWERRYHOUSE, A. R. and MILLER, A. A. (1930) 'Glaciation of Clun Forest, Radnor Forest and some adjoining districts', *Q. Jl geol. Soc. Lond.* **86**, 96–129.

GRINDLEY, H. E. (1904a) 'Further notes on ice action and ancient drainage of the River Wye', *Trans. Woolhope Nat. Fld Club*, pp. 336–8.

GRINDLEY, H. E. (1905a) 'Glaciation of the Wye valley', *Trans. Woolhope Nat. Fld Club*, pp. 163–4.

GRINDLEY, H. E. (1905b) 'Glacial dam at Llanvihangel', *Trans. Woolhope Nat. Fld Club*.

GRINDLEY, H. E. (1915) 'Mortimer's Cross and the Downton Gorge', *Trans. Woolhope Nat. Fld Club*, pp. 65–8.

GRINDLEY, H. E. (1917) 'Gravels of the lower Lugg and their relation to an earlier river system', *Trans. Woolhope Nat. Fld Club*, pp. 227–30.

GRINDLEY, H. E. (1918) 'Superficial deposits of the middle Wye', *Trans. Woolhope Nat. Fld Club*, pp. ii–viii.

GRINDLEY, H. E. (1923) 'A foraminiferous clay at Bredwardine, Herefordshire', *Geol. Mag.* **60**, 88–90.

GRINDLEY, H. E. (1926) 'Sectional report—Geology', *Trans. Woolhope Nat. Fld Club*, p. 195.

GRINDLEY, H. E. (1937) 'Note on the glaciation of the Teme valley', *Trans. Woolhope Nat. Fld Club*, pp. 94–5.

GRINDLEY, H. E. (1954) 'The Wye glacier', in *Herefordshire*. Woolhope Club Centenary Volume, Chapter 3.

HEY, R. W. (1959) 'Pleistocene deposits on the west side of the Malvern Hills', *Geol. Mag.* **96**, 403–17.

LEWIS, C. A. (1966) 'The Breconshire end-moraine', *Nature, Lond.* **212**, 1559–61.

LUCKMAN, B. H. (1966) 'Some aspects of the geomorphology of the Lugg and Arrow valleys', unpublished M.A. thesis, Manchester Univ.

MACKNEY, D. and BURNHAM, C. P. (1966) *The Soils of the Church Stretton district of Shropshire*, Mem. Soil Surv. England and Wales, Sheet 166, Harpenden.

MACKNEY, D. and BURNHAM, C. P. (1964) *The Soils of the West Midlands*. Bull. Soil Surv. Gt Br.

MEREWETHER, F. (1870) 'Drifts of the Woolhope area', *Trans. Woolhope Nat. Fld Club*, pp. 173–7.

MEREWETHER, F. (1877) 'Geology of the Woolhope drifts', *Trans. Woolhope Nat. Fld Club.*

MEREWETHER, F. (1878) 'Drifts of the neighbourhood (Mordiford)', *Trans. Woolhope Nat. Fld Club*, pp. 18–21.

MOORE, H. C. (1904) 'Drift in Herefordshire and evidence of land-ice', *Trans. Woolhope Nat. Fld Club*, pp. 330–5.

MURCHISON, R. I. (1834) On the gravel and alluvial deposits of those parts of the counties of Hereford, Salop and Worcester which consist of Old Red Sandstone with an account of the Puffstone or Travertin of Spout-house, and of the Southstone Rock near Tenbury. *Proc. Geol. Soc. Lond.* **2**, 77–8, also *Phil. Mag.*, ser. iii, **5**, 217–19.

MURCHISON, R. I. (1839) *The Silurian System*. London.

POCOCK, R. W. and WHITEHEAD, T. H. (1948) *British Regional Geology: The Welsh Borderland*, H.M.S.O. pp. vi and 82.

POCOCK, R. W., WHITEHEAD, T. H., WEDD, C. B. and ROBERTSON, T. (1938) *Shrewsbury District*. Mem. Geol. Surv. G.B., Sheet 152.

POCOCK, T. I. (1925) 'Terraces and drifts of the Welsh border and their relation to the drift of the English Midlands', *Z. Gletscherk*, **13**, 10–38.

POCOCK, T. I. (1940) 'Glacial drift and river terraces of the Herefordshire Wye', *Z. Gletscherk*. **27**, 98–117.

RICHARDSON, L. (1905) 'The Geology of Herefordshire', *Trans. Woolhope Nat. Fld Club*, pp. 1–68.

RICHARDSON, L. (1935) *The Wells and Springs of Herefordshire*. Mem. Geol. Surv.

SHOTTON, F. W. (1967) 'Age of the Irish Sea glaciation of the Midlands', *Nature, Lond.* **215**, 1366.

SYMONDS, P. B. (1924) 'Arthur's Cave and the Great Doward', *Trans. Woolhope Nat. Fld Club*, pp. 28–9.

SYMONDS, P. B. (1935) 'King Arthur's Cave', *Trans. Woolhope Nat. Fld Club*, pp. 132–5.

SYMONDS, W. S. (1869) A sketch of the proceedings of the Malvern Field Club from the Commencement to the close of 1868, *Trans. Malvern Fld Club*, pp. 5–37.

SYMONDS, W. S. (1872) *Records of the Rocks*. (*Notes on the Geology, Natural History and Antiquities of North and South Wales, Devon and Cornwall*). London.

SYMONDS, W. S. (1889) Notes on the geology of the botanical districts, in *Flora of Herefordshire* by W. H. Purchas and A. Ley. Hereford, pp. xxxvii, 549.

TOMLINSON, M. E. (1963) 'The Pleistocene chronology of the Midlands', *Proc. Geol. Ass.* **74**, 187–202.

TROTTER, F. M. and WELCH, F. B. A. (1960) *The Geology of the Area around Monmouth and Chepstow* (Sheets 233 and 250). Mem. Geol. Surv.

WILLS, L. J. (1938) 'The Pleistocene development of the Severn from Bridgnorth to the sea', *Q. Jl geol. Soc. Lond.* **94**, 161–242.

WRIGHT, J. (1905) 'Note on the grey and finely bedded clay at Bredwardine', *Trans. Woolhope Nat Fld Club*, pp. 167–8.

# CHAPTER NINE

# South-east and central South Wales

## D. Q. Bowen, B.Sc., Ph.D., F.G.S.

A diluvial origin (the biblical flood) was postulated for some of the 'drifts' of South Wales by William Buckland in 1823, but by the 1840s the former existence of glaciers was accepted widely and some years later Edgeworth David (1883) and Howard and Small (1901) had described features due to the former presence of glacier ice.

Between 1891 and 1920 a distinguished band of Geological Survey Officers, amongst whom can be numbered Strahan, Cantrill, H. H. Thomas, Dixon and O. T. Jones, mapped South Wales between Pontypool and St Brides Bay. Their mapping and descriptions of glacial phenomena still remain an outstanding contribution only now being supplemented and expanded by the resurvey of the coalfield and by Quaternary research workers, although mention of George's work in Gower (1932, 1933) in particular, together with that of Griffiths (1940), must be made. Strahan and his colleagues not only recognised the influence of glaciers nourished in central Wales, on the dipslope of Fforest Fawr and the Beacons, and the coalfield, but also of an ice-sheet which had its origins in Scotland and North Wales—the Irish Sea ice—which impinged upon the South Wales coastlands and which was coeval with the Welsh ice-sheet.

Following the earlier work of Wright (1914) who formalised the division of British glacial deposits into an Older Drift and a Newer Drift, Charlesworth (1929) outlined a South Wales end-moraine of Welsh ice during Newer Drift times when Irish Sea ice failed to reach the area as it had done during the Older Drift glaciation. A more recent view of multiple glaciation was put forward by Mitchell (1960) who favoured three glaciations, while in 1964 the monoglacial view was set out by Woodland and Evans.

### GLACIAL EROSION

The landscape today is not greatly unlike the preglacial one, glaciation having merely modified the detail of that time. Although the

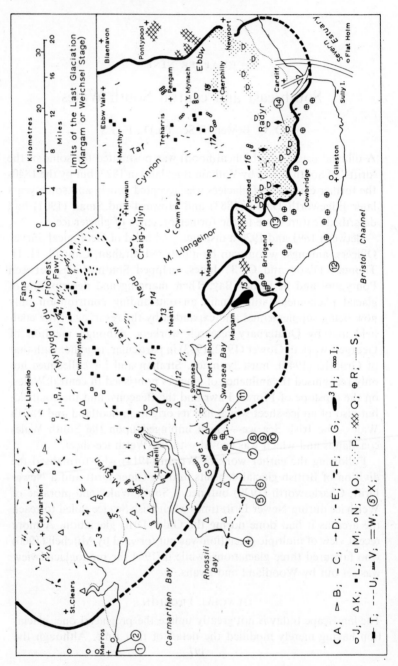

spectacular effects of Alpine sculpture are lacking, valleys were over-deepened and hillslopes oversteepened by glacial erosion. Oversteepened hillslopes contributed as a causal factor of postglacial landslips particularly in the central coalfield valleys (described for example by Woodland and Evans, 1964), while the rock floors of many valleys, now beneath a varying thickness of glacial deposits, are due in part to glacial erosion. No rock basins are known in South Wales.

The distinguishing marks of glacial erosion—roughening, polishing, bare grooved rock surfaces and roches moutonneés aligned in the direction of ice movement, are best seen on hard rocks such as the Basal Grit dipslopes of the north crop of the coalfield (Fig. 9.1). Here striations are ubiquitous while on weaker rocks it is likely that weathering would have erased them swiftly: a comparison of Fig. 9.1 with a geological map reveals their absence from all but the hardest rocks. On Carboniferous Limestone the life of striations is limited due to solutional weathering, but their occurrence near Ystradfellte (Strahan, Gibson and Cantrill, 1904) points to a recent glaciation. It is also significant that they do not occur outside the limit of the Last Glaciation (Fig. 9.1). If the till and blanket peat covering many areas were to be removed then the number of striations might well be supplemented. Used with caution (Flint, 1957), striations can assist in revealing regional ice movements as shown in Figs 9.1 and 9.3.

Without exception the corries in the area (Fig. 9.1) face between north-north-west and east: that is, where insolation was least effective because temperatures of the morning sun were lower than that of the

---

Fig. 9.1. South-east and central South Wales: selected Pleistocene features. Note: many symbols have necessarily been generalised and their exact location is deemed to be at their centres. Key: (A) corries. (B) roches moutoneés. (C) striation. (D) meltwater channel. (E) drumlin. (F) drumlins. (G) drumlinoid ridges. (H) moraines: (1) Sarnau. (2) Llanarthney. (3) Llandyfaelog. (4) Pontantwn. (5) Pont-newydd. (6) Ponthenry. (7) Machynys. (8) Waun gron. (9) Tirdonkin. (10) Pontlassau. (11) Glais. (12) Glandwr. (13) Aberdulais. (14) Clyne. (15) Margam-Pyle. (16) Talbot Green. (17) Treforest. (18) Bedwas. (I) esker. (J) kettle hole(s). (K) kamiform gravels. (L to O) erratic boulders. (L) O.R.S., Carb. Lmst. and Basal Grit. (M) Silurian. (N) Irish Sea. (O) Pennant Measures. (P) areas of ice stagnation. (Q) head. (R) palaeosol sites (after Crampton, 1964, 1966; Hooper and Hewgill, 1965). (S) Flandrian marine limit. (T) coastal solifluction terraces. (U) Interglacial raised beach and cliffs, occasional solifluction terraces. (V) till on raised beach. (W) buried channels. (X) sites referred to on Fig. 9.7

afternoon, and thus snow patches developed; on slopes leeward of extensive plateaux (Brown, 1960) where drifting snow was blown by the prevailing winds; and, again, as 'the present is the key to the past', in areas of greatest rainfall today. Initially they developed as snow patches lying in minor depressions but with increased growth and the development of active ice achieved corrie status. Some, such as Llyn-y-Fan-Fach and Llyn Fach, were occupied by corrie glaciers which formed moraines late in the Pleistocene, while others at lower elevations, such as Cwm Saerbren in the Rhondda Fawr, nourished large snow patches similar to those described by Watson (1966) in west Wales. The rock headwalls of Craig Clwyd Fechan and Craig Clwyd on the left bank of the Nedd overlook areas of local drift clearly distinguished from surrounding Brecknockshire drift. They nourished large snow patches during glaciation but failed to achieve corrie status before the main ice sheet overwhelmed them.

LAND-FORMS AND DEPOSITS OF GLACIAL DEPOSITION

Figure 9.2 shows the distribution of glacial and fluvioglacial deposits mapped by Strahan and his colleagues of the Geological Survey. Extensive areas are concealed by Flandrian (postglacial) marine and estuarine deposits between Newport and Cardiff, east of Swansea Bay, in Carmarthen Bay and the Burry estuary, as well as by alluvium in the river valleys.

Lodgement till was deposited between the ice and bedrock and not surprisingly is a very compact deposit. Its composition largely reflects that of local bedrock, notably in its colour. Hence the lodgement tills of the Millstone Grit and Coal Measures shale outcrops are clay rich, tenacious and of various shades of blue or blue-black, while the Lower Palaeozoic tills of the Towy valley are clay rich and yellow-blue in colour. Tills on the Old Red Sandstone, Millstone Grit and Pennant Measures assume a sandy or gravelly aspect, and such is the regional geology that all tills tend to become increasingly sandy or gravelly when traced from north to south. Indeed, fluvioglacial sand and gravel is difficult to separate from till, especially as all deposits near valley floors were subject to reworking and redeposition by meltwater during deglaciation. The attribute of colour frequently extends on to adjacent rock outcrops as in the upper Swansea valley, where red O.R.S. till oversteps the Carboniferous Limestone and Millstone Grit outcrops southwards, which is an indication of the direction of ice movement (Griffiths, 1940).

The arrangement of stones in a till, its fabric that is, shows that the majority of them have their long axes aligned in the direction of ice movement. Till fabric analysis in the Crosshands district, for example, shows that basal ice moved from north-east to south-west from the Amman valley across the north–south Loughor valley.

The lodgement till of ice which has crossed a sea area is usually very distinctive being calcareous, clay rich, blue or purple in colour and contains shells picked up from the sea-bed. Irish Sea ice penetrated the Ewenny valley to Pencoed where Storrie (Strahan and Cantrill 1904) found shells of *Cyprina islandica* in sand and Strahan and Cantrill described a lower Irish Sea till overlain by an upper

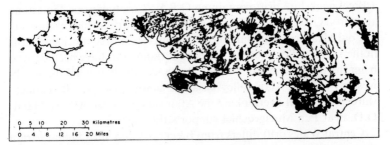

Fig. 9.2. Glacial and fluvioglacial deposits, after Strahan and his colleagues of the Geological Survey

Welsh till (see Fig. 9.7, 13 and footnote). At Ewenny shelly calcareous Irish Sea till is used for pottery making and the industry dates back to Roman times. The shells recorded by Prestwich (1892) from glacial gravel in Rhossili Bay (Strahan, 1907b) (Fig. 9.7, 3) were picked up from the floor of Carmarthen Bay and the Burry estuary and incorporated in outwash drainage from the ice margin of Welsh ice (Fig. 9.1).

Ablation till is composed of material carried in and on the ice, and, as its name suggests, was deposited when the ice melted. Thus it lacks the compaction of lodgement till and morphologically consists of hummocky surfaces common in many districts: west of Llanharan, between Pontypridd and Llantrisant and in the Vale of Glamorgan, where it is associated with fluvioglacial deposits. During its deposition considerable quantities of meltwater were available and the debris was frequently sorted and redeposited during and immediately after its initial deposition (Boulton, 1967; Sissons, 1967). Near

Rotherslade, in Gower (Fig. 9.7, 11), bedded ablation till, whose fabric shows a marked preferred orientation, was subject to redeposition partly by solifluction and partly by water as the imbricated gravel layers show (Bowen, 1966).

Together with striations the erratic content of glacial deposits is useful for deducing directions of ice movement (Fig. 9.3). When the provenance of an erratic is restricted to a particular outcrop it is known as an indicator. Useful indicators in South Wales are shown on Fig. 9.3, and it is interesting to note that many in the area are thought to come from Scotland (George, 1933).

According to their erratic assemblages the glacial deposits may be classified into provinces (Fig. 9.4). Edgeworth David (1883) recognised a Brecknockshire drift containing O.R.S., Carboniferous Limestone and Basal Grit erratics, and also a Glamorgan drift formed exclusively of Coal Measures material. The small outcrop of Brecknockshire drift in the Llynfi valley south of Maesteg is thought to have been derived from the upper layers of Vale of Neath ice which crossed the south-westerly-moving Afan valley basal ice south-eastwards. Brecknockshire ablation till overlooking the Afan valley at over 300 m (1000 ft) O.D. near Pen Moelgrochlef supports this view.

Central Wales drift differs from Brecknockshire drift in that part of its content comes from Lower Palaeozoic rocks beyond the O.R.S. outcrop. Hence the Silurian boulders at over 300 m (1000 ft) on Mynydd Bettws, south of the Amman valley, show that central Wales ice crossed the south-westerly moving Amman valley ice from north to south. This example of a basal ice shed was first recognised by Cantrill on Mynydd Sylen in Carmarthenshire (Strahan *et al.*, 1907).

The Irish Sea province is characterised by erratics from west Carmarthenshire, Pembrokeshire, North Wales and Scotland. On the coastal margin of west and north Pembrokeshire the till is calcareous and shelly but as the ice moved across the county the till assumed a 'land' facies (Chapter 10). Mixed Irish Sea and Welsh provinces occur in west Gower and the Ewenny valley and, at Pencoed, Storrie discovered eight igneous erratics which came from the Irish Sea province (Strahan and Cantrill, 1904). To a lesser extent the Ely valley deposits may be considered mixed, for an occasional erratic of westerly derivation has been discovered (Griffiths, 1940). George (1933) considered that the Gower mixed province was due to the commingling of Welsh and Irish Sea ice as did the Geological Survey before him, but the drift

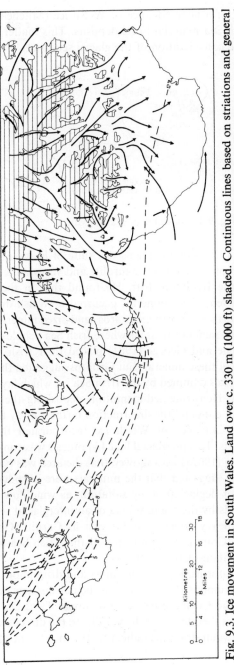

Fig. 9.3. Ice movement in South Wales. Land over c. 330 m (1000 ft) shaded. Continuous lines based on striations and general erratics. Broken lines—indicator fans (after Griffiths, 1940): (1) Llangadog rhyolite. (2) Llangynog rhyolite, agglomerate and diabase. (3) Builth Wells olivine dolerite. (4) Prescelly diabase, rhyolite and spotted slate. (5) Roch-Hayscastle andesite. (6) St David's granite. (7) Ramsey Island quartz albite porphyry. (8) St David's Head gabbro. (9) Clegyr agglomerate. (10) Llandeloy porphyrite. (11) Harlech (Barmouth) grit. (12) New Inn pyroxenic keratophyre. (T) Trilobites (*Asaphus* or *Ogygia*) from the Ffairfach Grit near Llandeilo

may be the result of a later advance of Welsh ice (Mitchell, 1960) incorporating Irish Sea material in its deposits. This matter is considered later in the consideration of the glacial sequence.

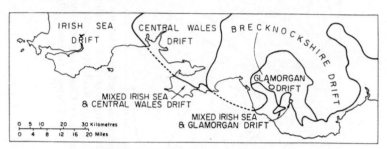

Fig. 9.4. Glacial deposits: provinces

Griffiths (1937, 1939, 1940) recognised a suite of heavy minerals, including, for example, andalusite, staurolite and vesuvianite, which were peculiar to the Irish Sea drift. He found that in the Burry estuary and lower Nedd these minerals occurred in Welsh drift and that when traced north and north-eastwards sand-sized minerals were replaced by silt-sized minerals in areas where Irish Sea pebble erratics were lacking. Pringle and George (1961) placed their Irish Sea ice margin at the limit of these minerals but Griffiths suggested that they were deposited in lakes dammed by the Irish Sea ice which lay across the coastal valleys, lacustrine sedimentation depositing the coarse fraction near the ice dam and the fine fraction farthest away from it. Then a subsequent advance of Welsh ice incorporated the lake deposits into its drifts. This proglacial lake hypothesis was considered untenable by Bowen (1967a) who showed that the area was occupied by stagnant ice and suggested that the minerals were introduced by meltwater, and after deglaciation, by saltation, as windblown dust, and in the coastlands by the rising waters of an interglacial sea. The discovery of such heavy minerals as far east as Cardiff (Griffiths, 1940) and beyond (Crampton, 1960) showed that Irish Sea ice had invaded the south Glamorgan coastlands.

Ice-moulded land-forms occur as drumlins and drumlinoid ridges of till (Fig. 9.1). Drumlins occur in two principal districts, around Hirwaun, and around Coelbren between the upper Tawe and Nedd valleys. In both cases their long axes lie parallel to the direction of ice movement, the Hirwaun drumlins indicating the movement of basal ice as it was deflected by the Craig-y-llyn ice towards the Cynon and Taf valleys. Drumlinoid ridges occur in adjacent areas (Fig. 9.1) and also conform in direction with that of ice movement. Elsewhere isolated drumlin mounds occur, as at Llangendeirne in the Gwendraeth Fach valley.

End-moraines in the area are all associated with Welsh ice. They range from the spectacular Glais moraine (Fig. 9.1, No. 11) and the subdued low moraine at Sarnau near St Clears, to the complex kamiform morainic belt between Margam and Pyle. The latter represents the limit of Welsh ice during the Last Glaciation (the Margam stage) while all the others represent readvances during deglaciation. Many block their valleys and have caused streams to flow along the valley side as at Aberdulais and Talbot Green.

### THE LANDFORMS AND DEPOSITS OF GLACIAL STREAMS

The influence of meltwater streams during glaciation was swiftly erased by advancing glaciers, but during deglaciation meltwater erosion and deposition was at its height. Deglaciation was not achieved merely by retreat of the ice margin, for ice-marginal recession is strictly a function of the infinitely more important process of ice-thinning. The pattern of downwasting was controlled by the underlying relief and as the ice thinned high ground projected through it as nunataks. In this way Mynydd Eglwysilian caused the separation of ice in the Caerphilly lowland from the main ice streams and, therefore, from the ice sources which provided its nourishment. The separated ice became stagnant (that is, it lost all forward motion) and a similar condition occurred south of the south crop of the coalfield between Pencoed and Newport, south of Townhill, Swansea, and in the synclinal Burry estuary (Fig. 9.1). Vast quantities of water were liberated by deglaciation and meltwater stream erosion was particularly effective in some areas as the numerous meltwater channels, also known as overflows, spillways or glacial drainage channels, show.

Submarginal channels run parallel and subparallel to the contours and their streams flowed in tunnels just inside the ice margin. Ice marginal channels are lacking because the crevassed condition of the marginal ice was incapable of supporting channel courses between the ice and valley sides. Examples of submarginal channels occur at Oystermouth cemetery near Mumbles, and at Wenallt, Bancyfelin. Such channels frequently cut the contours obliquely and where they run directly downslope are known as subglacial chutes (Mannerfelt,

1947); a fine example occurs at Merthyr, west of Carmarthen (map in Bowen, 1967a). Subglacial channels, cut beneath the ice, commonly have up and down or hump profiles, the short uphill portion having been cut by water flowing slightly uphill under hydrostatic pressure: good examples occur at Maesyprior (Fig. 9.5A), Upper Killay (588922) and Hensol Park (035785) in the Vale of Glamorgan (map

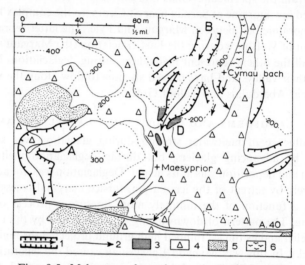

Fig. 9.5. Meltwater channels west of Carmarthen.
Key: (1) Large meltwater channel. (2) Small meltwater channel. (3) Former valley floor. (4) Till. (5) Sand and gravel. (6) Alluvium

in Bowen, 1967a). Some meltwater streams, flowing within, and upon the ice, were superimposed across spurs as the ice thinned (Price, 1960). Such spur channels (e.g. Fig. 9.5B and C) commence and end abruptly with no suggestion of continuity at either end for it was at those points that the meltwater stream left and re-entered the ice respectively after fashioning the channel in bedrock.

Meltwater channel systems (Bowen and Gregory, 1965) are rare in South Wales, but one example is found west of Carmarthen (Fig. 9.5). The earliest channels were superimposed across spurs from the ice (B and C) while the course of channel B was continued by an englacial (in the ice) stream across the Maesyprior valley to cut subglacial channel A with its up and down profile. Increasing ice-thinning led to the control of channel location by the underlying relief: subglacial channel D was formed later than B and C as it cuts below them, channel E lies at the valley side, and finally the valley floors were channelled.

At one time meltwater channels were associated with 'overflow' from supposed glacial lakes. Figure 9.6 shows the many such lakes which have been claimed in South Wales. With few exceptions, shorelines, bottom deposits and deltas are conspicuously lacking and even when they are claimed the features are open to alternative interpretation. It has been shown that the majority did not exist and that instead their areas were occupied by glacier ice which decayed by ice thinning (Bowen, 1967a). The one instance thought to be valid is Lake Teifi, but that is not to say that minor temporary lakes did not occur, and infrequent sections such as the laminated sediments in the Dulais valley, silts near Pentyrch, and Dafen, Llanelli, demonstrate their former existence. But the uncritical association of meltwater channels with overflow from supposed lakes must now be repudiated.

Depositional indications of meltwater streams occur principally in those areas where ice stagnation occurred. Eskers, the deposits of subglacial streams, occur east and west of Abertridwr (Fig. 9.1), and a fine example occurs east of Nelson (125955). North-west of Llanelli esker gravels are capped by ablation till which was deposited on final deglaciation and indicates the subglacial origin of the esker.

Kame and kettle surfaces occur at Llanilid near Pencoed, the kettles having been formed by the melting of buried blocks of ice within gravels; the numerous waterlogged hollows between Llanilid and Newport (aerial photographs revealed well over 100 of them) are all kettle holes produced in this extensive area of ice stagnation. Isolated kamiform gravel mounds (Fig. 9.1) may be the remnants of larger landforms no longer apparent, or individual kames formed in crevasses. Kame terraces, with some exceptions (Bowen and Gregory, 1966), are rare in South Wales, but some occur in the Vale of Glamorgan and many more may have been rendered unrecognisable by the slumping of their ice-contact faces.

Outwash formed by meltwater flowing from ice margins occurs in most valleys. Frequently it has been terraced by meltwater as well as by postglacial streams as in the Taf valley near Radyr (J. G. C. Anderson, 1960). North of Kidwelly in the Gwendraeth Fach valley a fine outwash train may be traced up-valley to the Llandyfaelog moraine, the ice margin of the day, while in the Thaw valley south of

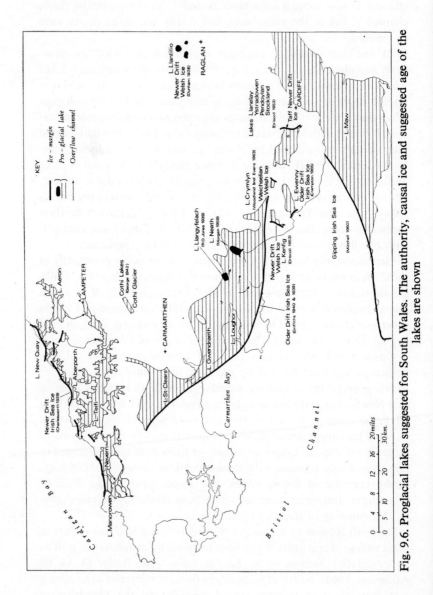

Fig. 9.6. Proglacial lakes suggested for South Wales. The authority, causal ice and suggested age of the lakes are shown

Cowbridge patches of formerly more extensive outwash occur near St Mary Church and Howe Mill.

## PERIGLACIAL DEPOSITS AND LAND-FORMS

Lying marginal to glaciated areas is the periglacial zone where a distinctive set of processes produces an equally distinctive set of land-forms and deposits. The principal process is the frost shattering of bare rock outcrops by cycles of freeze and thaw. Such mechanical weathering produces angular and subangular rock fragments which accumulate at the foot of hill-slopes and sea cliffs, sometimes filling valley bottoms, sometimes burying coastlines. These deposits, known as head, are common in South Wales and are frequently difficult to separate from glacial deposits, particularly when argillaceous in character. Head is especially extensive outside the limits of the last glaciation (Fig. 9.1) and mantles the slopes, for example, around Pontypool and Risca, while at Marros (Fig. 9.1) it has buried an old coastline which is only slowly being exhumed by the sea today (Fig. 9.9 H). The head deposits of Gower are exposed to advantage along the coast and as is shown later their stratigraphy is important for deducing the glacial sequence. Farther east, Cwm Nash and Cwm Marcross, west of St Donats, were filled by head which has been terraced by postglacial streams.

As well as responding to gravity, head was moved by solifluction. During summer the upper layers of permafrost thawed and became a highly viscous mobile layer capable of movement on all but the very gentlest of slopes. Downslope movement of this nature built up accumulations of head which mantled hill slopes and coastal cliffs as solifluction terraces (Fig. 9.9 F). Such coastal terraces when cut back by the sea have been mistaken for raised beaches (Steers, 1946), but the absence of marine deposits on their surfaces discounts such an origin. Inland solifluction terraces occur in the upper Tawe valley (Lewis, 1967) and elsewhere in upland South Wales (Taylor and Crampton, 1967). In Gower till was redeposited by solifluction (Bowen, 1966); this is discussed when the glacial sequence is considered.

Tors fashioned by mechanical weathering occur on Pennant Measures sandstones and are particularly well developed on the Rhondda Beds at Craig Ogwr at the head of the Ogwr valley. Sometimes isolated pillars up to 2·5 m (8 ft) high stand on hill slopes as, for example, on the left bank of the Cynon south of Mountain Ash. The upper slopes of that valley carry sandstone crags and pillars

8

which formed after glaciation for they would not have survived glacial erosion.

## PALAEOENVIRONMENTS DEDUCED FROM BIOLOGICAL AND PEDOLOGICAL DATA

### Flora

Botanical evidence is available from four sites: Bryn House, Swansea (621922), Cwmllynfell (738136) and Waen Ddu (185165) (Anderson, 1964), and Marros sands (Mitchell, 1967, personal communication).

At Marros sandy silty mud lying below head and on raised beach (the same horizons as the colluvial beds in Fig. 9.7, 2) yielded various seeds and fruits, moss stems and beetle fragments together with pollen. The fossil list (see p. 223) shows that trees and bushes were absent or rare but that there was a very rich herbaceous flora with a strong calcicole element. The environment was probably very similar to that of the Alleröd phase of the Late Glacial with open country-side, young soils and rich meadows. This environment was characteristic of late (Eemian) interglacial conditions.

Pollen analysis of deposits lying in a kettle hole at Bryn House, Swansea, enabled Dr Dorothy Anderson to recognise Pollen Zones I, II and III of the Late Glacial and Zone IV of the Postglacial (Fig. 9.8). The solifluction gravels of Zone I give way to the muds of Zone II and a climatic amelioration is indicated by the rise in birch (*Betula*) and decline in shrubs and grasses. During Zone II, the worldwide Alleröd phase, birch accounted for over 50 per cent of the pollen, but towards the end of the zone it declined and the shrubs and grasses rose correspondingly. Zone III had a cooler and wetter climate and birch still persisted, but as the climate improved the herb and shrub flora declined and the forest cover expanded. Zones Ia, Ib, Ic, II and III were also recognised in the Waen Ddu peat bog, but only Zones II and III occurred at Cwmllynfell. The Swansea site is particularly important because sedimentation probably began in the kettle hole as soon as it was formed by the melting of a buried block of ice in the gravels. Hence the Zone I deposits fix an upper date for the glacial

Fig. 9.7. The stratigraphy at some important sites. National grid references given for each site. Note: I am grateful to Messrs Ellis Gruffydd and D. Unwin for details of site 13. The lower till yielded no shells, no erratics and was non-calcereous. It may be a Welsh till. The overlying sandy clay may be frost wedged, but these may be desiccation cracks: in any event they mark a distinct hiatus

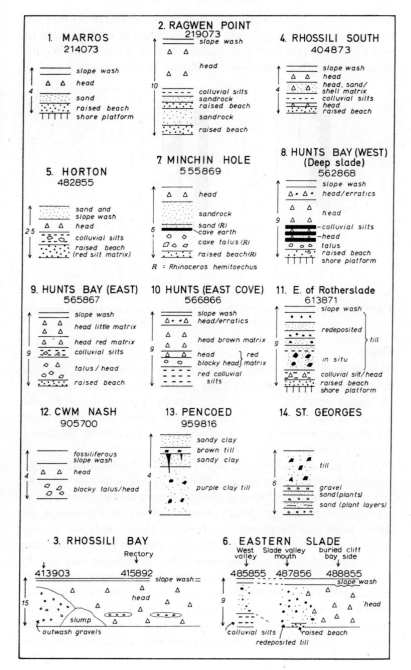

1. MARROS
214073

slope wash
head
sand
raised beach
shore platform

2. RAGWEN POINT
219073
slope wash

head

colluvial silts
sandrock
raised beach
sandrock

raised beach

4. RHOSSILI SOUTH
404873

slope wash
head
head, sand/
shell matrix
colluvial silts
head
raised beach

5. HORTON
482855

sand and
slope wash
head
colluvial silts
raised beach
(red silt matrix)

7. MINCHIN HOLE
555869

head

sandrock
sand (R)
cave earth
cave talus (R)
raised beach (R)

R = Rhinoceros hemitoechus

8. HUNTS BAY (WEST)
(Deep slade)
562868

slope wash
head/erratics

head

colluvial silts
head
talus
raised beach
shore platform

9. HUNTS BAY (EAST)
565867

slope wash
head little matrix
head red matrix
colluvial silts
talus / head
raised beach

10. HUNTS (EAST COVE)
566866

slope wash
head/erratics

head brown matrix

head        } red
blocky head} matrix
red colluvial
silts

11. E. of Rotherslade
613871
slope wash

redepposited        }
                    } till
in situ

colluvial silt/head
raised beach
shore platform

12. CWM NASH
905700

fossiliferous
slope wash
head

blocky talus/head

13. PENCOED
959816

sandy clay
brown till
sandy clay

purple clay till

14. ST. GEORGES

till

gravel
sand (plants)
sand (plant layers)

3. RHOSSILI BAY
Rectory

413903          415892
                      slope wash=

head

slump
outwash gravels

6. EASTERN SLADE
West   Slade valley   buried cliff
valley  mouth          bay side

485855   487856      488855
                        slope wash

head

colluvial silts    raised beach
redeposited till

gravels—about 15,000 years ago. It is probable, therefore, that the last glaciation of South Wales, the Margam stage, corresponded to well-known advance elsewhere, namely the Late Weichselian (Würm) which occurred between about 20,000 and 16,000 years ago.

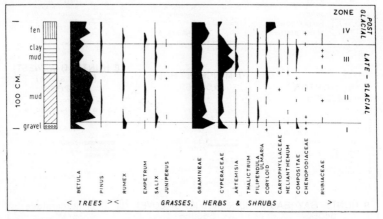

Fig. 9.8. Dr D. Anderson's pollen diagram for the Late Glacial deposits lying in a kettle hole at Bryn House, Swansea (1964)

*Fauna*

Marine shells in the raised beaches and in various glacial deposits have already been mentioned. Land shells recovered from sand lying on raised beach in Caswell Bay, namely *Cepaea nemoralis* and *Xerophila itala*, indicate a climate more genial than the present (Hinton, in George, 1932). They are to be referred to the raised beach or Minchin Hole interglacial.

At the seaward end of Cwm Nash (Fig. 9.7, 12) a fossiliferous slope wash overlies head. The entire sequence was originally believed to be marine and estuarine (Driscoll, 1953), but must now be interpreted as a series of slope deposits. Shells of land molluscs are abundant in the upper layer, and the presence of *Pomatias elegans* dates the assemblage as postglacial and after the climatic optimum (about 5000 years B.P.), while *Helix asperon* indicates a late prehistoric date (M. P. Kerney, 1967, personal communication). The slope wash is tufaceous and suggests formation by springs in a calcareous marsh closely surrounded by scrub or open woodland environment. This date suggests that the slope wash of the upper layer of most of the coastal sections (Fig. 9.7) and, indeed, much of the extensive hillwash inland (e.g.

Woodland and Evans, 1964) is postglacial, despite the recognition of loess affinities in the deposit at Port Eynon by Wirtz (1953). The high silt content of the Port Eynon deposit may have been derived from a true wind-blown loess.

The mammalian faunas of Gower have long been famous (Buckland, 1823) but despite the richness of the fauna most of it is of little value for no systematic studies have been made at any site, and faunal lists (Allen and Rutter, 1944, 1947) are meaningless as they bear no relation to the cave stratigraphy. Tundra species such as the now extinct woolly mammoth (*Elephas primigenius*), woolly rhinoceros (*Tichorhinus antiquitatis*), the ermine (*Mustela*) and open country wild horse (*Equus caballus*) have been found, while temperate or 'warm' species recovered are the woodland elephant (*Elephas antiquus*) the African hippopotamus (*H. amphibius*) and the grassland rhinoceros (*Dicorhinus hemitoechus*). Other species with no climatic preference include the lion (*Panthera leo*) the cave bear (*Ursus spelaeus*) and hyaena (*Crocuta*).

Fortunately the finds are related to stratigraphy at Minchin Hole and Bacon Hole. At Minchin Hole (Fig. 9.7, 7) *D. hemitoechus* was found in the *Patella* beach (Falconer, 1868), the talus layer and *Neritoides* beach (George, 1932), while *Elephas antiquus* occurred in a similar position to the *Neritoides* sand in Bacon Hole (Benson, 1852). This shows that all three basal layers in Minchin Hole are interglacial, a minor climatic deterioration being indicated by the talus layer which also yielded the loess steppe *Bison priscus*. Both in Bacon and Minchin Hole these warm layers are overlain by cold climate beds (head) and the simplest interpretation is to regard the warm beds as Last Interglacial and the cold beds as Last Glaciation periglacial deposits. Some support for this comes from Sutcliffe's suggestion that *D. hemitoechus* was an invader on a large scale during the Last Interglacial (Eemian) and that *Elephas antiquus* was also common (Shotton, 1962). The *Patella* and *Neritoides* raised beaches are thus thought to be Last Interglacial (Eemian) in age.

At Paviland cave Buckland (1823) discovered a human skeleton popularly known as the 'red lady of Paviland'. Subsequent investigation by Sollas (1913) in collaboration with French authorities identified the male skeleton as belonging to the Aurignacian culture which flourished during the Aurignacian interstadial between 50,000 and 23,000 years ago, and before the last major glacial advance. In an adjacent cave the Abbé Breuil and Sollas discovered the sole example of cave painting in Wales which consisted of crudely daubed red stripes. Recently skeletal remains of the 'red lady' have been dated by carbon 14 analysis to give an age of 18,460 ± 340 years B.P. (Oakley, 1968). The fact that Proto-Solutrean and Gravette implements (Upper Aurignacian of the older nomenclature, Oakley, 1964) are common in the Paviland collection lends substance to the view that the skeleton belonged to a level containing such an industry.

### PALAEOSOLS (FOSSIL SOILS) AND COLLUVIAL SILTS

Lying on various limestones are soils and remnants of soils which developed under conditions considerably warmer than the present day (Fig. 9.1). *Terra rossa* soils characteristic of hot dry Mediterranean climates have been found in Glamorgan, such as the soil profile on the southern slopes of Newton Down near Porthcawl (Crampton, 1966), and more commonly *terra fusca* soils characteristic of more temperate regions. Ball (1960) suggested that the red silty clay occurring over limestones in Gower and the Vale of Glamorgan was an interglacial relic lying south of the deposits of the Last Glaciation. Its absence within the glacial limit of the Margam stage (Fig. 9.1) supports this view. An accessible example, capping a small quarry, occurs near Colwinston, north of the A48 road (949765).

On Worms Head, Ball interpreted a red clay as a soil which developed under warmer conditions than the present. Later investigation showed that it was ubiquitous in Gower overlying the raised beaches but that it was not *in situ* (Bowen, 1966). It was thought at that time to have been redeposited by solifluction, but later was recognised as a colluvial deposit produced by soil erosion and sheet washing (Bowen, 1967b). Such deposits are widespread in the Mediterranean lands and are known by the French as *limon rouges*. Typically they occur in hollows, as on Stormy Down, and at the base of slopes where they attain considerable thicknesses (Fig. 9.7, 10). During their deposition at the base of coastal cliffs, when interglacial were changing to cold conditions, sea-level was falling slowly and was exposing an ever-increasing sea-floor on which sand was subject to deflation (Fig. 9.9 E). Hence many sections of the *limon rouges* contain sand grains derived in this way.

### THE CHANGING SEA-LEVEL

In South Wales several glacio-eustatic movements of sea-level occurred in response to stages of glacial advance, when sea-level was low, and interglacial stages, when sea-level was high. The low sea-level of the last cold stage is indicated by the rock floors of coastal valleys which were fashioned by rejuvenated streams in response to a low base level, for during the Last Glaciation sea level has been estimated to have been as low as 130 m (425 ft) below the present (Donn, Farrand and Ewing, 1962). These channels were later filled with glacial and postglacial deposits. Hence the buried channel of the Tawe at one point lies at 44·5 m (146 ft) below O.D. (O. T. Jones, 1942), the Taff and Rhymney buried channels lie at 15 m (50 ft) below O.D. (J. G. C. Anderson and Blundell, 1965), and the Burry estuary at least 30 m (100 ft) below O.D. (Strahan, 1907a).

High interglacial sea-levels are represented by raised beaches and shore platforms, while doubtful representatives of the former occur at Pendine and Porthkerry as a plastering of sandrock against the cliffs. The status of sea-levels indicated by high-level platforms (Driscoll, 1958; Bowen, 1964) is uncertain despite correlation of the '600-foot (183 m) platform' with the Early Pleistocene (Brown, 1960), for evidence from St Erth (Mitchell, 1967) suggests that the Early Pleistocene sea-level was not above 60 m (200 ft) O.D.

Raised beaches rest at various heights between 1·5 and 16 m (5 to 55 ft) O.D. on shore platforms whose inner margins meet their contemporary cliff lines. Several facies of beach occur: pebbles, shells and shingle, or any combination of these. Generally it is cemented to form a tough conglomerate but sometimes, as at Horton (Fig. 9.7, 5), it is uncemented at 9·5 m (31 ft) O.D. Occasionally a certain amount of angular material (cliff fall) is contained in its upper layers and this probably indicates colder conditions. The most ubiquitous raised beach was named the *Patella* beach, after the limpet shell of that name (George, 1932), but unfortunately the shell assemblage it contains has a very extensive time range which gives it no chronological value. Equally the shells are too diversified in ecological tolerance to provide climatic data for conditions under which the beach accumulated. As well as local pebbles the beaches contain erratic material from the coalfield as well as from the Irish Sea, and George speculated on the possibility that these had been rafted in on ice floes and that the beach accumulated in cold conditions. Another view is that they were derived from glacial deposits and incorporated in an interglacial beach (Zeuner, 1959; Mitchell, 1960): the balance of evidence now favours this view. It has already been argued that the mammalian remains point to an interglacial age for the two beaches.

At the entrance to Minchin Hole occurs an important section (Fig. 9.7, 7) first described by George (1932) and later by Bowen (1966). Separated from the *Patella* beach by a layer of talus is a sand deposit which yielded species of *Neritoides* after which George named this beach (11·25 m: 37 ft). George claimed to recognise the *Neritoides* beach as the upper member of a threefold stratigraphy of (i) *Patella* beach, (ii) head, (iii) shells, at Shirecoombe, but Bowen (1966) argued that the upper two members described by George are a commingling of *Patella* beach and head.

The Heatherslade beach at 0·46 m (1·5 ft) O.D. was discovered and described by George (1932) and may only be seen during low-water mark at Southgate. It occurs on the 'modern' shore platform, whose inner margin lies at 0·15 m (0·5 ft) O.D., and covers about 24 m² (250 sq. yd). It differs from the *Patella* beach in the greater size of its pebbles and its enhanced erratic content.

The raised beaches are frequently covered by blown sand, sometimes loose, sometimes cemented hard as sandrock or aeolianite, in areas where sand does not occur on the foreshore today. This material accumulated as sea-level fell at the close of the interglacial exposing increased areas of the sea-floor, from where the sand was obtained by deflation.

The melting of the world's glaciers after the Last Glaciation was responsible for the most recent marine transgression, the Flandrian. It reached its high-water mark of about 4·5 m (15 ft) O.D. about 5000 years ago. This shoreline is best called a marine limit rather than a raised beach and consists of marine and estuarine muds and clays abutting landwards against till and various gravels between Newport and Cardiff, while between Laugharne and Pendine, at Pwll, Bynea, and east of Swansea Bay it lies at the foot of what are now abandoned sea cliffs originally fashioned during the raised beach interglacial. The postglacial slope evolution of the Pendine cliffs has been studied by Savigear (1953).

### PLEISTOCENE CHRONOLOGY

The chronology of Pleistocene events in South Wales, is conveniently summarised in Table 9.1.

There is little evidence for the Pleistocene development of South Wales prior to the deposition of the so-called 'Pencoed' glacial material, although circumstantial evidence from north Devon suggests that the 'preglacial' shore platforms of Gower antedate the

TABLE 9.1. *Pleistocene chronology in South Wales*

| Local stage | Events | Climate | Correlation |
|---|---|---|---|
| Postglacial | 4·5 m (15 ft) marine limit<br>Slope washes | *t* | Flandrian |
| MARGAM | Late Glacial — Zone III, Zone II {Bryn House deposits / Swansea}, Zone I, Valley end-moraines, Margam Glaciation | *c*<br>*c*<br>*g* | Younger Dryas<br>Alleröd<br>Older Dryas |
|  | PAVILAND INTERSTADIAL |  | Aurignacian Interstadial (Upton Warren Interstadial Complex) |
|  | Early cold interval<br>head formation<br>sheet-washing phase<br>erosion of buried channels | *c* | WEICHSELIAN |
| MINCHIN HOLE INTERGLACIAL | *Terra rossa* and *terra fusca* soil formation — Horton beach, minor climatic deterioration, Minchin Hole beach | *t* | Eemian (Ipswich) |
| PENCOED | Irish Sea Glaciation | *c*<br>*g* | Saale or Riss (Gipping) |
|  | Shore-platform and cliff formation. Beach material probably preserved but cannot be separately identified from the Eemian beaches | *t* | Holstein (Hoxnian) |

*t*, temperate;   *c*, cold;   *g*, glacial deposits known.

8*

Pencoed Irish Sea Glaciation, and they may be of Hoxnian Interglacial age or older.

### The Pencoed (*Saale or Riss*) cold stage*

Irish Sea and Welsh ice-sheets completely covered the area during this
stage. The former met the latter around St Clears, was held off by the
Cefn Bryn upland in Gower, but crossed the south Glamorgan coastlands to Cardiff. Whenever it came into contact with Welsh ice it
deflected the latter south-eastwards (Griffiths, 1940). Investigations
in south-west England have confirmed that the Bristol Channel was
packed by Irish Sea ice at this time (Stephens, 1966; Mitchell and
Orme, 1967).

Evidence for the Irish Sea ice-sheet in Glamorgan is sparse, and
Strahan and Cantrill (1902/1912) wrote of the low-lying Mesozoic
areas in the south that it was 'practically certain that they were never
covered with glacial drift'. Moreover, they noted that the supposed
Cretaceous flints in the Ely valley only make their appearance where
the Lias begins to contain chert. Yet, the evidence of erratics at
Pencoed (Strahan) and farther east (Griffiths, 1940) points to glaciation from the west. Recently Crampton suggested that Lias cobbles in
the southern Vale are glacial and that heavy minerals of the Irish Sea
drift are widespread (1966). Figure 9.2 shows that the southern Vale is
not unlike south Pembrokeshire in terms of its drift cover and it is not
unreasonable to assume that Irish Sea drift has been removed by
erosion: certainly it has never been suggested that the south Pembrokeshire plateau was not glaciated. The southern Vale merits
detailed reinvestigation.

The mixed drift in the Ewenny valley is the product of a late re-
expansion of Welsh ice during this stage, but the west Gower province is more enigmatical and is considered with the Margam cold
stage.

No undoubted deposits of the coeval Welsh ice can be identified
and the Irish Sea deposits west of Carmarthen Bay have been severely
denuded: it is significant that the porous gravels have survived while
the finer tills have been removed. Remaining deposits have been
severely disturbed by frost action during the Margam cold stage
when they lay in the periglacial area and 5·5 m (18 ft) frost wedges

---

* This is named after the lower till described by Strahan and Cantrill (1904) at
Pencoed. It was called the Fremington stage by Mitchell (1960) but as Fremington
is in north Devon the designation is inappropriate in South Wales.

occur in a gravel quarry at Llandre (093203) while intense cryoturbation has affected the upper gravels. Figure 9.2 demonstrates clearly the contrast between the highly dissected drifts of this older stage and the more extensive drifts of the later Margam stage (Fig. 9.1).

*The Minchin Hole (Eemian or Last) interglacial stage*

Hippopotamus, the zonal fossil of the Eemian Interglacial (King, 1955), together with other mammalia frequent in that interglacial (*Elephas antiquus* and *Dicorhinus hemitoechus*) show that temperatures were higher than the present day. *Terra rossa* and *terra fusca* soils lying on limestone plateau in Glamorgan also point to warmer conditions with distinct Mediterranean affinities. These palaeosols are younger than the Irish Sea glaciation as they lie on ground traversed by that ice-sheet.

As sea-level rose from its low stand during the Pencoed cold stage it destroyed Irish Sea and Welsh glacial deposits, some of the erratics of which found their way to the raised beaches which were deposited on wave-cut shore platforms lying marginal to the interglacial cliff line (Fig. 9.9 B). It is possible, however, that the general coastal configuration had earlier origins in the Hoxnian interglacial (pre-Pencoed stage) as the giant erratics beneath raised beach in north Devon (Stephens, 1966 and Chapter 11) suggest. Even so, the raised beach deposits of South Wales and the detail of their coastline was the product of the Last Interglacial (Eemian) sea, although it is possible that patches of indistinguishable Hoxnian Interglacial raised beach remain, (Bowen, 1969a).

Following the deposition of the *Patella* beach, a minor climatic deterioration is recorded by the talus layer separating the *Patella* and *Neritoides* beaches at Minchin Hole (Fig. 9.7, 7 and 9.9 C), and elsewhere by the talus in the upper layers of the raised beach (e.g. Fig. 9.7, 8). It was during this minor fall in sea-level that the *Patella* beach was cemented into a tough conglomerate. The *Heatherslade* beach may have been deposited at the same time as the *Neritoides* sand (Bowen, 1966) but may equally be postglacial (George, 1932). But at Horton and Eastern Slade (Fig. 9.7, 5) the red colluvial silt matrix of the beach shows that it accumulated late in the interglacial (*Neritoides* interval) immediately preceding the sheet-washing phase of the earliest part of the succeeding cold Margam stage (Fig. 9.9 D, E). At Ragwen Point the upper beach may be the equivalent of the *Neritoides* sand (Fig. 9.7, 2).

## The Margam (*Weichsel or Würm*) *cold stage**

The most complete record of the last cold stage occurs in the periglacial zone of that time (Fig. 9.1). The stage opened with sea-level falling and with the interglacial soils being subject to soil erosion caused by a reduced vegetation cover due to increasing cold, and re-

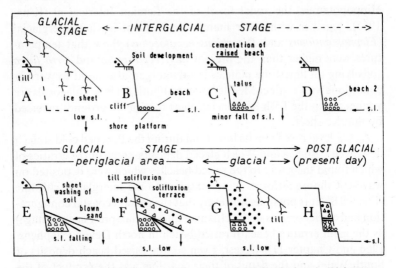

Fig. 9.9. A model of coastal evolution in South Wales during the Pleistocene (see text). Note: G only occurred between Langland Bay and Mumbles. H represents the existing polycyclic profile as seen at many sites (cf. Fig. 9.7)

deposition by sheet-washing as colluvial silts which rest on raised beaches or the overlying talus layers (Fig. 9.9 E). A similar sheet-washing phase is known in Europe at the beginning of the Last Glaciation (Rohdenburg, 1965). Increasing cold produced extensive head formation and the burial of the coastal zone beneath solifluction debris (Fig. 9.9 F). The lower layers of head (sometimes all the head) in Gower have a red silty matrix derived from the interglacial soil, and at Rhossili (Fig. 9.7, 4) deflation from an increasingly exposed sea-floor as sea-level fell (as in Fig. 9.9 E) provided a sandy shingle matrix for the head which at this site is formed of fossil fragments

* Known as the Newer Drift by Charlesworth (1929) and the Llandaff stage by Mitchell (1960). Llandaff, however, is not a morainic locality nor is it at the margin of glaciation, hence the use of Margam as a stage name is more appropriate.

from the Lower Caninia zone of the Carboniferous Limestone forming the local cliffs (George, 1940).

At Eastern Slade, Pencoed stage till from around Hangman's Cross
was redeposited by solifluction down the Eastern Slade valley, and
the smaller valley farther west, to accumulate on the foreshore on top
of colluvial silts. Its solifluction origin is confirmed by the marked
preferred orientation of its pebbles down valley, its restriction to the
line of the valley, and because it passes laterally into periglacial
head on either side of the seaward end of the valley where till was un-
available for solifluction and material was derived from the interglacial cliffs instead (Fig. 9.7, 6, Bowen, 1969b).

Gravel erratics within head or as discrete lenses or layers passing
laterally into head occur towards the top of many sections (Fig. 9.7,
10). The earlier (and lowest) erratic free head layers were derived from
the local cliffs, but extensive solifluction, due to increasing cold,
moved gravel erratics from the plateau area immediately inland to be
deposited later (and therefore higher) than the locally produced
earliest lower layers.

All the deposits overlying the raised beaches at Marros and
Ragwen Point are unquestionably periglacial slope deposits derived
from shale and sandstone outcrops in the interglacial cliff behind.
The absence of even a single erratic further demonstrates their nonglacial origin, for there is a considerable lithological diversity inland
in the direction from which ice came during the earlier Pencoed stage
and from which it would have come had it reached this area during
the Margam cold stage. The Marros and Ragwen raised beaches are,
therefore, younger than the Pencoed glacial advance, but older than
the Margam cold stage periglacial deposits which rest upon them.

These coastal sections record the entire cold stage, but from about
50,000 to 23,000 years ago a climatic amelioration occurred, the
*Paviland interstadial*, when Aurignacian man lived in Gower. The
interstadial may be represented by the hiatus in the head stratigraphy
at Hunts Bay (Fig. 9.7, 10) between the lower layers with a red silt
matrix and the upper layers with a very limited chocolate brown
matrix, but it is not possible to state this with any certainty.

At the same time as the upper head layers were accumulating
glaciation occurred inland, the Margam advance, which, on the basis
of the date assigned to its upper limit by the Zone I deposits at
Swansea (Anderson, 1964), is correlated with the continental Late
Weichselian or Upper Würm glaciation (Woldstedt, 1967). Its

maximum occurred about 17,000 years ago and till and ablation till of this advance only rest on the raised beaches between Langland and Mumbles (Fig. 9.7, 11, Fig. 9.9 G). In and east of Swansea Bay the ice limit is essentially that envisaged by Charlesworth (1929) and outside its margin extensive head accumulation took place, while it is only in this periglacial zone that the interglacial soil escaped glacial erosion. East of Pencoed the limit of glacial and fluvioglacial deposits is clearly defined on the St Nicholas ridge which held the ice in the Ely valley, while between Newport and Cardiff the glaciers reached just offshore where gravels are now covered by Flandrian sediments. At Llanilid the limit is marked by a fine assemblage of kames and kettles thought to be the product of Irish Sea ice by Woodland and Evans (1964) because of the red colour of the local tills. Such colouring, however, was obtained from the Trias outlier at Llanharan and the landforms are thus the product of Welsh ice. Moreover kettle holes of the Pencoed (Saale–Riss–Gipping) glaciation are not known in Britain (R. G. West, 1966, personal communication).

The complex kamiform Pyle–Margam moraine represents the limit of the Tawe–Nedd piedmont ice which expanded into Swansea Bay, while west of Swansea the ice failed to surmount the high ground of Cefn-Bryn and Llanmadoc Hill, but a lobe of ice may have crossed Ryers Down and the Fairyhill water gap to sit in west Gower where till is extensive showing no signs of weathering while constructional surfaces developed on gravels occur (Fig. 9.1). Outwash from the ice margin, which may have stood where Whitford Point now lies, occurs in Rhossili Bay and on Worms Head. From the Gwendraeths and Towy valleys ice entered the head of Carmarthen Bay while in the west it reached the St Clears–Laugharne area: exceptionally fresh meltwater channels occur north of the A40 road (e.g. Fig. 9.5).

It is not possible to say whether this was an ice-sheet or a predominantly valley glaciation. Woodland and Evans (1964) suggested that much of the coalfield may have been above the ice but in the west a considerable thickness of central Wales ice spilled over the Carmarthenshire scarplands and carried Silurian boulders to over 300 m (1000 ft) on Mynydd Bettws. Likewise the boulders on Mynydd Eglwysilian cannot be disassociated with this glaciation unless good evidence to the contrary is available. The balance of evidence, therefore, is in favour of ice-sheet conditions with possibly some of the higher coalfield areas being ice free.

Deglaciation was punctuated by phases of readvance when the

South Wales valley end-moraines were formed (Fig. 9.1), the most spectacular example being that at Glais. It is not possible to correlate these from valley to valley, all that can be said is that they occurred in times of re-expansion during general deglaciation.

Even during deglaciation head continued to form, especially in areas of favourable relief as for example at the foot of Rhossili Down where it covered the earlier outwash. Inland, head mantles hillslopes, merging with till so that they are frequently difficult to distinguish. Tor formation occurred and extensive scree slopes formed.

A late corrie glaciation probably occurred in the higher corries during Zone I of the Late Glacial, while Zone III was probably characterised by snow patch formation.

Sea-level rose during deglaciation, being known as the Flandrian transgression which reached its upper limit of about 4·5 m (15 ft) O.D. about 5000 years ago, and radiocarbon dating of peats formed at sea level 6500 years ago has shown that the South Wales coast has been stable since that time (Churchill, 1965), any isostatic recovery after depression beneath the ice having taken place by that date.

## Appendix

### FOSSIL LISTS*

FROM MARROS

Seeds and fruits etc.: *Gramineae, Cyperaceae, Juncus bufonius, J. effusus, J. acutiflorus/articulatus, Ranunculus flammula, R. reptans, R. linqua, R. aquatilis* s.l., *Montia fontana* agg., *Caryophyllaceae, Stellaria cf. alsine, Callitriche* sp.

Pollen: *Gramineae, Cyperaceae, Compositae, Pinus, Salix, Betula* (including *B. nana* type), *Hippophae, Alnus, Carpinus, Picea, Taxus, Tilia, Tsuga. Empetrum, Calluna. Artemisia, Caryophyllaceae, Camppolygonum, Potentilla, Ranunculaceae, Rumex, Sanquisorba, Thalictrum,* and *Umbelliferae. Caltha, Lemna, Littorella* and *Potamoqeton.*

FROM CWM NASH†

*Pomatias elegans, Helix asperon, Helix nemotalis, Helicella virgatc, Lymnaea trunculata, Oxychiluscellarius, Discus rotundatus.*

* Contributed by Professor G. F. Mitchell, Trinity College, Dublin; Miss Alison M. Loader and Miss R. Andrew, Cambridge University.

† Contributed by Dr M. P. Kerney, Imperial College, London.

REFERENCES

ALLEN, E. E. and RUTTER, J. G. (1944) 'A survey of the Gower caves with an account of recent excavations, Pt. 1', *Proc. Swansea sci. Fld Nat. Soc.* 2, 221–46. *Ibid.* Pt. 2 (1947), 2, 263–90.

ANDERSON, D. (1964) 'Data for Late Glacial and Post Glacial history in South Wales', unpublished Ph.D. thesis, University of Wales.

ANDERSON, J. G. C. (1960) *Geology around the University Towns: The Cardiff District.* Geologists' Association Guide No. 16.

ANDERSON, J. G. C. and BLUNDELL, C. R. K. (1965) 'The sub-drift surface and buried valleys of the Cardiff district', *Proc. Geol. Ass.* 76, 367–78.

BALL, D. F. (1960) 'Relic-soil on limestone in South Wales', *Nature, Lond.* 187 (4736), 497–8.

BENSON, S. (1852) 'Account of the cave deposits of Bacon Hole', *Rep. Swansea Sci. Soc.*, 10.

BOULTON, G. S. (1967) 'The development of a complex supraglacial moraine at the margin of Sørbreen, Nyfriesland, Vestspitsbergen', *Jl. Glaciol.* 6, 717–36.

BOWEN, D. Q. (1964) 'The denudation chronology of central South Wales', *Abs. Papers*, 20th *Int. Geogr. Congr. London*, p. 120.

BOWEN, D. Q. (1966) 'Dating Pleistocene events in south-west Wales', *Nature, Lond.* 211 (5048), 475–6.

BOWEN, D. Q. (1967a) 'On the supposed ice-dammed lakes of South Wales', *Trans. Cardiff Nat. Soc.* 93 (1964–66), 4–17.

BOWEN, D. Q. (1967b) 'The geomorphic relationships of a palaeosol in South Wales', paper read to the British Soil Science Society.

BOWEN, D. Q. and GREGORY, K. J. (1965) 'A glacial drainage system near Fishguard, Pembrokeshire', *Proc. Geol. Ass.* 74, 275–82.

BOWEN, D. Q. and GREGORY, K. J. (1966) 'Fluvioglacial deposits between Newport, Pembs. and Cardigan', *Occ. Pub. Br. Geomorph. Res. Gp*, 2, 25–8.

BOWEN, D. Q. (1969a) 'A re-evaluation of the coastal Pleistocene succession in south-west Britain,' *Résumés des Com. VIII INQUA Congress*, Paris, 183.

BOWEN, D. Q. (1969b) in *Coastal Pleistocene deposits in Wales*, Quat. Res. Ass. H'book, 12–16.

BROWN, E. H. (1960) *The Relief and Drainage of Wales.* Cardiff.

BUCKLAND, W. (1823) *Reliquiae Diluvianae.*

CANTRILL, T. C., DIXON, E. E. L., THOMAS, H. H. and JONES, O. T. (1916) *The Geology of the South Wales Coalfield. Part 7. The Country around Milford.* Mem. Geol. Surv. Gr. Br. H.M.S.O.

CHARLESWORTH, J. K. (1929) 'The South Wales end-moraine', *Q. Jl geol. Soc. Lond.* 85, 335–58.

CHURCHILL, D. M. (1965) 'The displacement of deposits formed at sea-level 6500 years ago in southern Britain', *Quaternaria*, 239–49.

CRAMPTON, C. B. (1960) 'Analysis of heavy minerals in the Carboniferous Limestone, Millstone Grit and soils derived from certain glacial gravels of Glamorgan and Monmouthshire', *Trans. Cardiff Nat. Soc.* 87, 13–22.

CRAMPTON, C. B. (1961) 'An interpretation of the micromineralogy of certain Glamorgan soils: the influence of ice and wind', *Jl. Soil Sci.* **12**, 158–71.

CRAMPTON, C. B. (1964) 'Certain aspects of soils developed on calcareous parent materials in South Wales', *Trans. Cardiff Nat. Soc.* **91**, 4–16.

CRAMPTON, C. B. (1965) *Guide to Excursions British Society of Soil Science*, Swansea 1965.

CRAMPTON, C. B. (1966) 'Certain effects of glacial events in the Vale of Glamorgan, South Wales', *Jl. Glaciol.* **6**, 261–6.

CRAMPTON, C. B. and TAYLOR, J. A. (1967) 'Solifluction terraces in South Wales', *Biul. Peryglac.* **16**, 15–36.

DAVID, J. E. (1883) 'On the evidence of glacial action in south Brecknockshire and east Glamorganshire', *Q. Jl geol. Soc. Lond.* **39**, 39–54.

DIXON, E. E. L. (1921) *The Geology of the South Wales Coalfield. Part 8. The country around Pembroke and Tenby.* Mem. Geol. Surv. Gr. Br. H.M.S.O.

DONN, W. L., FARRAND, W. R. and EWING, M. (1962) 'Pleistocene ice volumes and sea-level lowering', *Jl. Geol.* **70**, 206–15.

DRISCOLL, E. M. (1953) 'Some aspects of the geomorphology of the Vale of Glamorgan', unpublished M.Sc. thesis, University of London.

DRISCOLL, E. M. (1958) 'The denudation chronology of the Vale of Glamorgan', *Trans. Inst. Br. Geogr.* **25**, 45–57.

DUNHAM, K. C. (1938) *Summary of Progress, Geological Survey.* H.M.S.O.

FALCONER, H. (1868) *Palaeontological Memoirs and Notes of the late Hugh Falconer.* C. Murchison.

FLINT, R. F. (1957) *Glacial and Pleistocene Geology.* New York, Wiley.

GEORGE, T. N. (1932) 'The Quaternary beaches of Gower', *Proc. Geol. Ass.* **43**, 291–324.

GEORGE, T. N. (1933) 'The glacial deposits of Gower', *Geol. Mag.* **70**, 208–32.

GEORGE, T. N. (1940) 'The structure of Gower', *Q. Jl geol. Soc. Lond.* **96**, 131–98.

GEORGE, T. N. (1942) 'The development of the Towy and Upper Usk drainage pattern', *Q. Jl geol. Soc. Lond.* **98**, 89–137.

GRIFFITHS, J. C. (1937) 'The glacial deposits between the River Tawe and the River Towy', unpublished Ph.D. thesis, University of Wales.

GRIFFITHS, J. C. (1939) 'The mineralogy of the glacial deposits of the region between the rivers Neath and Towy, South Wales', *Proc. Geol. Ass.* **50**, 433–62.

GRIFFITHS, J. C. (1940) 'The glacial deposits west of the Taff', unpublished Ph.D. thesis, University of London.

HOOPER, L. J. and HEWGILL, D. (1965) in Crampton, 1965.

HOWARD, F. T. and SMALL, E. W. (1901) 'Notes on ice action in South Wales', *Trans. Cardiff Nat. Soc.* **32**, 44–8.

JONES, O. T. (1942) 'The buried channel of the Tawe valley near Ynystawe, Glamorganshire', *Q. Jl geol. Soc. Lond.* **98**, 61–88.

JONES, R. O. (1939) 'The evolution of the Neath–Tawe drainage system South Wales', *Proc. Geol. Ass.* **50**, 530–66.

KING, W. B. R. (1955) 'The Pleistocene epoch in England', *Q. Jl geol. Soc. Lond.* 111, 187–208.

LEWIS, C. A. (1967) 'The Breconshire end-moraine', *Nature, Lond.* 212 (5070), 1559–61.

MANNERFELT, C. M. (1947) 'Nagra glacialmorfologiska formelement', *Geog. Annlr*, 27, 3–329.

MITCHELL, G. F. (1960) 'The Pleistocene history of the Irish Sea', *Advmt Sci.* 17, 313–25.

MITCHELL, G. F. (1967) 'Preliminary results of excavations at St Erth, Cornwall', paper read to the Brit. Ass. Adv. Sci.

MITCHELL, G. F. and ORME, A. R. (1967) 'The Pleistocene deposits of the Isles of Scilly', *Q. Jl geol. Soc. Lond.* 123, 59–92.

OAKLEY, K. P. (1964) *Frameworks for Dating Fossil Man.* London.

OAKLEY, K. P. (1968) 'The date of the "Red Lady" of Paviland', *Antiquity*, 42, 306–7.

PRESTWICH, J. (1892) 'The Raised Beaches and "Head" or Rubble Drift of the South of England', *Q. Jl geol. Soc. Lond.* 48, 263–343.

PRICE, R. J. (1960) 'Glacial meltwater channels in the upper Tweed basin', *Geogr. Jl.* 126, 483–9.

PRINGLE, J. and GEORGE, T. N. (1961) *British Regional Geology. South Wales.* H.M.S.O. (1st edn 1948.)

ROHDENBURG, H. (1965) 'Untersuchungen zur pleistozänen Formung am Beispiel der Westabdachung des Göttinger Waldes', *Giessener Geographische Schriften.* 7, 5–76.

SAVIGEAR, R. A. G. (1953) 'Some observations on slope development in South Wales', *Trans. Inst. Brit. Geog.* 18, 31–52.

SHOTTON, F. W. (1962) 'The physical background of Britain in the Pleistocene', *Advmt Sci.* 19, 193–206.

SISSONS, J. B. (1967) *The Evolution of Scotland's Scenery.* Oliver & Boyd.

SOLLAS, W. J. (1913) 'Paviland Cave: an Aurignacian station in Wales', *Jour. Roy. Anthrop. Inst.* 43, 325.

STEERS, J. A. (1946) *The Coastline of England and Wales.* Cambridge University Press. 2nd edn 1964.

STEPHENS, N. (1966) 'Some Pleistocene deposits in north Devon', *Biul. Peryglac.* 15, 103.

STRAHAN, A. (1907a) *The Geology of the South Wales Coalfield. Part 8. The country around Swansea.* Mem. Geol. Surv. Gr. Br. H.M.S.O.

STRAHAN, A. (1907b) *Ibid. Part 9. The country around West Gower and Pembrey.*

STRAHAN, A. (1909) *Ibid. Part 1. The country around Newport.* (2nd edn.)

STRAHAN, A. and CANTRILL, T. C. (1902) *Ibid. Part 3. The country around Cardiff.* (2nd edn 1912.)

STRAHAN, A. and CANTRILL, T. C. (1904) *Ibid. Part 6. The country around Bridgend.*

STRAHAN, A. and GIBSON, W. (1900) *Ibid. Part 2. The country around Abergavenny* (2nd edn 1927 by T. Robertson).

STRAHAN, A., TIDDEMAN, R. H. and GIBSON, W. (1903) *Ibid. Part 4. The country around Pontypridd and Maesteg* (2nd ed 1917; 3rd edn, see under Woodland and Evans).

STRAHAN, A., GIBSON, W. and CANTRILL, T. C. (1904) *Ibid. Part 5. The country around Merthyr Tydfil* (2nd edn, 1931, by T. Robertson).

STRAHAN, A., CANTRILL, T. C., DIXON, E. E. L. and THOMAS, H. H. (1907) *Ibid. Part 7. The country around Ammanford.*

STRAHAN, A., CANTRILL, T. C., DIXON, E. E. L. and THOMAS, H. H. (1909) *Ibid. Part 10. The country around Carmarthen.*

STRAHAN, A., CANTRILL, T. C., DIXON, E. E. L., THOMAS, H. H. and JONES, O. T. (1914) *Ibid. Part 11. The country around Haverfordwest.*

TAYLOR, J. A. and CRAMPTON, C. B. (1967) 'Solifluction terraces in South Wales', *Biul. Peryglac.* **16**, 15–36.

WATSON, E. (1966) 'Two nivation cirques near Aberystwyth, Wales', *Biul. Peryglac.* **15**, 79–101.

WIRTZ, D. (1953) 'Zur Stratigraphie des Pleistocäns in Westen der Britischen Inseln', *Neues Jahr. Geol. u. Pal.* **96**, 267–303.

WOLDSTEDT, P. (1967) 'The Quaternary of Germany', in *The Geologic Systems: The Quaternary*, ed. K. Rankama, Vol. 2. Interscience.

WOODLAND, A. W. and EVANS, W. B. (1964) *The Geology of the South Wales Coalfield. Part 4. The country around Pontypridd and Maesteg.* Mem. Geol. Surv. Gr. Br. H.M.S.O.

WRIGHT, W. B. (1914) *The Quaternary Ice Age.* London.

ZEUNER, F. E. (1959) *The Pleistocene Period.* 2nd edn. Hutchinson.

# CHAPTER TEN

# Pembrokeshire

Brian S. John, D.Phil., M.A.

The Pembrokeshire peninsula is thrust south-westwards into the oceanic approaches to Great Britain. It is remote from the centres of local Welsh glaciation, and as a consequence has experienced a less complex pattern of glacial events than some of the other areas considered in this book. Whilst the adjacent Teifi and Towy valleys have suffered intermittent local valley glaciation, there is little evidence in Pembrokeshire for the conflict of Welsh ice and Irish Sea ice as seen on other parts of the western seaboard of Wales. There may have been local glaciers in the Prescelly Mountains at some stages of the Pleistocene, but the glacial landforms and deposits described in the county appear to be related almost entirely to the processes of glaciation by the Irish Sea glacier. As such, they are relatively simple and readily permit geomorphological and stratigraphic interpretation.

The landscape of Pembrokeshire is related orographically to the coastal plateaux of Lleyn and Anglesey. There is an irregular series of platform remnants sloping from 120 m (c. 400 ft) in the north-east to 30 m (c. 100 ft) in the Castlemartin peninsula (Miller, 1937). Above these platforms there are more irregular surfaces on the flanks of the Prescelly Mountains, which rise through gentle undulations to 520 m (1760 ft) in the north-east of the county. To the west of the mountains the Treffgarne ridge points a long finger of moorland from the foothills to the coast of St Bride's Bay, while several rocky monadnocks stand above the coastal platforms of Dewisland and Pen Caer. In aspect, north Pembrokeshire is wild and treeless except for occasional small woods which shelter on the slopes of steep-sided valleys. Around the lower drainage basins of the Eastern Cleddau and Western Cleddau the landscape is more deeply dissected, but most of the interfluves lie beneath an altitude of 100 m (300 ft), and extensive erosion surfaces can still be easily discerned (Burton, 1952). The major feature of southern Pembrokeshire is the drowned river system of Milford Haven, with tidal inlets as far inland as Haverfordwest and

Canaston Bridge. South of Milford Haven the Castlemartin peninsula bears comparison with Dewisland in the overall appearance of its flat, relatively treeless landscape; however, it lacks the upstanding monadnocks of the St David's area, and the coastal flats at Flimston and Bosherston are more perfectly preserved than any in the north of the county.

Geologically the county is complex (Pringle and George, 1948). The north is composed largely of Ordovician and Silurian sedimentary rocks, with igneous intrusives and extrusives especially in the Prescelly Mountains, Pen Caer, and around the western part of the Treffgarne ridge. Dewisland is as distinctive geologically as it is orographically, for here there is a major eroded anticline exposing Pre-Cambrian igneous rocks and Cambrian sedimentaries.

South of Haverfordwest and Narberth more recent rocks are exposed. A belt of Coal Measures runs from St Bride's Bay to Carmarthen Bay. There is an extensive outcrop of Old Red Sandstone centred around Milford Haven, and much of the spectacular cliff scenery of the south coast of the Castlemartin peninsula is carved in Carboniferous limestone. The county's structural trends are Caledonian in the area of Lower Palaeozoic rocks, and Armorican in the south. In spite of the geological diversity of the county, the landscape reflects bedrock and structure only in some details; for example, the major inlets on the south shore of Milford Haven are related to the exploitation of beds of the 'lower limestone shale' of Lower Carboniferous age, while the upstanding monadnocks and rocky headlands of north Pembrokeshire coincide with exposures of Ordovician intrusive rocks.

Any study of glaciation in Pembrokeshire must be intimately related to the county's major morphological features. In the first place, the extensive flat erosion surfaces are only thinly covered with drift; inland morainic and fluvioglacial landforms are rare, and drift exposures infrequent. On the other hand the rocky indented coastline has innumerable small coves which have been plugged with till and other glacial deposits. In these localities comprehensive drift sections can be observed, and of necessity the greater part of this chapter is based on interpretations of coastal drift stratigraphy as revealed in these sections. Indeed, past authors such as Prestwich (1892), Jehu (1904), Leach (1911) and Griffiths (1940) have all been forced to use the coastal drift stratigraphy to interpret the sequence of Pleistocene events.

Erratics and drift characteristics were the first elements of the glacial landscape to excite attention in this area. As a result of the importance of the county's solid geology, geological investigation commenced at an early stage, and the specialist literature before 1900 contained scattered references to the glaciation of Pembrokeshire. But it was not until 1904, with the work of T. J. Jehu, that any comprehensive examination of the glacial deposits was undertaken. Jehu worked mainly in the north of the county, and in the course of his skilful and comprehensive analysis of the drifts he proposed the tripartite succession (Lower boulder clay, Middle sands and gravels, Upper boulder clay) which was in fashion at that time. His differentiation of a lower or Irish Sea till from an upper till or 'rubbly-drift' has been used by many subsequent authors in discussions of the Pleistocene chronology of west Wales. Jehu agreed with the suggestion of Hicks (1885) that the Irish Sea glacier had overridden most of north Pembrokeshire from north-west to south-east. A further advance came in 1911, when Leach showed from Porth-clais (742237) that solifluction deposits and till overlie the local representatives of the raised beach series. Thus by 1911 the five major components of the Pembrokeshire coastal sections had been described: namely raised beach gravels, solifluction deposits, Irish Sea till, fluvioglacial sands and gravels, and local till or rubbly-drift. Glacial morphology as such was not examined in any detail until Charlesworth (1929) analysed some of the features of north Pembrokeshire. He recognised a complex series of 'overflow' channels on the north-western flanks of the Prescelly Mountains, supposedly formed in association with a belt of hummocky fluvioglacial material on the coast around Fishguard which he termed the 'South Wales end-moraine'. He related this end-moraine to the last (Newer Drift) glaciation, and referred all the supposedly denuded drifts further south in the county to the Older Drift glaciation. Thus his work supported the thesis of two glaciations which had arisen from Jehu's stratigraphic studies. Although much has been written on the glaciation of Pembrokeshire since 1929, little fresh fieldwork has been undertaken and much of the discussion has been preoccupied with glacial chronology. The story of glaciation told by Pringle and George (1948) is based essentially on the work of Jehu and Charlesworth, and is still accepted as authoritative by many people.

It is worth examining some of the long-accepted hypotheses with the aid of specific examples. In the following pages there is an examination of the glacial features of Pembrokeshire, both from a stratigraphic and morphological point of view. Four coastal sections in north Pembrokeshire are described, and three from south Pembrokeshire. Following these descriptions, an attempt is made to correlate the events represented at each locality, and then to assess the reliability of earlier schemes of Pleistocene chronology in the area. The meltwater channels of north Pembrokeshire are described and interpreted, and finally an independent sequence of events is presented with the help of analyses of organic material and radiocarbon age determinations.

DRIFT STRATIGRAPHY

### North Pembrokeshire

At Aber-mawr (883347), close to the base of the Pen Caer peninsula, occurs possibly the most useful drift section in Pembrokeshire. Above the present storm-beach at the head of the bay is a drift cliff up to 13 m (40 ft) high in which both Irish Sea till and Jehu's problematical 'rubbly-drift' are displayed (Fig. 10.1A). The lower part of the drift cliff is composed of a variety of facies of head, including two blocky head horizons and one horizon of bedrock flakes with erratic pebbles. There is at least one weathering horizon towards the base of the head sequence. Overlying the head and truncating its pseudo-stratification is a wedge of massive purple Irish Sea till up to 2·5 m (8 ft) thick. This till is tenaceous and plastic, although there are occasional thin lenses of sand and silt in its upper layers. It is highly calcareous except within 0·5 m (18 inches) of its upper surface and within 0·25 m (9 inches) of its base. Like most exposures of Irish Sea till in the area, it contains abundant fragments of marine mollusca and pieces of carbonised wood, as well as a variety of erratics from northern sources (Jehu, 1904). The till is overlain by up to 4·5 m (15 ft) of outwash sands and gravels which display many complex bedding structures, apparently indicative of deposition in a dead-ice environment. In places discontinuous patches of weathered Irish Sea till are interbedded with sands and gravels.

Overlying the purple till in the highest part of the section, and itself overlain by only 0·75 m (2 ft) of sandy loam and blown sand, is a deposit described by Jehu as 'a rubbly clay, full of boulders of all sizes, most of which are angular and sub-angular, and derived from rocks in the locality' (p. 75). He assigned the deposit to his Upper boulder clay, and Synge (1963) has also interpreted the deposit as a true till. However, in view of the predominance of local rock types and the downslope orientation of the great majority of included rock fragments, the deposit is more likely to be an upper head which has incorporated some glacial and fluvioglacial material.

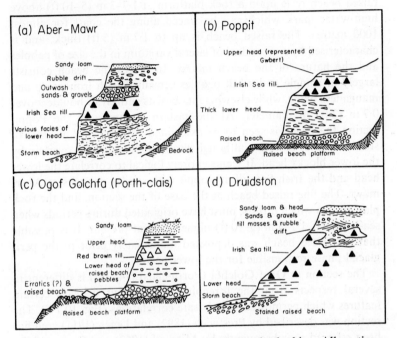

Fig. 10.1. Coastal drift sections in north Pembrokeshire. All sections generalised and not to scale

The only true glacial deposits at Aber-mawr, therefore, are the Irish Sea till and its associated outwash; the glacial drifts are underlain by a thick sequence of periglacial slope deposits and overlain by a thinner head horizon, apparently indicating that the Irish Sea glaciation was preceded by a prolonged periglacial phase and succeeded by a shorter periglacial phase.

At Poppit (145490), on the Teifi estuary, is a drift exposure which is similar in many respects (Fig. 10.1 B). Again the dominant components of the drift sequence are a thick, blocky lower head and an overlying purple shelly Irish Sea till; the head is at least 6 m (20 ft) thick in the western part of the exposure, while the Irish Sea till attains a maximum thickness of at least 2 m (7 ft). To date, no upper head has been found above the Irish Sea till at this site; however, it exists at the mouth of the Teifi estuary, and it may well be present on the higher, inaccessible part of the drift cliff. The most interesting element of the Poppit succession occurs at the base of the drift exposure, where a fine raised beach rests upon a rock platform at 1·7–3 m (5–10 ft) above high-water mark which can be traced along the coast for almost 1000 metres. The raised beach is up to 1·7 m (5 ft) thick, and is characterised by a great deal of lateral variation in the size of pebbles and the nature of the beach matrix. In places the beach consists largely of shingle and small pebbles cemented with iron oxide and manganese oxide, while elsewhere it consists entirely of boulders over 0·7 m (2 ft) in diameter. At the eastern extremity of the Poppit section the beach is seen to be associated with up to 2 m (7 ft) of stratified sands and silts, although elsewhere it is directly overlain by the lower blocky head. It seems entirely logical to correlate the lower head and the Irish Sea till at Poppit with similar deposits at Aber-mawr. The fine raised beach at the base of the section, and the rock platform on which it rests, must have originated during periods when sea-level was 1·7–3 m (5–10 ft) higher than at present. It is possible that the raised beach was deposited immediately prior to the peri-glacial phase responsible for the lower head.

The section at Ogof Golchfa (Porth-clais) (742237) is different in several respects from the foregoing section, but displays certain features which permit reasonably safe correlations (Fig. 10.1 C). The section was described by Prestwich (1892) and Leach (1911), and has been referred to subsequently by Mitchell (1962) and Synge (1963). The drift deposits at Ogof Golchfa rest upon a series of rock plat-forms varying in height between 4·5–11 m (15–35 ft) O.D. on a small headland just to the west of Porth-clais harbour. The platform is striated in places, and supports a number of boulders which have been referred by Leach to a cold (possibly glacial) period which preceded the deposition of overlying beach gravels. However, there seems little evidence in favour of this, for most of the boulders are of local origin, and the only provable far-travelled erratics at the site appear to rest on the raised beach, and could easily have been emplaced by slump-ing from the till higher in the section. It is nevertheless true that there are erratic pebbles in the raised beach itself; these may have been derived from pre-existing glacial deposits in the area, as suggested by Mitchell (1960). The beach attains a maximum thickness of *c.* 1 m (3 ft) on the inaccessible western part of the exposure, but at one time it must have been of considerable thickness, for the 2·7 m (8 ft) of overlying lower head consists almost entirely of soliflucted beach pebbles. Above this head there is up to 2·7 m (8 ft) of red-brown non-calcareous till which consists of a variety of local and foreign striated rock fragments and boulders in a sandy and gravelly matrix. This till has been referred to by Mitchell (1962) and Synge (1963) as a local Welsh till, presumably laid down by Welsh ice which was contemporaneous with, or slightly later than, the Irish Sea ice in this area. On the other hand the erratic suite of the till, and the striations on the platform surface, indicate that the ice which overrode this site moved towards the south-east. This is in accord with the directions of ice-movement recognised by Hicks (1885) and Jehu (1904) for the Irish Sea glacier, and it seems more likely that the Ogof Golchfa till is a local 'land facies' of the Irish Sea till. This is apparently confirmed by the distribution of calcareous Irish Sea till around the Pembrokeshire coasts, for it is found only where the Irish Sea ice moved onshore. Where it moved offshore, as on the south side of the St David's peninsula, only stony and sandy local tills were laid down. Above the till is a thin upper head of local Cambrian sandstone and shale fragments, and capping the section is about 0·7 m (2 ft) of sandy loam. On the basis of the whole section it is possible, therefore, to construct a chronology for Ogof Golchfa which contains the early and late elements of the chronology represented at Poppit and Aber-mawr respectively: a stage of marine platform erosion was followed by the deposition of a storm beach, and later by an early prolonged period of solifluction, a glaciation from the north-west, and a late periglacial phase of limited duration.

At Druidston (862173), on the west-facing shore of St Bride's Bay, there is a section in which true Irish Sea till is again represented (Cantrill *et al.*, 1916; Bowen, 1966). At the head of Druidston Haven a deeply cut rock channel is completely plugged with drift deposits (Fig. 10.1D). At the base of the section cemented and stained raised beach shingle is occasionally revealed beneath the level of the present storm-beach. This is overlain by *c.* 2 m (6 ft) of blocky quartzite head which is also stained with iron oxide, and then by up to 15 m (50 ft) of Irish Sea till. This till exposure is the most extensive in Pembrokeshire, and is being actively denuded today by the processes of gullying and sludging. The till is highly calcareous, and contains northern erratics, marine mollusc fragments, and pieces of carbonised wood. In its upper layers it is decalcified and gleyed to a depth of *c.* 1·3 m

(4½ ft), and there is another thin decalcified reddish layer at the base of the till. The upper layers of the till contain a high proportion of sand and gravel, and it grades upwards into a variable deposit of rubble-drift which incorporates patches of bedded outwash sands, layers of till, and soliflucted bedrock fragments and glacial drift. In its stratigraphic position and its lithology, the rubble-drift is comparable to the rubble-drift at Aber-mawr. As at Aber-mawr also, it is overlain by 0·3 m (12 inches) of sandy loam with included head fragments. From the stratigraphic association of Irish Sea till with discontinuous masses of contorted outwash deposits, it seems possible that the upper layers of the till were deposited in a dead-ice environment (John, 1965a). Fabric analyses favour this interpretation, and there is no reliable evidence in support of Bowen's (1966) contention that the whole till mass at Druidston has been soliflucted. In most of its characteristics, therefore, the Druidston section is comparable to the sections at Poppit and Aber-mawr. A period of raised beach deposition was followed by a periglacial phase and then by an Irish Sea glaciation. The wasting of the Irish Sea ice, and the following periglacial phase, are represented at this site by the variable rubble-drift above the till.

At other localities on the north Pembrokeshire coast similar drift successions are encountered. On the western coast of St David's peninsula the well-known Whitesands sections (733273) again show that the Irish Sea till is underlain by head and overlain by a variable rubble-drift of local rock fragments, erratics, and lenses of sand and gravel (Jehu, 1904; Green, 1911). In the drift cliffs in the southern part of Whitesands beach there is no Irish Sea till. However, the thickness of the main (lower) head is confirmed, and it is underlain by a stained red sand and by a cemented basal deposit which incorporates erratic boulders, raised beach pebbles and head fragments resting on a raised beach platform. Above the main head at this site is a thick cemented 'sandrock', a deposit recorded nowhere else in North Pembrokeshire. At Porthmelgan (728279), close to St David's Head, there is a drift succession which is closely comparable to the successions at Druidston and Gwbert (163495), while small sections at Parrog (050398) on the Afon Nevern estuary also reveal an 'ideal' succession. There are few sections where real anomalies occur, although at Pen Dal-aderyn (717233) at the southern end of Ramsey Sound there is no till, no fluvioglacial deposits, and no rubble-drift. A drift cliff up to 3·7 m (12 ft) high reveals at least six successive head facies, the lowest resting on a fine raised beach platform. Again, there is an anomaly at Caerbwdy (765243) on the south side of the St David's peninsula, where a variety of fluvioglacial deposits is seen *beneath* 3 m (10 ft) of stony local till.

## South Pembrokeshire

There are few well-preserved drift sections in south Pembrokeshire, and from those which can be studied the stratigraphic relationship of the drift succession with that of north Pembrokeshire is not immediately apparent. Calcareous Irish Sea till is conspicuous by its absence from the Castlemartin peninsula, and the drifts as a whole are more closely related to those of the Bristol Channel coasts than those of the Irish Sea.

However, at Mullock Bridge (811080), close to the north shore of Milford Haven, there is a drift sequence which can be related with some confidence to the sequence around St Bride's Bay (Fig. 10.2A). A large gravel-pit has been cut into the surface of a kame terrace on the side of the Gann valley, exposing a thick sequence of fine white basal sands and coarser gravels with marine mollusc fragments and erratics from the north-west. These materials are of fluvioglacial origin, and display complex cross-bedding and slumping and contortion structures compatible with deposition among masses of dead ice. Beneath the basal sands there is a thin layer of laminated silt, while here and there in the pit a blocky head of Devonian red marl fragments is seen at the base of the succession. At the top of the succession there is up to 2 m (6 ft) of torrential gravels, overlain by about 0·7 m (24 inches) of sandy loam. On the outer flank of the terrace the uppermost deposit is a wedge of gravels which have been redeposited by solifluction. In places the terrace supports a stony local till, beneath which the outwash gravels appear to be contorted; and another important feature of the terrace is a finely developed fossil ice-wedge which passes from the torrential gravels into the shelly gravels beneath. The deposits in the pit suggest that the following sequence of events may have occurred. Initially there was a periglacial phase of head formation, followed by a glaciation from the north and west by the Irish Sea glacier. The outwash deposits and till patches were laid down during the wastage of this glacier, and there was later solifluction and ice-wedge formation during a short periglacial phase. As elsewhere, the sandy loam on the terrace surface may indicate a final phase of aeolian deposition. This sequence of events is similar in all respects to the chronology proposed for north Pembrokeshire.

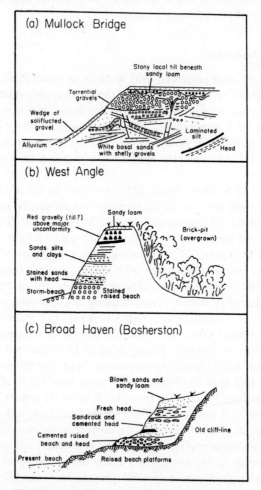

Fig. 10.2. Coastal drift sections in south Pembrokeshire. All sections generalised and not to scale

At West Angle (853032), at the mouth of Milford Haven and on its southern shore, there is a series of sections which are similar in some respects to those further north. At the head of the main bay is a cliff section close to the brick-pit described by Dixon (1921). Just above the level of the present storm-beach there is a stained raised beach about 0·5 m (18 inches) thick, which is overlain by a sandy deposit with raised beach pebbles and head (Fig. 10.2B). This passes upwards into a series of white sands and silts, light grey silts, and then dark grey clay. In the sands and silts there are wood fragments, small pieces of organic matter, and some fragments of marine mollusca. It is possible that this suite of deposits represents a marine transgression, with the lower sandy facies representing the initial rise of sea-level and the upper laminated clay representing a later stage of tidal estuarine conditions (John, 1968). The character of pollen from the sands and silts is compatible with an estuarine environment.

Above this suite there is an unconformity, overlain by up to 3·7 m (12 ft) of rich red sand and gravel, which occasionally assumes the character of a gravelly local till. This apparently glacial deposit is not overlain by head in the section exposed above the beach, but Dixon (1921) has recorded head above the equivalent deposit closer to the old cliff line of the bay in the adjacent brick-pit. Capping the whole drift succession is the ubiquitous sandy loam.

While the suite of sands, silts and clays at West Angle is a unique feature, it is quite possible that the raised beach is the equivalent of the raised beaches of north Pembrokeshire. Furthermore, the reddish glacial deposits overlain by head and sandy loam may also be equated with the upper members of the drift succession further north. The apparent transgression at West Angle may be related to the warmer stage responsible for the raised beach, for the evidence from Ogof Golchfa and elsewhere shows that sea-level at some time during the 'raised beach' stage was high enough to construct storm-beaches at over 9 m (30 ft) O.D. This may indicate a maximum sea-level of at least 4·5 m (15 ft) O.D.

Finally a drift succession at Broad Haven, near Bosherston on the south Pembrokeshire coast, is worthy of mention (Fig. 10.2C). On the cliffs adjacent to the outlet of Bosherston Pools there are several raised beach platform remnants cut across Carboniferous limestone at altitudes of 1–4·5 m (3–15 ft) above high-water mark. The upper platform remnants are badly dissected by gullies; however, these gullies often contain masses of solidly cemented shelly raised beach deposits, which are also found in isolated patches on the lower platform surfaces. In places the raised beach is seen to grade upwards into a cemented blocky limestone head or into roughly stratified sandrock up to 2 m (6 ft) thick. On the easternmost rock platform there are thick masses of cemented raised beach pebbles and head on the platform surface, often roofing over gullies and small caves. This head appears to underlie the sandrock, although the latter deposit is also overlain by another cemented blocky head. In addition there is a non-cemented head up to 1 m (3 ft) thick above the cemented deposits on the westernmost platform remnant. Thus there appear to be three head deposits at Broad Haven; the lowest is intimately associated with the raised beach, while another cemented head postdates the sandrock and pre-dates an apparently recent fresh head. At the top of the sections there is up to 2 m (6 ft) of blown sands and sandy loams. In many respects the drift succession at Broad Haven is similar to successions from Gower and the south-west peninsula, especially in view of the lack of till at the site. The cemented raised beach, the lower head deposits and the sandrock may be equivalent in age to the raised beaches and main head of north Pembrokeshire, while the non-cemented upper head and the blown sands may be the equivalents of the deposits at the top of sections further north. In this case the Irish Sea glaciation may have occurred between the uppermost cemented deposit and the base of the uncemented upper head. In view of the lack of glacial deposits in the vicinity it is impossible to estimate at this stage whether the Irish Sea ice overrode this locality.

### Relative dating of the Pembrokeshire drifts

There are many clues in the foregoing descriptions which indicate a relatively uniform sequence of Pleistocene events in Pembrokeshire. Most of the correlations between deposits at different localities have already been mentioned but they can now be expressed in diagrammatic form as in Fig. 10.3. The raised beaches (3) are widespread around the Pembrokeshire coasts at a variety of altitudes beneath 11 m (35 ft) O.D., and it is reasonable to assign them to an interglacial period of fluctuating sea-levels. The raised beaches described on the foregoing pages may have been deposited at different stages of this interglacial, and it is possible that a marine transgression is represented by the deposits at West Angle. To avoid confusion later in this chapter, the raised beaches are here referred to the 'Poppit interglacial', named after the excellent exposure in the Teifi estuary. The lower or main head (4) is generally a thick deposit which may indicate a prolonged periglacial phase. It overlies the raised beach and is therefore younger, and by the same stratigraphic principle it must be

25 A subglacial meltwater channel at Maesyprior near Carmarthen. Note the hump on the channel long profile. The channel (A in fig. 9.5) makes a right-angled turn to the south (right-hand side) in the middle foreground.

26 A view of the cliffed solifluction terrace between Oxwich Point and Eastern Slade. Landwards the partially buried interglacial cliff may be seen, while seawards the interglacial shore zone has been exhumed by the sea from beneath solifluction debris.

27 A tor developed on Rhondda Beds on Craig Ogwr at the head of the Ogwr valley. (Immediately south of the A4061 road may be seen an isolated pillar of sandstone on the west-facing slope of Craig-y-gelli.)

28 Raised (*Patella*) beach at Seven Slades, Gower. The beach has been cemented into a tough conglomerate and contains shell fragments, pebbles of Carboniferous Limestone, Old Red Sandstone, and Coal Measures Sandstones. Note that the amount of angular material increases towards the top of the deposit.

29 A fossil ice wedge and traces of cryoturbation in the torrential gravels, Mullock Bridge gravel-pit.

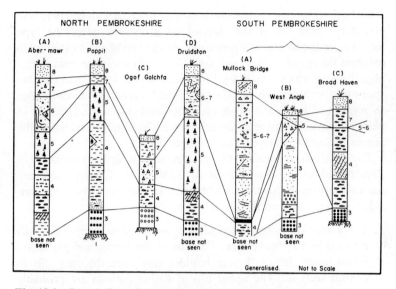

Fig. 10.3. Correlation of selected stratigraphic columns for the Pembroke-shire drifts (see text for explanation)

older than the overlying glacial drift. From the stratigraphic columns it seems that there is only one suite of glacial deposits in the county. This suite comprises calcareous Irish Sea tills (5), local stony tills (5) laid down as the 'land facies' of the Irish Sea till, and fluvioglacial

TABLE 10.1. *Drift stratigraphy and Pleistocene events in west Wales*

| Drifts | Possible events |
|---|---|
| 8. Sandy loams and blown sand | Temperate conditions—aeolian and hillwash deposition |
| 7. Rubble-drift and upper head | Short periglacial phase |
| 6. Fluvioglacial sands and gravels ⎫<br>5. Irish Sea till and local tills ⎭ | Irish Sea (Dewisland) glaciation |
| 4. Thick lower head | Prolonged periglacial phase |
| 3. Raised beach | Poppit interglacial—sea-level up to 9 m (30 ft) O.D.? |
| 2. Erratics | Glaciation? |
| 1. (Rock platform) | Interglacial(s)—sea-levels up to 15 m (50 ft) O.D.? |

sands and gravels (6). The deposits may be referred with safety to one glaciation, here called the 'Dewisland glaciation' after the type localities in the St David's peninsula. At the top of the stratigraphic columns the upper head (7) and sandy loams (8) may date from a periglacial phase which was less severe and less prolonged than that responsible for the main head; the upper head is generally much thinner than that which formed before the Dewisland glaciation, although in places the availability of easily solifiucted glacial materials led to the accumulation of relatively thick 'rubble-drifts'.

The conclusions based on the interpretation of drift stratigraphy in Pembrokeshire are summarised in Table 10.1. Clearly there are several features of this table which provoke comment when one considers earlier views on the glaciation of Pembrokeshire.

1. There appears to be no foundation for the classic tripartite drift division in Pembrokeshire. The tills of the Dewisland glaciation (constituting Jehu's 'Lower boulder clay') are the only tills in coastal sections. The upper tills of earlier authors are probably solifluction deposits which are the equivalents of the upper head, although they may sometimes include masses of glacial drift.

2. The tills of the Dewisland glaciation are found at least as far south as West Angle. Furthermore, there are no unique glacial deposits on the north Pembrokeshire coast within Professor Charlesworth's 'South Wales end-moraine'. It seems, therefore, that the feature does not mark the limit of the Newer Drift glaciation. The most recent glaciation of Pembrokeshire was the Dewisland glaciation, during which the Irish Sea ice extended well to the south of Charlesworth's limit.

3. Although there is only one suite of coherent glacial deposits in Pembrokeshire, there are some indications that there may have been an 'early glaciation' prior to the Poppit interglacial. At Abermawr, Ogof Golchfa and Whitesands erratics have been recorded either in association with raised beach deposits or in the lower head. Leach (1911) and Stephens (1966) have favoured the idea that such erratics may be ice-rafted during cold periods, but as suggested by Mitchell (1960) they are best explained by relating them to the destruction of pre-existing glacial deposits. Whichever hypothesis is correct the erratics must be indicators of a stage of glacial climate somewhere in western Britain, even if the ice of this early glaciation did not override Pembrokeshire.

## GLACIAL MELTWATER CHANNELS

Apart from a possible corrie and several nivation hollows in the Prescelly Mountains, the only noticeable landforms of glacial erosion in Pembrokeshire are its meltwater channels. These are spectacular elements of the landscape wherever they occur, although they are concentrated for the most part in the north of the county. These channels contribute a great deal towards our understanding of the Pleistocene period in Pembrokeshire, and provide especially valuable information about the course of the postulated early glaciation.

### Channels in Dewisland

In Dewisland there are many deeply cut channels on both the north and south coasts of the peninsula. The larger channels are occasionally 30 m (100 ft) deep, and are incised into the coastal platforms; examples are Merry Vale, the Solfach and St Elvis valleys, and the Abereiddy and Aber-mawr valleys, all of which support small misfit streams at the present day. The latter two valleys are especially well marked. They are steep-sided and flat-floored, and are elements of diverging channel systems which all but isolate Barry 'Island' and Mynydd Morfa from the mainland. Between Solva and Newgale there are deeply cut 'cwms' on the coast at a number of localities, and shallow dry channels are seen at St Justinian (724255), Carneddgwion (745287), and Pwll Crochan (823323). In addition there are distinct one-sided benches on the north coast at Pen Clegyr (769298) and Carreg-yr-afr (757290).

Previously the deeply cut channels have been ascribed to subaerial stream rejuvenation (Cox, 1930; Burton, 1952). While many of them may indeed be subaerial in origin, the size of the largest ones points to erosion by considerable volumes of meltwater. At Porthmelgan and St Justinian there are rock channels with humped long-profiles, and at Morfa (880331) and Barry Island Farm (814317) adjacent channels are linked by apparent meltwater routes across cols. Cwm Mawr and Crow Cwm near Newgale (847224) are sizeable features too large to have been cut by simple headward erosion by their misfit streams. In Merry Vale (741243), close to the point where the road from St David's enters the channel, an old valley spur supports a complex arcuate channel which has two outlets (Fig. 10.4). The channel is up to 6 m (20 ft) deep, has a humped long profile, and its intake and outlet points hang up to 7·5 m (25 ft) above the present valley floor.

Thus many of the features of the Dewisland coastal channels are difficult to explain by the processes of 'normal' stream erosion. They have several of the diagnostic features of glacial meltwater channels (Sissons, 1960, 1961), and indeed much evidence seems to indicate that there has been subglacial erosion by meltwater at some stage of valley development.

Fig. 10.4. Merry Vale and the arcuate channel across the spur close to Porthclais, Pembrokeshire

## The Gwaun–Jordanston system of channels

This channel system, on the north-western flanks of the Prescelly Mountains, has long been recognised as a legacy of glaciation. The present-day streams of the area flow for the most part in steep-sided and flat-floored rock channels which form an interconnected system (Fig. 10.5), although few of them are used for the whole of their length by one stream. Just as the channels are cut through pre-existing cols, so the present watersheds of the area occur on the flat channel floors (Fig. 10.6). The largest channel is that of the Gwaun (030340), which is over 14 km (8 miles) long and which passes between the hill masses of Carningli and Mynydd Prescelly to connect the drainage basins of Fishguard Bay and Newport Bay. Close to the western end of the channel there is an intricate pattern where the Cwmonnen channel joins the Gwaun from the north-east and where three major exits from the Gwaun channel swing towards the south and west. The greatest complexity is seen in that part of the channel system which culminates in the Scleddau channel (in the extreme west), while the most spectacular feature of all is the Nant-y-bugail

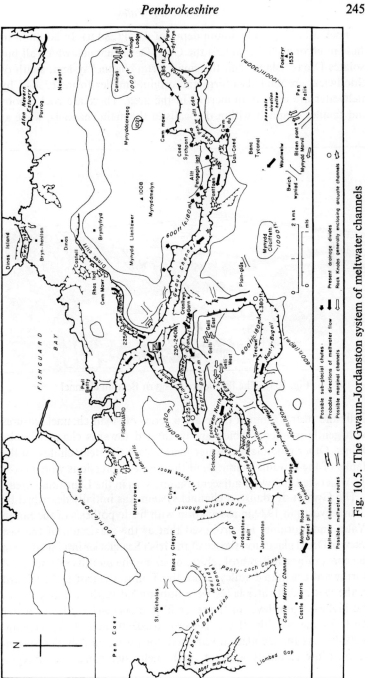

Fig. 10.5. The Gwaun-Jordanston system of meltwater channels

channel, attaining a maximum depth of 85 m (280 ft). In addition to these interconnected channels the Cwm Dewi channel, which all but isolates Dinas 'Island' and the small Drim channel system, south of Goodwick, are related in form and origin. Smaller forms eroded by meltwater are equally numerous in the area. There are rock knobs and smaller channels within the walls of the main channels, while at

Fig. 10.6. The flat-floored Esgyrn Bottom channel

Gelli and near Carningli there are a series of small channel features with gentle gradients which are orientated towards channel intake points. Other possible meltwater features are the steep chutes on the flanks of the Gwaun channel and the gentle depressions which cross the watersheds between adjacent channels, as in the Llanstinan area.

The Gwaun–Jordanston channel system was initially described by Charlesworth in 1929 in association with his hypothesis of the South Wales end-moraine. He proposed that at the maximum of the last (Newer Drift) glaciation the ice of the Irish Sea glacier impinged upon the coast of north Pembrokeshire, blocking the mouths of a series of northward-orientated valleys from the Teifi in the east to Fishguard in the west. Meltwater lakes were impounded in these valleys, with the highest water-level in 'Lake Teifi' and successively lower water levels to the west in 'Lake Nevern' (070390) and 'Lake Gwaun' (030340). As the ice-margin retreated northwards a complex system of overflow channels was cut, with meltwater sometimes flowing westward along the ice-margin and sometimes over pre-existing cols between the drainage basins (Fig. 10.7). The Gwaun–Jordanston system was thought to have been cut at a relatively late stage of deglaciation by meltwaters overflowing from Lake Gwaun via a series of successively lower cols into the Western Cleddau basin. Since ice completely blocked the drainage basin of Fishguard Bay at this time, a 'Lake Manorowen' (940370) was thought to have been impounded in the extreme west; this lake also overflowed southwards via the Jordanston channel into the Western Cleddau basin. Thus the greater part of the meltwater from the glacial lake series is thought to have flowed into Milford Haven via the Western Cleddau and the spectacular Treffgarne Gorge.

Since Charlesworth first proposed this glacial lake hypothesis, it has been accepted with various modifications, by Griffiths (1940), M. Jones (1946), and O. T. Jones (1965). The hypothesis has received widespread recognition as a result of its inclusion in the *South Wales Regional Geology Handbook* (Pringle and George, 1948), and Charlesworth has restated his views in 1963.

It has been recognised in recent years that there are several features of the Gwaun–Jordanston channel system which the glacial lake hypothesis cannot explain. Among these the following may be quoted:

(*a*) Several inconsistencies appear in Charlesworth's lake overflow hypothesis if one attempts a simple reconstruction of preglacial relief in the area. For example, Lake Nevern supposedly had a surface level of *c*. 116 m (380 ft) O.D., and overflowed a low col at about this altitude into the Gwaun valley. However, a reconstruction of the preglacial watershed in this area shows that the col was probably at an altitude of *c*. 180 m (600 ft) O.D. Again, Lake Gwaun, with a surface level below 116 m (380 ft), supposedly overflowed by a lower col into the Nant-y-bugail channel; but a reconstruction of this col shows that it had an altitude of *c*. 200 m (650 ft) O.D. Further problems arise when one attempts to reconstruct the altitudes of preglacial cols in the Esgyrn Bottom–Criney Brook area.

(*b*) At least three major channels in the area are not accounted for in Charlesworth's hypothesis. These are the Blaenffos channel (195368) (intake at *c*. 170 m (560 ft) O.D.) and the Crymmych channel (190335) (intake at *c*. 240 m: 720 ft O.D.) to the south of the Teifi basin, and the Llanwern channel within the Gwaun–Jordanston system. Further, the group of channels around Escalwen originates in a

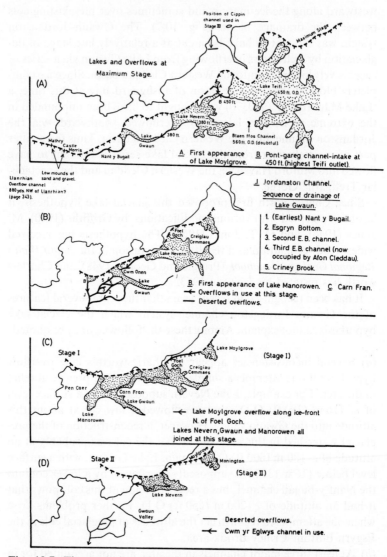

Fig. 10.7. The evolution of glacial lakes and overflow channels in the Fishguard area according to Charlesworth (1929)

group of shallow depressions at an altitude of 160 m (525 ft) O.D. on the hillside above Gelli; none of these small channels could have been used as an overflow route by a Gwaun lake with a maximum surface altitude of under 116 m (380 ft) O.D., and since the channels are not cut into preglacial cols there could have been no overflow at this point even from a lake with a much higher surface level.

(*c*) Detailed field mapping of the drifts within the confines of 'Lake Manorowen' has revealed no lacustrine deposits and no fluvioglacial deposits which can have been laid down deltaically. There is no field evidence that this lake ever existed. Similarly the depositional evidence in favour of Lake Nevern appears scanty.

(*d*) As noted by Bowen and Gregory (1965), the humped long profiles of the channels and their steep longitudinal gradients are difficult to account for by subaerial overflows of meltwater. In the case of the Cwmonnen channel, for example, meltwater would have had to flow uphill for a distance of about 1·5 km (1 mile) in its passage from Lake Nevern to Lake Gwaun.

(*e*) Joy (1963) has pointed out that in the Gwaun channel and elsewhere there has been virtually no change in the position of the watershed between preglacial times and the present day, even though its altitude has been lowered by over 62 m (200 ft). As indicated in Fig. 10.8, the exploitation of a col by subaerial meltwater must be accompanied by headward retreat of the col; in the case of the Gwaun and Nant-y-bugail channels the cols have not only remained stationary, but have had long channel sections eroded upstream of them, at levels which should have been submerged beneath the waters of Lakes Nevern and Gwaun.

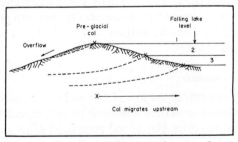

Fig. 10.8. Postulated sequence of events when a meltwater lake overflows a pre-glacial col

It seems, therefore, that the glacial lake hypothesis can no longer be considered reliable in this part of north Pembrokeshire. As an alternative it may be suggested that the Gwaun–Jordanston channel system was cut by meltwater flowing subglacially and marginally in

9*

association with the wastage of an ice-sheet which inundated this whole area. The humped long profiles of the channels and their anastomosing pattern are diagnostic features of subglacial channel systems (Derbyshire, 1958), as are the discordant junctions of distributory channels and the rock knobs and arcuate channels on the floor of the Gwaun valley and elsewhere. The steep gullies on the north wall of the Gwaun channel may be subglacial chutes (Bowen and Gregory, 1965) while possible marginal or submarginal channels and terraces occur at Gelli, on the flanks of Carningli (065372), and at Ty-canol (090370). There are other degraded channels and benches which may have been cut by meltwaters flowing along an ice-margin above Dinas cliff, on the northern slopes of Mynydd Prescelly at Banc Llwydlos (090327), beneath Foelfeddau (100327) and beneath Mynydd-bach (108329), at Carnbica (128329), and further east on the slopes of Foeldrygarn (162336). (Some of these benches may be controlled in part by structure and bedrock lithology.)

Bowen and Gregory (1965) have attempted a reconstruction of the stages of meltwater erosion in this area; however, in view of the complexity of the features described above the present author would prefer to mention only three broad stages:

1. An initial stage of subglacial channel erosion with meltwaters flowing generally towards the west and south-west.
2. A secondary stage of subglacial erosion, with the Gwaun channel incised beneath the level of the main distributary channels.
3. A stage when the ice-surface in the area lay beneath an altitude of 300 m (1000 ft) with the cutting of marginal and submarginal channels and terraces.

*The age of the channels*

Several clues concerning the age of the meltwater channels can be gained on examination of their relationships with superficial deposits. At several localities thick glacial and periglacial drifts are seen to be banked within the walls of channels. At Porthmelgan, for example, lower head with an associated silt band is found beneath Irish Sea till within the walls of a humped meltwater channel, which curves across St David's Head. At Aber-mawr the thick lower head shows every indication of being banked against the steep wall of a meltwater channel, and the rest of the drift fill similarly masks the channel wall for some hundreds of metres inland. Again, at Druidston, West Angle and West Dale coastal sections reveal that both lower head and raised beach deposits underlie the true glacial drifts within the walls of apparent meltwater channels. In the Gwaun–Jordanston channel system evidence of this type is not so clearly displayed, largely as a result of the lack of good coastal exposures. However, the presence of thick Irish Sea till plugging the intake of the Cwmonnen channel (Jehu, 1904) apparently indicates that this channel, at least, was in existence prior to the depositional phase of the Dewisland glaciation.

In view of the evidence from Dewisland it seems that a phase of beach deposition and a subaerial phase of head accumulation intervened between the cutting of these channels and the Dewisland glaciation, which was responsible for the introduction of the Irish Sea till and its associated drifts. Thus it seems that the meltwater channels and the Irish Sea drifts must be related to different glaciations.* Following the discussion on relative dating earlier in the chapter, it seems entirely logical to assign the meltwater channels to the 'early glaciation' (or glaciations) which preceded the deposition of the raised beaches.

This dating is apparently supported by the orientation of the Gwaun–Jordanston channel system. It has been stated above that the Irish Sea ice of the Dewisland glaciation overrode north Pembrokeshire from north-west towards south-east. The maximum ice-gradient must have followed this orientation. It has been shown by Mannerfelt (1945) and other workers that the orientation of subglacial meltwater channels, while controlled in part by bedrock relief, is controlled above all by the orientation of the ice surface gradient. Therefore subglacial meltwater channels of the Dewisland glaciation would be expected to run from approximately north-west to south-east. On the other hand the large channel system in question has an overall orientation from north-east towards south-west, with several component channels running east to west or even south-east to north-west. It would be extremely difficult to relate these components to the Dewisland glaciation, and it seems more satisfactory to relate them to an early glaciation, when there may have been an ice-gradient sloping from north-east to south-west on the flank of the Prescelly Mountains.

One fact appears to militate against this dating: namely the 'fresh' appearance of the channels and the excellent preservation of steep

* Bowen (1966) has argued that some of the Pembrokeshire tills are solifucted and thus in secondary position. If this is so, then both channels and drifts could be claimed to date from the same (last) glaciation. However, fabric analyses indicate that the drifts are largely *in situ* within the channel walls.

walls in many cases unmasked by accumulations of head and till. However, it is not only possible but likely that the channels were used for a second time during the Dewisland glaciation, with considerable modification in places. Indeed, channel-cutting episodes (2) and (3) referred to above may be related to the last glaciation and meltwater action may have been responsible for the steepening of denuded valley walls and the removal of much accumulated periglacial debris within the channels. Even the freshest channels may have been eroded initially during the early glaciation.

## TOWARDS AN ABSOLUTE CHRONOLOGY

In continental Europe, it has long been accepted that there have been at least three glaciations. In ascending order of age, these are termed Mindel (Elster), Riss (Saale) and Würm (Weichsel). Inevitably, past workers have attempted to relate the glaciations of west Wales not only to the traditional British scheme of Older Drift and Newer Drift glaciations, but also to the pattern of continental glaciation. But because of the lack of reliable criteria for dating, discussion has been based on such qualitative factors as the 'freshness' of hummocky drift surfaces and the depth of weathering in calcareous till. Consequently strong differences of opinion have emerged, with authors representing two distinct schools of thought.

1. Those who consider that Pembrokeshire was not affected by Irish Sea ice during the Würm or Last Glaciation. Mitchell (1960) and Stephens and Synge (1966) estimate that the drifts of the county are heavily denuded and probably of Riss age. They consider that the Irish Sea ice of the Würm glaciation did not impinge upon the Pembrokeshire coast, but that the margin passed well to the north, from Lleyn to Wexford (Fig. 10.9).

2. Those who believe that Pembrokeshire was glaciated, at least in part, during the Würm. As noted earlier, Charlesworth (1929) considered that the north coastal area was glaciated during the Last Glaciation. He was followed in this view, with minor modifications, by Wirtz (1953). Also Bowen and Gregory (1965), while discounting the existence of the South Wales end-moraine, consider that Würm ice did cross the Pembrokeshire coast to the west of the Teifi estuary. On the other hand Zeuner (1959) believed that all of the Pembrokeshire drifts should be dated to the Last Glaciation, although he seemed to accept two phases of the glaciation.

In the foregoing pages it is suggested that the only suite of glacial deposits in Pembrokeshire belongs to the Dewisland glaciation. Clearly, therefore, these deposits must be dated entirely to the Riss glaciation or entirely to the Würm. If they are of Riss age, it is sur-

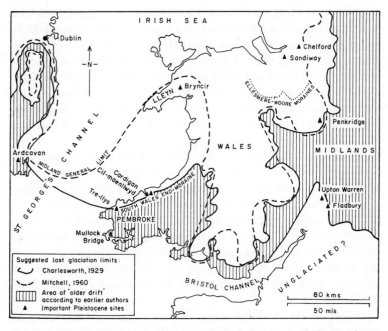

Fig. 10.9. Suggested limits of the 'Last' glaciation around the southern Irish Sea basin

prising that they have not been more completely modified by the processes of subaerial denudation and weathering during the succeeding interglacial and glacial periods; it has been noted above that strongly calcareous till may still be found within 1·6 m (5 ft) of the surface in places, while in the highly pervious gravels at Mullock Bridge complete mollusc valves can still be found within 1 m (3 ft) of the surface. Also, the relatively thin upper head seems to indicate a short time lapse since the Dewisland glaciation. It seems more likely that the Pembrokeshire drifts should be dated to the Würm glaciation, with the meltwater channels and early erratics assigned to the Riss.

This hypothesis can be tested in several ways in Pembrokeshire. Luckily organic remains are relatively frequent in the glacial drifts, and it has proved possible to use these remains to add some precision to the dating of the Dewisland glaciation and related events.

*Faunal analysis of marine mollusca*

Well-preserved fragments of marine mollusca are found in both Irish Sea till and glacial outwash sands and gravels, and it is reasonable to assume that these have been dredged by overriding ice from pre-existing beach and sea-bed deposits in St George's Channel and the southern Irish Sea basin. Although some shells have been perfectly preserved during transport and redeposition, most are comminuted and abraded to such an extent that the recognition of species is some-what difficult. However, the identification of these mollusca is a worthwhile exercise, on the following grounds:

(*a*) Specific mollusc species can be used with reasonable certainty as 'cold-water' or 'warm-water' indicators (Baden-Powell, 1955); with qualifications these indicators may suggest what climatic conditions were prevalent at some stage prior to the onset of glaciation.

(*b*) While some species became extinct during the Pleistocene, other molluscs arrived in western waters relatively late. By bearing in mind the extinction and arrival dates for such species (in so far as they are known) it may be possible to estimate the age of the glacial deposits in which they are now incorporated (Baden-Powell, 1967).

Unfortunately there are many difficulties involved in dating by means of derived marine mollusca. In the first place the species dis-covered in a glacial deposit are likely to have been derived from many localities with different environmental conditions; therefore the species can under no circumstances represent 'communities' in the biological sense (Craig, 1964). Secondly, the species are likely to have been dredged from deposits of different ages on the sea-floor (McMillan, 1964). Again, the species found in a glacial deposit may be only those species which have managed to survive transport by ice; in part this may reflect chance circumstances (e.g. transport in a frozen block of silt or clay) or, more likely, the resistant nature of some valves assisting their survival (Thompson and Worsley, 1966). And yet in spite of these difficulties marine mollusca were widely used, with some success, around the turn of the century for the zoning of the glacial deposits of eastern England. More recently McMillan (1964) has undertaken a comprehensive analysis of the marine mollusca in the glacial deposits of Wexford in south-east Ireland, although she does not attempt to use the mollusca as media for dating. In Pembrokeshire marine mollusc fragments have been collected from fluvioglacial outwash deposits at two sites: Tre-llys (898349) near the north Pembrokeshire coast, and Mullock Bridge (811080) near the mouth of Milford Haven. The former site is in the sand and gravel complex which was thought by Charlesworth to constitute the 'South Wales end-moraine', and the latter is well to the south, in the area of supposed Older Drift. In addition, marine mollusc fragments were collected from Irish Sea till at Druidston (862173) and Abermawr (883347). Following collection, the fragments were identified with the help of Mr D. F. W. Baden-Powell. The records for each site are shown on Table 10.2, together with assemblages mentioned by past workers from north Pembrokeshire and the Teifi estuary.

The most significant point to emerge from this table is that the mollusc species have come from different environments. Some of them, like *Trophonopsis clathratus* and *Chlamys islandica*, are indicators of cold climatic conditions; while others, like *Mytilus edulis*, *Littorina littorea*, and *Nucella lapillus*, are common in Pembrokeshire waters today (Crothers, 1966). Thus neither the Mullock Bridge nor Tre-llys assemblages can be considered as representative mollusc communities; the molluscs lived at different depths, at different times, and probably in different geographical locations prior to the Dewisland glaciation. The same may be said of the smaller assemblages collected from the Irish Sea till at the three localities on the table. This may seem to be a somewhat disappointing interpretation of the mollusc assemblages, but two important points do emerge.

1. On the basis of the contained marine mollusca, it would be impossible to distinguish any age differences between any of the localities examined. Thus it is quite possible that the Irish Sea till and the outwash deposits are the same age—as has already been suggested on purely stratigraphic grounds. Again it seems that the outwash deposits at Tre-llys and Mullock Bridge are the same age, reinforcing the impression that the South Wales end-moraine does not mark a major glacial limit.
2. The character of the total assemblage of molluscs seems to indicate a Late Pleistocene age. No extinct species have been discovered so far in the Pembrokeshire deposits. On the other hand *Turritella*

*communis* (thought by Baden-Powell in 1955 to have developed relatively late in the Pleistocene) is the only *Turritella* species represented. These facts, together with the occurrence of many other present-day Pembrokeshire species in the gravels, invite some speculation that part of the assemblage may have lived as recently as the Last Interglacial.

*Pollen analysis*

As yet, little pollen analysis has been undertaken in west Wales except on deposits of probable postglacial age (Godwin and Mitchell, 1938). This is largely because no coherent deposits of peat, soil or gyttja* have been discovered in close association with the periglacial or glacial drifts of the county. Field workers have been reluctant to examine head, till and outwash sands and gravels largely because such materials are unpromising for the preservation of pollen in large enough quantities for worthwhile analysis. However, some recent continental literature has shown that pollen and other organic remains can be preserved in glacial deposits and can yield evidence concerning climatic conditions at some stage prior to glaciation (see, for example, van Gijzel *et al.*, 1960; Andersen, 1965; Hansen, 1965). A much more reliable indication of climatic conditions can of course be gained from pollen discovered *in situ* in interglacial or interstadial deposits; much of the accepted British glacial/interglacial sequences is based on this type of evidence, although it must be admitted that pollen diagrams can only be used as indicators of specific nonglacial stages or intervals if careful stratigraphic studies are taken into account for each site investigated (Crabtree, 1968; West, 1963; Watts, 1967).

Preliminary pollen analyses have kindly been undertaken by Dr I. G. Simmons for two sites in Pembrokeshire where organic materials have been discovered in association with drifts. At West Angle samples of sand, gravel and silt were collected from the raised beach and the overlying deposits which are thought to have represented an interglacial transgression (see p. 239). Results showed that the deposits were laid down in an interglacial environment, although which interglacial is not known (John, 1968).

At Cil-maenllwyd (203482), in the Banc-y-Warren (202482) complex of outwash sands and gravels overlooking the town of Cardigan,

---

* Gyttja is a muddy deposit with a high organic content, normally formed by sedimentary accumulation beneath low-water mark (West, 1968, p. 52).

a sample of sand was collected from a horizon which included organic material in the side of a small gravel-pit (John, 1967). The sands and gravels were laid down in their entirety by the meltwater of the wasting Irish Sea ice-sheet, and it was felt that all the organic material in this sample would be derived and thus subject to intensive mixing. And so it proved. The sample was found to hold a surprisingly large amount of well-preserved pollen, with *Pinus* the dominant constituent (77 per cent of total tree pollen) followed by a type resembling *Fraxinus* and then *Quercus*. Other pollen included *Corylus*, *Tilia cordata*, *Carpinus*, and *Alnus*, with some *Gramineae* and *Cyperaceae*. Spores of Filicales were frequent, with 21 per cent of total tree pollen. Tree pollen accounted for 74 per cent of the 280 pollen grains and spores counted, and it may be assumed that the environment at some stage prior to glaciation was largely forested. This accords with the common occurrence of wood fragments in the sands at Cil-maenllwyd. However, detailed examination of the Cil-maenllwyd samples showed that they contained some reworked Carboniferous and Mesozoic spores and pollen (A. Hibbert, personal communication), indicating that the organic remains in the pit have come from a variety of deposits of different ages. The dating of glacial deposits by means of derived pollen clearly involves the same risks as those pointed out on the foregoing pages for dating by means of derived marine mollusca.

Thus the analysis of pollen samples from the Pembrokeshire Pleistocene deposits has yielded little concrete evidence concerning their age. However, this investigation does suggest that further analyses of pollen from glacial drifts should be undertaken. The West Angle sands and silts are almost certainly of interglacial age, although more detailed sampling and analysis at the site is needed before they can be assigned with certainty to any specific interglacial.

*Radiocarbon dating*

Radiocarbon dating, although as subject as any other method to its special limitations, has been employed with some success in west Wales for the dating of glacial deposits. This dating has proved possible because of the relatively large amounts of organic material to be found in the drifts at a variety of sites.

The dates so far obtained are summarised in Table 10.3. All the samples were collected from glacial drifts of the Dewisland glaciation, and the organic materials used were wood fragments, peaty organic mud, and marine mollusc fragments. Of these, the latter material, collected from Mullock Bridge and Tre-llys for age determinations 3 and 4, is the least reliable; but in spite of the criticisms of these dates by Bowen (1966) and Shotton (1967a) there is no reason to disbelieve them. Indeed, they may be treated with some confidence because they are closely similar to dates 1 and 2, obtained from similar stratigraphic positions and from samples thought by the author to be inherently reliable.*

TABLE 10.3. *Some radiocarbon age determinations for south-west Wales*

1. Banc-y-Warren, Cardigans. (SN 202482) Peaty organic mud from fluvioglacial outwash
   $31,800 {}^{+1400}_{-1200}$ years B.P. (I-2559) (Brown *et al.*, 1967)
2. Cil-maenllwyd, Cardigans. (SN 203482) Wood fragments from fluvioglacial outwash
   $33,750 {}^{+2500}_{-1900}$ years B.P. (I-2564) (John, 1967)
3. Tre-llys, Pembs (SM 898349) Marine mollusc fragments from fluvioglacial outwash
   $37,310 {}^{+1515}_{-1275}$ years B.P. (NPL-97) (John, 1965)
4. Mullock Bridge, Pembs (SM 811080) Marine mollusc fragments from fluvioglacial outwash
   $37,960 {}^{+1700}_{-1400}$ years B.P. (NPL-80) (John, 1965)
5. Druidston, Pembs (SM 862173) Marine mollusc fragments from calcareous Irish Sea till
   > 36,300 years B.P. (I-1687)
6. Aber-mawr, Pembs (SM 883347) Wood fragments from calcareous Irish Sea till
   > 40,300 years B.P. (NPL-98)
7. Aber-mawr II, Pembs (SM 883347) Wood fragments from calcareous Irish Sea till
   > 54,300 years B.P. (GrN-5281)

The dates on the table fall very clearly into two categories. Samples 1–4 lie in the radiocarbon age range of approximately 30,000 years B.P.–40,000 years B.P., within the span of the Middle Würm of Gross (1964) and the Interpleniglacial of the Last Glaciation referred to by van der Hammen *et al.* (1967). According to these authors this period was one of fluctuating interstadial conditions. Thus, by correlation the organic remains in the drifts of the Dewisland glaciation appear to date from the interstadial phase which separated the Early Würm from the Late or Main Würm. The Dewisland glaciation itself, during

* G. S. Boulton (1968) is not of this opinion, and argues that the Cil-maenllwyd sample may be contaminated.

which the organic remains were incorporated in glacial drift, must have occurred after this interstadial—in the Main Würm.

The second group of samples was too old for satisfactory radiocarbon age determinations at the laboratories concerned. The quoted dates cannot be used for any chronological assessments. All these samples were obtained from Irish Sea till, and in view of their great age it might be supposed that the till is older than the outwash deposits which yielded the dateable samples 1–4 (Shotton, 1967a). However, it has already been pointed out that the marine mollusc fragments from Druidston constitute a similar assemblage to that of the outwash deposits, and it is considered that a further age determination on mollusca from the site could yield a date similar to those for Tre-llys and Mullock Bridge. The two attempts to date coniferous wood fragments from the Irish Sea till at Aber-mawr demonstrate the risks involved in using derived material for radiocarbon samples. Most of the wood fragments from the till appear to be Tertiary lignite, and even some of the freshest fragments collected belong to no recognisable present-day species of conifer in northern Europe (personal communication from the Director of the Royal Botanic Gardens, Kew). While the dated fragments could be of Last Interglacial (Eemian) age, there is a strong possibility that they might be pre-Quaternary. In spite of this there seems no reason as yet to reconsider the assessment of the age of the glacial drifts. The radiocarbon evidence set out on the table does seem to indicate that there was an interstadial phase of possibly boreal climate during the Middle Würm in the Irish Sea area, and that the last inundation of Irish Sea ice occurred during the Main Würm (John, 1965b).

### CONCLUSION

From the foregoing pages it appears that the landscape of Pembrokeshire owes relatively little to the climatic fluctuations of the late Pleistocene. The richest legacy of glaciation appears in the form of the meltwater channels of north Pembrokeshire, while the glacial drifts of the county are insignificant as landscape forms, even in the area of Charlesworth's South Wales end-moraine. However, while the geomorphologist in Pembrokeshire may be denied the 'sublime contemplation' afforded by the spectacular glacial landforms of Snowdonia, he can at least employ his techniques profitably and enjoyably in the elucidation of glacial chronology. Like a detective he can collect his clues, construct his hypothesis, and test it in a variety of ways.

From the clues set out in the first two sections of this chapter it is suggested that the older hypotheses concerning the glaciation of Pembrokeshire are unreliable. The glacial features of Pembrokeshire can be related instead to two glaciations, with the meltwater channels dateable to the Early Glaciation and all of the true glacial drifts to the Dewisland glaciation.* It is suggested, largely on stratigraphic grounds, that the former glaciation was the equivalent of the continental Riss, and the latter the equivalent of the Main Würm.

TABLE 10.4 *Drift stratigraphy and Late Pleistocene events in west Wales*

| *Drifts* | *Suggested sequence of events* | *Continental chronology* |
|---|---|---|
| Sandy loams and blown sand | Aeolian deposition with temperature fluctuations | Holocene |
| Rubble-drift and upper head | Short periglacial phase | Zone III (Younger Dryas) Cold<br>Zone II (Allerød) Warmer<br>Zone I (Older Dryas) Cold |
| Fluvioglacial sands and gravels<br><br>Irish Sea till and local tills | Irish Sea (Dewisland) Glaciation | Main Würm Glaciation (25,000–16,000 years B.P.) |
| ? | Some weathering? | Paudorf Interstadial (28,000 years B.P.)<br>Middle Würm Inter-stadials (50,000–30,000 years B.P.) |
| Thick lower (main) head | Prolonged periglacial phase with fluctuations | Early Würm periglacial stages (70,000–50,000 years B.P.) |
| Raised beach with erratics | Interglacial—sea levels up to 9 m (30 ft) O.D. | Eemian Interglacial ( > 70,000 years B.P.) |
| (Rock platform) | Interglacial—sea levels up to 15 m (50 ft) O.D. | ? |

* A major unanswered problem concerns the *extent* of the Main Würm glaciation of Pembrokeshire. The lack of coherent glacial drifts in south Pembrokeshire has been mentioned on p. 237, and is taken by Bowen (1967) to indicate that the south and east of the county remained ice-free at this time. The present author is inclined to believe otherwise, although detailed field investigations are needed in the basin of the Eastern Cleddau and on the Castlemartin peninsula before this view can be substantiated.

The testing of this hypothesis produces a favourable convergence of evidence, as indicated in the foregoing paragraphs. It appears to the author, therefore, that the glacial chronology of Pembrokeshire provides a very close parallel with that of continental Europe, as summarised in Table 10.4. Similar correlations, on a more restricted scale, have already been suggested for other parts of Britain. At the top of the table the correlation of the upper head of Pembrokeshire with the Late Glacial seems reasonable in the light of related investigations; Watson (1965) and Kerney (1963) have presented evidence for considerable solifluction in the Late Glacial in Cardiganshire and Kent, while Worsley (1966) has demonstrated that the climate at this time was cold enough for ice-wedge development in the Cheshire–Shropshire basin.

Concerning full glacial events, Penny (1964) and John (1965b) have suggested before that the last glaciation of the south-western British Isles was the equivalent of the continental Main Würm (Weichsel) glaciation. On the basis of radiocarbon dates Boulton and Worsley (1965) agree with this dating for the last Irish Sea glaciation of the Cheshire–Shropshire basin, and Shotton (1967b) for the Midlands. Brown *et al.* (1967) and Saunders (1968) have presented supporting evidence from Wales. There is a similar convergence of evidence concerning events prior to this glaciation, and Coope, in association with Shotton and Strachan (1961) and with Sands (1966), has been analysing evidence which appears to confirm the existence of a Middle Würm interstadial complex in the Midlands. This interstadial (now called the Upton Warren Interstadial) is supported by radiocarbon dates from Fladbury, Upton Warren and elsewhere which are closely similar to the dates from Tre-llys and Mullock Bridge in Pembrokeshire. It must be admitted, however, that no stratigraphic horizons which can be assigned with any certainty to the Middle Würm have yet been discovered in Pembrokeshire. They may well exist as weathering horizons in the lower head at Aber-mawr, Pen-deudraeth and elsewhere, although pollen-analytical evidence is needed to confirm this.

There is little independent evidence from Britain for the events of the Early Würm. The view of a widespread Early Würm glaciation, postulated, for example, by Simpson and West (1958), Godwin (1960) and Shotton (1967b), no longer seems reliable in the light of recent evidence, and it is more likely that the period was one of severe periglacial conditions punctuated by at least one interstadial. It is likely that the lower head of Pembrokeshire is primarily the stratigraphic equivalent of the Early Würm coversands of the Netherlands (Zagwijn, 1961), although the upper facies of the head may have accumulated during phases of the Middle Würm immediately prior to the Dewisland glaciation.

Lower in the table the correlations become more debatable. There may be some doubt, for example, about the correlation of the raised beaches of the Poppit interglacial with the Eemian interglacial. However, the author considers that this can be defended on more than stratigraphic grounds (John, 1968), and Zeuner argued strongly in favour of this dating many years ago. Likewise, there is no direct supporting evidence for the correlation of the Early Glaciation of Pembrokeshire with the continental Riss. However, the Riss glaciation of northern Europe was both more intensive and more extensive than the succeeding Würm, and we may speculate that similar circumstances prevailed in western Britain. Of course, the meltwater channels and erratics of Pembrokeshire may in reality date from more than *one* early glaciation.

This story of glacial and nonglacial events in Pembrokeshire may be open to many criticisms. To the author it seems the most reasonable story in the light of the evidence currently available. Given the speed of present-day advance in the field of glacial geomorphology, it may not survive for more than a few years. But at least it will serve as another working hypothesis.

### REFERENCES

ANDERSEN, S. T. (1965) 'Pollen analysis and till stratigraphy at Lindø, Denmark', *Geol. Soc. Am. Sp. Paper* **84**, 65–78.

BADEN-POWELL, D. F. W. (1955) 'The correlation of the Pliocene and Pleistocene marine beds of Britain and the Mediterranean', *Proc. Geol. Ass. Lond.* **66** (4), 271–92.

BADEN-POWELL, D. F. W. (1967) 'On the marine mollusca of the Nar Valley Clay and their relation to the Palaeolithic sequence', *Trans. Norf. and Norwich Nat. Soc.* **21** (1), 32–42.

BOULTON, G. S. (1968) 'A Middle Würm interstadial in south-west Wales', *Geol. Mag.* **105** (2), (letter), 190–1.

BOULTON, G. S. and WORSLEY, P. (1965) 'Late Weichselian glaciation in the Cheshire–Shropshire basin', *Nature, Lond.* **207**, 704–6.

BOWEN, D. Q. (1966) 'Dating Pleistocene events in south-west Wales', *Nature, Lond.* **211** (5048), 475–6.

BOWEN, D. Q. (1967) 'On the supposed ice dammed lakes of South Wales', *Trans. Cardiff Nat. Soc.* **93**, (1964-66), 4–17.

BOWEN, D. Q. and GREGORY, K. J. (1965) 'A glacial drainage system near Fishguard, Pembrokeshire', *Proc. Geol. Ass. Lond.* **76**, 275–81.

BROWN, M. J. F. *et al.* (1967) 'A new radiocarbon date for Wales', *Nature, Lond.* **213**, 1220–1.

BURTON, B. W. (1952) 'A contribution to the geomorphology of Pembrokeshire', unpublished M.Sc. thesis, Univ. of London.

CANTRILL, T. C. *et al.* (1916) *The Geology of the South Wales Coalfield. Part XII, The Country around Milford.* Mem. Geol. Surv. Gr. Br., p. 154.

CHARLESWORTH, J. K. (1929) 'The South Wales end-moraine', *Quart. Jl geol. Soc. Lond.* **85** (3), 335–58.

CHARLESWORTH, J. K. (1963) 'Some observations on the Irish Pleistocene', *Proc. Roy. Irish Acad.* B, **62**, 295–322.

COOPE, G. R. and SANDS, C. H. S. (1966) 'Insect faunas of the last glaciation from the Tame valley, Warwickshire', *Proc. Roy. Soc.* B, **244**, 379–421.

COOPE, G. R., SHOTTON, F. W. and STRACHAN, I. (1961) 'A Late Pleistocene fauna and flora from Upton Warren, Worcestershire', *Phil. Trans. Roy. Soc.* **244**, 379–421.

COX, A. H. *et al.* (1930) 'The geology of the St David's district, Pembrokeshire', *Proc. Geol. Assoc.* **41**, 241–73.

CRABTREE, K. (1968) 'Pollen analysis', *Science Progress*, **56**, 83–101.

CRAIG, G. Y. (1964) 'An ecological approach to the study of fossil marine invertebrates' in A. E. M. Nairn, ed., *Problems in Palaeoclimatology*, Interscience, p. 583.

CROTHERS, J. H. (1966) 'Dale Fort marine fauna', *Field Studies*, **2** (supplement), 169 pp.

DERBYSHIRE, E. (1958) 'The identification and classification of glacial drainage channels from aerial photographs', *Geogr. Annlr.* **40**, p. 188.

DIXON, E. E. L. (1906) *Summary of progress for 1905*, Mem. Geol. Surv. Gr. Br. H.M.S.O., p. 70.

DIXON, E. E. L. (1907) *Summary of progress for 1906*, Mem. Geol. Surv. Gr. Br. H.M.S.O., p. 63.

DIXON, E. E. L. (1921) *The Geology of the South Wales Coalfield. Pt. XIII. The Country around Pembroke and Tenby.* Mem. Geol. Surv. Gr. Br. H.M.S.O., p. 189.

GIJZEL, P. VAN, OVERWHEEL, C. J. and VEENSTRA, H. J. (1960) 'Geological investigations of boulder-clay of eastern Groningen', *Leidse Geol. Meded.* **24**, 721–59.

GODWIN, H. (1960) 'Radiocarbon dating and Quaternary history in Britain', *Proc. Roy. Soc.* B, **153**, 287–320.

GODWIN, H. (1962) *Nature, Lond.* **195** (4845), (letter), 984.

GODWIN, H. and MITCHELL, G. F. (1938) 'Stratigraphy and development of two raised bogs near Tregaron, Cardiganshire', *New. Phytol.* **39**, 425–54.

GREEN, J. F. N. (1911) 'The geology of the district around St David's, Pembrokeshire', *Proc. Geol. Ass. Lond.* **22**, 121–41.

GRIFFITHS, J. C. (1940) 'The glacial deposits west of the Taff, South Wales', unpublished Ph.D. thesis, Univ. of London.

GROSS, H. (1964) 'Das Mittelwürm in Mitteleuropa und angrenzenden Gebeiten', *Eisz. und Geg.* **15**, 187–98.

HAMMEN, T. VAN DER *et al.* (1967) 'Stratigraphy, climatic succession and radiocarbon dating of the last glacial in the Netherlands', *Geol. en Mijnb.* **46** (3), 79–95.

HANSEN, S. (1965) 'The Quaternary of Denmark', in K. Rankama, ed. *The Quaternary*, vol. 1, Interscience Publishers, pp. 1–90.

HICKS, H. (1885) In T. G. Bonney, 'On the so-called Diorite of Little Knott (Cumberland), with further remarks on the occurrence of Picrite in Wales', *Quart. Jl geol. Soc. Lond.* **41**, 511–22.

JEHU, T. J. (1904) 'The glacial deposits of northern Pembrokeshire', *Trans. Roy. Soc. Edin.* **41** (1), 53–87.

JOHN, B. S. (1965a) 'Aspects of the glaciation and superficial deposits of Pembrokeshire', unpublished D.Phil. thesis, Oxford University. 2 vols.

JOHN, B. S. (1965b) 'A possible Main Würm glaciation in west Pembrokeshire', *Nature, Lond.* **207** (4997), 622–3.

JOHN, B. S. (1967) 'Further evidence for a Middle Würm Interstadial and a Main Würm glaciation of south-west Wales', *Geol. Mag.* **104** (6), 630–3.

JOHN, B. S. (1968) 'Age of raised beach deposits of south-western Britain', *Nature, Lond.* **218** (5142), 665–7.

JONES, M. (1946) 'The development of the Teifi drainage system', unpublished M.Sc. thesis, University of Wales.

JONES, O. T. (1965) 'The glacial and post-glacial history of the lower Teifi valley', *Quart. Jl geol. Soc. Lond.* **121**, 247–81.

JOY, J. (1963) 'Glacial drainage features in north Pembrokeshire', unpublished undergraduate dissertation, Univ. of Southampton.

KERNEY, M. P. (1963) 'Late-glacial deposits on the chalk of south-east England', *Phil. Trans. Roy. Soc.* B, **246**, 203–54.

LEACH, A. L. (1911) 'On the relation of the glacial drift to the raised beach near Porth Clais, St David's'., *Geol. Mag.* **8**, 462–466.

MANNERFELT, C. M. (1945) 'Några glacialmorfologiska formelement', *Geogr. Annlr.* **27**, 1–239.

MCMILLAN, N. F. (1964) 'The mollusca of the Wexford Gravels (Pleistocene) S.E. Ireland', *Proc. Roy. Irish Acad.* **63**B (15), 265–89.

MILLER, A. A. (1937) 'The 600-foot platform in Carmarthenshire and Pembrokeshire', *Geogr Jl* **90**, 148–59.

MITCHELL, G. F. (1960) 'The Pleistocene history of the Irish Sea', *Advmt Sci.* **68**, 313–25.

MITCHELL, G. F. (1962) 'Summer field meeting in Wales and Ireland', *Proc. Geol. Ass. Lond.* **73**, 197–213.

PENNY, L. F. (1964) 'A review of the Last Glaciation in Great Britain', *Proc. Yorks. geol. Soc.* **34** (4), 387–411.

PRESTWICH, J. (1892) 'The raised beaches and "head" or rubble-drift of the south of England', *Q. Jl geol. Soc. Lond.* **48**, 263–343.

PRINGLE, J. and GEORGE, T. N. (1948) *South Wales: British Regional Geol.* H.M.S.O. London.

SAUNDERS, G. (1968) 'Glaciation of possible Scottish readvance age in north-west Wales', *Nature, Lond.* **218** (5136), 76–8.

SHOTTON, F. W. (1967a) 'The problems and contributions of methods of absolute dating within the Pleistocene period', *Quart. Jl geol. Soc. Lond.* **122**, 357–83.

SHOTTON, F. W. (1967b) 'Age of the Irish Sea glaciation of the Midlands', *Nature, Lond.* **215**, 1366.

SIMPSON, I. M. and WEST, R. G. (1958) 'On the stratigraphy and palaeobotany of a late-Pleistocene organic deposit at Chelford, Cheshire', *New Phytol.* **57**, 239–50.

SISSONS, J. B. (1960) 'Some aspects of glacial drainage channels in Britain, Part I', *Scott. geogr. Mag.* **76**, 131–46.

SISSONS, J. B. (1961) 'Some aspects of glacial drainage channels in Britain, Part II', *Scott. geogr. Mag.* **77**, 15–36.

STEPHENS, N. (1966) 'Some Pleistocene deposits in north Devon', *Biul. Peryglac.* **15**, 103–14.

STEPHENS, N. and SYNGE, F. M. (1966) 'Pleistocene shorelines', in *Essays in Geomorphology*, ed. G. H. Dury. Heinemann Educational, pp. 1–51.

SYNGE, F. M. (1963) 'A correlation between the drifts of south-east Ireland and those of Wales', *Ir. Geogr.* **4** (5), 360–6.

THOMPSON, D. B. and WORSLEY, P. (1966) 'A late Pleistocene marine mollusca from the drifts of the Cheshire Plain', *Geol. J.* **5**, 197–207.

WATSON, E. (1965) 'Periglacial structure in the Aberystwyth region of central Wales', *Proc. Geol. Ass. Lond.* **76**, 443.

WATTS, W. A. (1967) 'Interglacial deposits at Kildromin Townland, near Herbertstown, Co. Limerick', *Proc. Roy. Ir. Acad.* **65**B (15), 339–48.

WEST, R. G. (1963) 'Problems of the British Quaternary', *Proc. Geol. Ass. Lond.* **74** (2), 147–86.

WEST, R. G. (1968) *Pleistocene Geology and Biology*. Longmans.

WILLIAMS, K. E. (1927) 'The glacial drifts of western Cardiganshire', *Geol. Mag.* **64**, 205–27.

WIRTZ, D. (1953) 'Zur Stratigraphie des Pleistozäns im Westen der Britischen Inseln', *Neues Jb. Geol.* p. 267–303.

WORSLEY, P. (1966) 'Fossil frost-wedge polygons at Congleton, Cheshire, England', *Geogr. Annlr* **48**A, 211–19.

ZAGWIJN, W. H. (1961) 'Vegetation, climate, and radiocarbon datings in the Late Pleistocene of the Netherlands', *Meded. Geol. Sticht.* **14**, 15–45.

ZEUNER, F. E. (1959) *The Pleistocene Period*. 2nd edn. Hutchinson.

# CHAPTER ELEVEN

# The West Country and Southern Ireland

Nicholas Stephens, M.Sc., Ph.D.

During the various cold and temperate periods of the Quaternary an extremely varied series of deposits was laid down, including marine sand, shells, and shingle (now forming raised beaches), boulder clays, periglacial deposits (often called 'head' in the literature), and river terrace gravels, sometimes containing vertebrate remains and human artefacts. In addition, large erratic boulders have been recorded from the Devon and Cornish coasts, their presence having been accounted for by agencies such as regional ice-advances, pack-ice and icebergs (Charlesworth, 1957; Mitchell, 1960; Stephens, 1966). Holocene marine alluvium fills much of the Severn estuary and extensive deposits of postglacial peats occupy large areas of the Somerset Levels, where, in places, they are interbedded with marine clays.

The interpretation and correlation of these deposits presents many difficulties, mainly on account of the lack of datable fossil material, and the intermittent development of good sections along tens of miles of coastline. Notable contributions to the study of the Pleistocene at inland sites in south-west England have been made by Waters (1960, 1961, 1962, 1964, 1965), Brunsden (1964), and Orme (1960, 1962, 1964) who has examined areas in south Devon between Dartmoor and the sea. Moreover, fieldwork has shown that there are some close comparisons between the superficial deposits in south-west England and South Wales, and in southern Ireland (Stephens, 1966a and b). G. F. Mitchell (1960, 1962), and Synge (1963) have already made attempts to correlate the Pleistocene deposits on either side of the southern Irish Sea, and Mitchell and Orme (1967) have published an account of the Pleistocene geomorphology of the Scilly Isles. It is therefore the aim of this account to provide a fresh appraisal of the available evidence from Somerset, Devon and Cornwall, and to comment on the possibilities, and the difficulties, of achieving a satisfactory chronology of Quaternary events in this area, and in the south of Ireland and South Wales.

Although the age of the southern limits of the Pleistocene ice-sheets (Fig. 11.1) may be disputed it is clear that most of the area comprising the lower Severn valley and nearly all of the south-west peninsula of Somerset, Devon and Cornwall and part of South Wales lay within a periglacial zone during the glacial periods. It is not surprising, therefore, to find an abundance of relic periglacial deposits and features well displayed in the present landscape. The thick head deposits of coastal sections, the presence of fossil ice-wedges and involutions in superficial materials, the altiplanation terraces, block fields and tors of some upland areas, including Dartmoor and Bodmin Moor, and the two-storied cliffs (bevelled and hog's-backed), all testify to the action of freeze-thaw processes. It is evident that erosion in interglacial and postglacial periods has been quite incapable of obliterating a relic periglacial landscape of wide extent, where convex slopes (zone of wastage) and concave slopes (zone of deposition) dominate the landscape (Te Punga, 1957). Williams (1965) considers that much of Devon and Cornwall was relatively free of permafrost during the last glacial period, except on high ground, but the Bristol Channel coast east of Barnstaple Bay, and the lower Severn valley are shown to have had extensive permafrost. The line of division of Fig. 11.1 between extensive and discontinuous permafrost is adapted from Williams's map (1965). In the descriptions which follow it may well seem necessary to consider a greater area as having had active layers over permafrost on the evidence of the existence of fossil ice-wedges. The critical mean annual temperature for extensive continuous permafrost has been given as $-6°C$ (Péwé, 1964). Even if certain areas lay outside this critical limit in the extreme south-west of Cornwall, nevertheless an extremely rigorous periglacial climate must have occurred during the maximum of the last glacial period and, one presumes, during earlier cold periods.

Multiple layers of soliflucted head occur in some coastal sections, and their interpretation presents difficulties (cf. Watson, 1965), while on Dartmoor two and possibly three periods of cold climate have been recognised by Waters (1964) as responsible for gelifraction of the sound bedrock and the geliturbation of already weathered material. These cryergic processes have given to the high moorland many of its salient physiographic features, the clitters, valley-side buttresses, terraces of rubble-drift, and smooth slopes on the granite outcrops, as well as altiplanation terraces on the aureole rocks and have contributed to the shaping of the tors (Palmer and Nielson, 1960). There

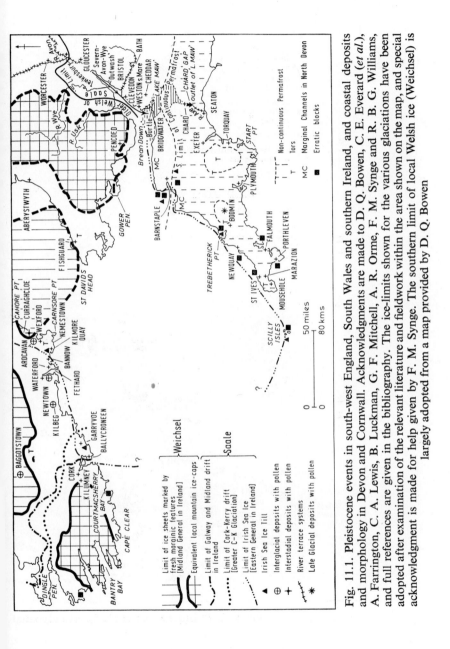

Fig. 11.1. Pleistocene events in south-west England, South Wales and southern Ireland, and coastal deposits and morphology in Devon and Cornwall. Acknowledgments are made to D. Q. Bowen, C. E. Everard (*et al.*), A. Farrington, C. A. Lewis, B. Luckman, G. F. Mitchell, A. R. Orme, F. M. Synge and R. B. G. Williams, and full references are given in the bibliography. The ice-limits shown for the various glaciations have been adopted after examination of the relevant literature and fieldwork within the area shown on the map, and special acknowledgment is made for help given by F. M. Synge. The southern limit of local Welsh ice (Weichsel) is largely adopted from a map provided by D. Q. Bowen

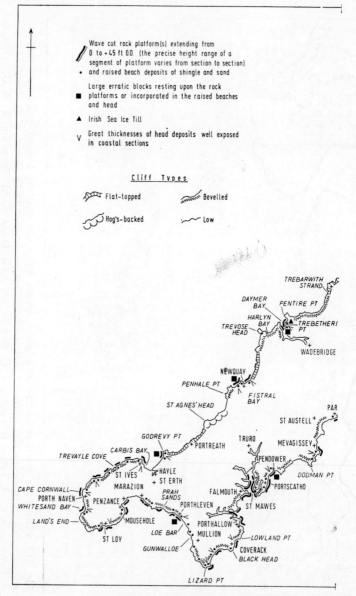

Fig. 11.2.  Pleistocene marine platforms and coastal

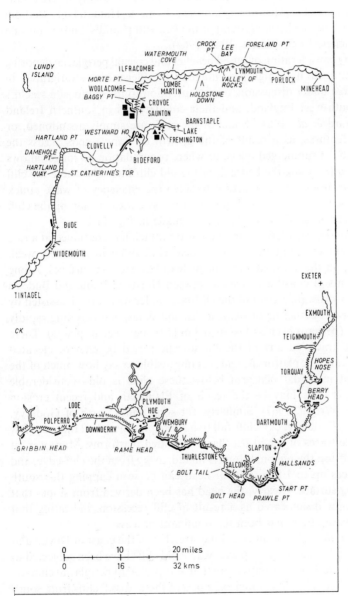

deposits, and cliff types in Devon and Cornwall

10

can be little doubt that cryergic action has been equally severe on the other uplands—Mendips, Quantocks, Exmoor, Haldon Hills, Bodmin Moor, and the higher ground of the Penwith (Land's End) peninsula (Guilcher, 1950; Waters, 1961, 1962).

Any consideration of raised beaches and coastal periglacial deposits, which make up a high percentage of the Pleistocene stratigraphy in the area under discussion, directs attention to the composite seacliffs of south-west England, and their counterparts in southern Ireland (Everard *et al.*, 1964). Some of the cliffs are simply abandoned, or relic features, as a result of the growth of shingle spits across the mouths of submerged inlets, or where progradation no longer allows the sea to attack the bedrock of the old cliff line. Elsewhere the cliff profiles reflect the interaction between the efficiency of wave attack and the intensity of periglacial cryergic processes in shaping the cliff.

Three broad divisions have been made in Fig. 11.2:

(*a*) *Flat-topped cliffs* occur where wave attack has continued at a rate sufficient to maintain a near-vertical cliff, often in relatively weak rocks, on the seaward edges of plateau-like erosion surfaces. Along segments of coastline, such as between Hartland Point and Bude in north Cornwall, retreat of the cliff-face under wave attack assisted by subaerial weathering of weak shales and slates, is proceeding rapidly today (perhaps even as fast as 0·3 m (1 ft) per year in places). There are no traces of any of the Pleistocene raised beaches or elevated wave-cut rock platforms, and it is impossible to say how much of the present intertidal platform below these cliffs is old—considerable portions of it may be the result of Flandrian and recent erosion (L. W. Wright, 1967). Similarly, for some miles north of Godrevy, Cornwall, steep rocky cliffs fall to a shore platform, and there are few reliable traces of raised beaches or elevated platforms. Moreover, at the cliff-top the regional slope is inland, away from the cliff-edge, and in places up to 1·8 m (6 ft) of head can be seen capping this south sloping surface. Clearly, the head has been derived from slopes that have now disappeared as a result of cliff recession, indicating that here wave attack has been the dominant process.

(*b*) *Bevelled cliffs* occur over long stretches of the coast of Devon and Cornwall, indicating the places where old cliffs have been subjected to cryergic action of perhaps more than one period of periglacial climate. Orme (1962) has illustrated examples of these 'fossil' cliffs from south Devon, where occasionally buttresses and cliff-tors stand out from head-strewn slopes. Good examples occur between Dartmouth and

30 Freshwater Gut, Croyde Bay, North Devon. The large erratic boulder of garnetiferous hypersthene-bearing granulite of gneissose type is seen to the left of Mr A. W. Smith (holding the geological hammer) and resting upon a wave-planated platform in Pilton slates: the huge boulder is estimated to weigh 50,000 Kg (48 tons) approximately, and is firmly wedged at the back of the 7.6m (25 ft) platform, against the rock cliff which leads upwards to the 15 m (50 ft) platform; it is overlain partially by the main head. The smaller round boulder *below* the hammer is a local rock type.

31 Garryvoe, Co. Cork. The old rock cliff on the right of the photograph is blanketed by beach deposits, lower head and glacial till (Irish Sea Eastern General or Ballycroneen till), and the surface of the 'terrace' is composed of a thin layer of upper head over deeply weathered till. The cliff is formed almost entirely in superficial deposits resting upon an elevated rock platform; the latter has been eroded and replaced by a wave-cut surface extending below high water mark, but it is not certain how much of this platform is 'younger' than the superficial deposits seen in the cliff sections (cf. Westward Ho!, and see A. Farrington, 'The early-glacial raised beach in Co. Cork', *Sci. Proc. Roy. Dub. Soc.*, **2,** 1966).

32 Fethard, Co. Wexford. The elevated rock platform can be seen in the cliff face, where a measuring scale represents 1.2 m (4 ft). The platform is overlain by beach gravels with associated angular blocks (head?), the latter can be seen at the base of the measuring scale; above there are 3.7–4.9 m (12–16 ft) of inclined beds of sand with interbedded lenses and thin layers of angular stones and slivers of slate rock. The sand is believed to be blown sand which accumulated against the old rock cliffs (now hidden) at a time of low sea-level, and when climatic conditions permitted frost-shattering of the rock. The upper 1.2–1.8 m (4–6ft) of the section is composed of deeply weathered and cryoturbated till of inland origin. This till is regarded as contemporaneous with the Irish Sea till (=Eastern General or Ballycroneen till=Fremington till) seen elsewhere in coastal sections (e.g. Kilmore Quay; Mitchell, 1962; Synge, 1963, 1964b). Consequently the beach gravels are considered to pre-date the Irish Sea till which is regarded as Saale in age. This section has many similarities with those recorded in Barnstaple Bay, North Devon (see Stephens and Synge, 1966, Plates 4 and 5, facing page 21), except that in North Devon the till is usually replaced by a massive head deposit.

33 The photograph shows frost-erected cobbles of the Eemian (?) beach, on the foreshore at Westward Ho!, North Devon. The beach cobbles and the head in which they are embedded pass below post-glacial sediments of clay and peat (see fig. 11.6b).

Bolt Tail, where sometimes the base of the old cliff is protected by head. Elsewhere (near Start Point) renewed trimming has occurred, and at Slapton the head has been removed, and the base of the old cliff re-etched, and cut back, by wave attack.

(*c*) *Hog's-backed cliffs* are really a special form of bevelled cliff, found best developed where wave-truncated cliffs form the lower element of whale-backed ridges running parallel or sub-parallel to the coast. They are specially well developed between Coombe Martin and Foreland Point, north Devon. Cotton (1951) has referred to these cliffs as 'two-tier cliffs', and the long (often convex) subaerial slope which forms the upper segment of profile (Fig. 11.2) can be regarded as a periglacially altered 'fossil' cliffline, at one or more former levels. There is considerable variation in the lower part of the profile; for example, on the south-west side of Baggy Point, and at Saunton, raised beach deposits and head form a conspicuous 'apron' or terrace of deposits below a marked convexity of profile developed on rock carrying only a thin skin of head; whereas, on the north-east side of Baggy Point sheer rocky cliffs fall away below the convexity, a type of profile repeated for mile after mile east of Ilfracombe (Fig. 11.2).

All three cliff types can be seen in southern Ireland, where similar periglacial processes have operated, but where glacial tills blanket parts of the solid rock, or together with raised beach gravels and head, make up the entire cliff above various wave-cut rock platforms. Flat-topped cliffs are recorded by Orme (1962) on Old Red Sandstone south-east of the Drum Hills, Co. Waterford; bevelled cliffs are commonly developed and hog's-backed cliffs are also known in Co. Waterford on resistant Palaeozoic rock. Conspicuous examples of 'fossil' cliffs have been recorded by Orme (1964) and Farrington (1966) where head and till obliterate the base of older rock cliffs.

The precise relationship of the wave-cut platforms, and the raised beaches with erratics, to the buried rock channels of the coastal estuaries is unknown at present (Farrington, 1959; McFarlane, 1955). However, the existence of rock platforms at the back of large bays (Barnstaple Bay, Tor Bay, Courtmacsherry Bay), and in some of the larger inlets and rias (Cork Harbour, Waterford, Camel estuary below Wadebridge, and Fal estuary between Falmouth and St Mawes) indicates that the present coastline was, in large measure, delineated by marine erosion in early Pleistocene times; and that in places the erosion followed a period of valley incision and deepening to levels well below the present base-level of the rivers.

10

The Bristol Channel and Severn estuary have been described as a drowned river valley (North, 1964) and there is no doubt that positive and negative movements have occurred many times since this major physiographic feature was formed. Kidson (1964) reviews the more recent literature concerning the origin of this large drowned depression, within the confines of which elevated beaches and buried channels in rock testify to significant fluctuations of sea-level. A considerable amount of geological information has now been obtained for the floor of the Bristol Channel and south-west approaches of the British Isles (Donovan, 1961; Stride, 1962, 1963; Whittard and Bradshaw, 1965). Preliminary considerations of these data suggest that it may have considerable value in the unravelling of the Pleistocene fluctuations of sea-level.

Thus the Pleistocene and Holocene periods have witnessed the final etching of a land surface largely blocked-out and moulded in pre-Quaternary times. So far as the coasts are concerned, where some of the best exposures of the varied deposits occur, the coastal morphology reflects closely the storm wave, macrotidal environment (tidal range varies from about 4·6 m (15 ft) at Land's End, to 14 m (46 ft) at the mouth of the Bristol Avon) outlined by Davies (1964). The height range of ancient beaches and wave-cut platforms must be expected to vary just as much as their modern counterparts, underlining the need for extreme care in making purely altimetric correlations of similar deposits (Stephens and Synge, 1966). Mitchell (1960, 1962) has published similar correlations to that shown in Table 11.1 and West (1963) has given some alternative interpretations.

NORTH DEVON

*Barnstaple Bay*

The sequence of Pleistocene deposits which overlie a series of low wave-cut rock platforms in the coastal sections around Barnstaple Bay have been described by Sedgwick and Murchison (1840), Maw (1864), Hughes (1887), Dewey (1913), and more recently by Mitchell (1960) and Stephens (1966b).

Between Saunton and Baggy Point (Fig. 11.3) there are traces of three distinct platforms—height ranges 0–6 m (0–20 ft) O.D.; 5·5–7·6 m (18–25 ft) O.D. and 10·7–15 m (35–50 ft) O.D.—although one is seldom able to see all three platforms in profile, one above the other. The lowest platform passes below modern high-water mark and hence below modern beach level. But at Bloody Basin and Middleborough its great age and exhumed nature are demonstrated by the Pleistocene beach and head deposits cemented upon it, which are now being eroded by the sea. The two higher platforms are best seen between Middleborough and Pencil Rock, and at Freshwater Gut a 50 800 kg (50 ton) 'giant' erratic block of granulite gneiss 1·38 m³ (10 × 7 × 7 cubic ft) is situated near the upper limit of the 5·5–7·6 m

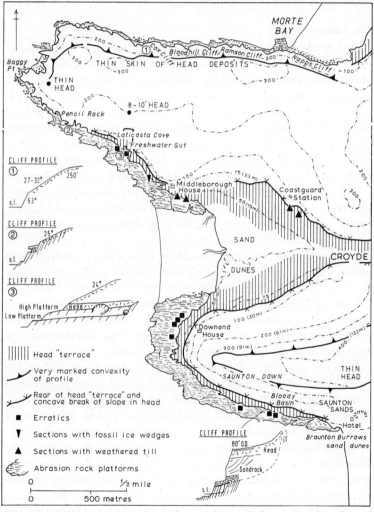

Fig. 11.3 Pleistocene deposits and coastal morphology near Croyde and Saunton, north Devon

(18–25 ft) platform, where it notches the higher platform (Taylor, 1956).

The enigma posed by these rock platforms is that they are all relic features by virtue of the suite of Pleistocene deposits which rest upon them. While some late-Pleistocene and Holocene modification of the lower platforms has taken place it is certain that much wave energy has been directed to the stripping of the superficial deposits and exhumation of old rock surfaces.

## The erratics

Near Croyde and Saunton numerous erratic boulders rest on wave-cut rock platforms (Fig. 11.3) (Taylor, 1956). At Freshwater Gut main (lower) head seals the large erratic, and at Saunton a large pink gneissose granite boulder is buried by sandrock, which is the upper part of the raised beach sequence, and also sealed below main head (see p. 278). These erratics compare in size with the Porthleven erratic (Flett and Hill, 1912; Stephens and Synge, 1966), and there seems to be no doubt whatever that they were delivered on to pre-existing rock platforms, and were later sealed by a variety of deposits.

The problem whether these large erratics were moved into position by a regional ice-sheet (Mindel/Elster or Lowestoft stage of Mitchell, 1960), or by floating icebergs in the early Pleistocene, has not been solved satisfactorily, except that the widespread distribution of erratics on the Bristol Channel, southern Ireland and English Channel coasts, and on the French coast (Tricart, 1956) would appear to support the latter hypothesis (Mitchell, 1965; Stephens, 1966b). In Devon and Cornwall, and in Ireland, the very large erratics appear to be confined to a narrow zone along the coast, below 9 m (30 ft) O.D., within the reach of storm waves at the present time, but their true age relationship to the raised beach is not yet known.

Flint is a common constituent of many modern beaches along the Bristol Channel and English Channel coasts, and it is present also in some of the raised beach deposits, in the Fremington boulder clay, and some head and outwash gravel deposits. At first sight it might appear to be a diagnostic 'erratic', only brought to the area by ice-carriage. But the modern beaches have probably received flint as the Flandrian transgression rolled material landwards during the post-glacial rise of sea-level. Submarine outcrops of flint-bearing Cretaceous rocks are known in the English Channel, but this explanation is less certain for the north coast of Cornwall and Devon (Whittard and Bradshaw, 1965), where southward moving ice-sheets (Fig. 11.1) might have carried flint. There is, however, another explanation for the presence of some of the flint. At Orleigh Court, a few miles south-west of Bideford, at about 106–122 m (350–400 ft) O.D. there is some 7·6 m (25 ft) of flint nodules, chert pebbles and sand (I. Rogers and Simpson, 1937). The deposit resembles other Pliocene outliers in Cornwall, and is not unlike the deposits capping the Haldon Hills near Exeter; it is certainly not a glacial deposit but a remanié of a once larger outcrop which may have contributed material to many different deposits.

### The raised beach deposits

At the Bloody Basin near Saunton (Fig. 11.3) a section shows the beach shingle, 2 m (7 ft) in places, and overlying sandrock (cemented shelly sands = aeolianite) resting upon a rock platform which is covered at times of spring tides. The sandrock may be up to 9 m (30 ft) thick and is always overlain by the main head. The upper or younger head can be seen in roadside cuts (near Saunton Sands Hotel, Fig. 11.3) where fresh, angular, shattered rock rests upon bedrock. This very coarse blocky layer of upper head, which lacks the high content of sandy matrix of the lower, main head, rarely reaches the top of the present cliff, in marked contrast to the main head which forms much of the cliff profile.

Near Saunton the horizontal bedding of the sandrock is often striking for the first 1·8–3 m (6–10 ft) upwards from the base, thence changing to gently sloping beds of sand with fine fragments of slate head dispersed through the sand in places (cf. sections at Fethard, Co. Wexford, in Mitchell, 1962, Plate 4; and see Stephens and Synge, 1966). The shell content of the upper sand rock (Hughes, 1887) and comparison with the bedding in modern dunes suggests a windblown deposit or series of fossil dunes.

At Pencil Rock the rock platform is at 13·7 m (45 ft) O.D. The raised beach shingle extends to 18·3 m (60 ft) O.D., and the overlying sandrock to well over 30·5 m (100 ft) O.D. Fragments of sandrock were recovered from the head on the cliff-path (Fig. 11.3). The shingle beach is 1–1·5 m (4–5 ft) thick and composed of well-rounded pebbles and cobbles, 0·17 m (6–8 inches) long, with occasional larger boulders. The matrix is made of coarse sand, shale slivers, small pebbles, shells and shell fragments—all of temperate species. The sandrock consists of cemented sand, shell fragments, and pebbles, with pieces of shale scattered through it in a haphazard manner: it has a well-bedded appearance, and in places has broken up into large cube-shaped blocks. Just as at some sections in Cornwall (e.g. Prah Sands, Porthleven and Pendower), and in southern Ireland (Farrington, 1966), the raised beach contained many angular blocks, which may be described tentatively as an old head deposit. This ancient head may have been contemporaneous with the ice-advance responsible for the distribution of the large erratics and therefore considerably older than the *transgressive* raised beach. On the other hand we may also regard the raised beach as a *regressive* feature, formed as sea-level began to fall during the onset of the cold conditions which accounted for the main head on the coastal slope; although it is recognised that ordinary cliff falls could account for the angular debris included in the beach, which might have been quickly buried by rapid beach accumulation. The subsequent exposure of a sandy strand allowed wind-transported material to accumulate to a great height, while head was forming. This would account for the layers and lenses of head found in the sandrock at all levels. Eventually the accumulation of head became dominant and a great 'terrace' formed seawards of the old cliffline, completely covering the sandrock, beach gravels, 'giant' erratics and a series of rock platforms.

*The main head*

The main head is a solifluction deposit of great thickness, which is regarded as having been formed under periglacial climatic conditions during the Saale glacial period. This is suggested because in south-east Ireland there is no such enormous deposit of head which can be associated with the advance of the last glaciation ice (= Midland General in Ireland) to the southern Ireland end-moraine (Synge, 1963) (Fig. 11.1). Outside this end-moraine the head deposits, and disturbance of the older tills rarely exceeds 2 m (6 ft). While the age of the southern Ireland end-moraine depends in some measure upon the somewhat unsatisfactory Ardcavan site (Mitchell, 1960), this moraine marks the outermost limit of fresh, relatively unweathered drift. The main head on the other hand is associated with the purple, shelly, calcareous Irish Sea (Eastern General or Ballycroneen) till in south-east Ireland (Farrington, 1944), deposited by Irish Sea ice moving across the eastern and southern coastlands.

At Garryvoe, Co. Waterford, the calcareous Irish Sea till rests on main head, which in turn overlies raised beach shingle containing erratics. All the deposits rest on a rock platform, providing a close analogy with north Devon. Where the Irish Sea (Ballycroneen) till has been subjected to weathering it is decalcified to depths of 2·4–4·5 m (8–14 ft), in complete contrast to the tills north of the southern Ireland end-moraine. Furthermore, at the Kilbeg and Newtown Great Interglacial (= Hoxnian or Holstein) sites (Watts, 1959), the till resting upon the peats is derived from the north (Munster General Glaciation) and was almost certainly responsible for excluding the Irish Sea (Ballycroneen) ice from these sites (Mitchell, 1960, 1962). The circumstantial evidence is strong, but not conclusive. If this contemporaneity is accepted, allied to the highly contrasting depth of weathering of the 'older' and 'younger' tills, and if the difference of degree of head formation is considered, then a Saalian age for the main head seems not unreasonable.

In south-west England, as in southern Ireland, the main head consists of blocks of all sizes of local slate, sandstone and quartz, set in a sandy matrix but without erratic pebbles, except occasionally as a surface find. Where it has not been disturbed by later frost action the rock fragments have a preferred orientation downslope, often lying at low angles to the horizontal (0–10°) and projecting out of the face of the cliff (Kirby, 1967). The material has moved downslope as a kind of sludge and spread out as a great apron or solifluction terrace at the foot of the coastal slope (e.g. Saunton Down, Baggy Point). It must be emphasised that there is no evidence to suggest that the terrace-like surfaces of the solifluction deposits owe their form to wave action, for no marine deposits have been found to cap the main head in south-west England, South Wales or southern Ireland (M. A. Arber, 1960; Stephens, 1961b, 1966b).

The main head is the thickest of the solifluction deposits, and at certain points in Devon and Cornwall (e.g. Croyde Bay, Godrevy) it can be seen to be highly weathered (due partly perhaps to the accumulation of previously weathered material?), and disturbed by cryoturbation, frost cracks, wedges and convolutions which extend down 1·5–1·8 m (5 or 6 ft) below the surface. This is interpreted as meaning that the head had ceased movement when renewed periglacial activity churned the surface of the deposit. This could have taken place during the Saale cold period, or resulted from renewed freeze-thaw action during the Weichsel cold period (cf. Bryant, 1966).

Between the hotel at Middleborough and Freshwater Gut (Fig. 11.3) the surface of the head deposits flattens out to a shelf of almost negligible slope (< 3°) at the cliff edge. An upper head or slope wash (deposit 2, Fig. 11.4B) overlies 0·6–1·0 m (2–3 ft) of disturbed head, and the slate fragments are erected on end instead of lying flat with their long axes dipping gently seawards. Frost action is invoked to explain this cryoturbation of the head, where fossil ice wedges occur

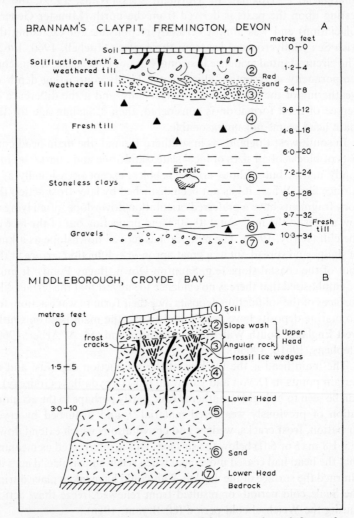

Fig. 11.4. (A) Composite section of the east and south faces of Brannam's Claypit, Fremington, north Devon. (B) Cliff-section near Middleborough House (see Fig. 11.2), Croyde Bay, north Devon

(Fig. 11.4B). The wedges appear to be confined to a 0·9 m (3 ft) layer of angular slate fragments (mostly 2 inches long) overlying a more sandy horizon (0·3–0·6 m: 1–2 ft), below which is more head, consisting of a mass of fine slate rubble set in a sandy matrix, the whole being well-weathered.

Two possible interpretations of the full section at Middleborough House have been suggested (Stephens, 1966b; Waters, 1966). Layers 3, 4, 5, 6 and 7 have been attributed to the Weichsel cold period, representing the waxing cold phase of the Last Glaciation as the ice advanced; layer 3 (= ice wedges) the coldest and driest part of the cold period, and layer 2 a return to less cold solifluction conditions during the waning of the glaciation. The interpretation of such sections in multiple layers of head is crucial in south-west England and southern Ireland, although perhaps too much should not be based upon one section. At Middleborough layer 3 consists of relatively fresh rock fragments, and contains less clay-mineral material than the layers 4, 5, 6 and 7, which are more weathered (cf. the two head deposits at Godrevy). While therefore, it is tempting to interpret the whole set of deposits as belonging to the last glacial period, this is still unsatisfactory. Layer 3 closely resembles the shattered rock in the quarry near Saunton, and along the road to Croyde Bay, where rock strata have been distorted down the coastal slope. Here the head is clearly 'young', but nowhere is there any sign of the lower head material, the latter is found only in the great 'terraces' of head. Consequently, it is believed that there are two separate head deposits present. The horizon containing the fossil ice wedges indicates that the material in which they occur had already ceased to move significantly when the wedges were formed, whereas only continuous mass movement could account for the great thickness and form of the lower (main) head. Another distinct phase of severe cold climatic conditions is therefore suggested to account for the presence of the wedges (Williams, 1965) and for the finer head (layer 2) which moved downslope to seal them off. The upper heads (coarse layer + wedges + stony layer) may therefore represent the total depth of the 'active layer' over permafrost during the Last Glaciation (Weichsel), and the sandier head below may be older (Saale age?) (Williams, 1965).

*The Fremington boulder clay*

The Fremington boulder clay (Maw, 1864; Mitchell, 1960; Stephens, 1966b) extends along the depression from Fremington to Lake and a

10*

small outlier occurs at Clampit, situated on a lower rock ridge which is orientated east to west, from Hele to Penhill (Fig. 11.5). A narrow tidal creek known as Fremington Pill occupies part of the western end of the depression between the rock ridges, and a wave-cut rock platform is seen around the western end of Hele–Penhill ridge. It is particularly well exposed in the railway cutting near Fremington Station, and from a height of 10 m (33 ft) O.D. it sinks northward and disappears below the modern beach before the old Lime Kiln is reached. Beach deposits rest upon the rock platform on either side of the tidal creek, and in the railway cutting. The beach shingle forms hummocky ground near Combrew Farm (where the surface is at about 16 m (50 ft) O.D.); it is overlain by weathered till in places near Fremington Station (Dewey, 1913) and by fresh, unweathered boulder clay in the clay pits.

When unweathered the Fremington boulder clays (layers 4 and 6, Fig. 11.4A), and the included mass of stoneless clays (layer 5) are highly calcareous, contain shell fragments, pieces of lignite, erratics and striated stones, and from analyses appear indistinguishable from the Irish Sea till, known as 'Eastern General' or 'Ballycroneen till' in Ireland (Stephens, 1966b). A composite section from the east and south faces of Brannam's Pit is reproduced in Fig. 11. 4A.

Mottled stony clay containing erratics (= weathered till) and solifluction 'earth' (layer 2) overlie the boulder clay. Deep frost cracks or pipes, filled with a sandy-clay paste which gives a non-calcareous reaction, extend some 1·5 m (5 ft) down into the stony clay, but no fossil ice-wedges have been detected; many of the angular rock fragments are inclined at vertical or high angle positions which may be interpreted as resulting from severe frost heaving and reorientation. A distinct red-coloured layer of gritty-sand about 0·6 m (2 ft) thick separates the upper stony clay (layer 2) from 1·2 m (4 ft) of red weathered (decalcified) till (layer 3) in places. Thus the total depth of the weathered and disturbed material overlying the fresh till is from 2·4 to 3·6 m (8 to 12 ft), a depth comparable with that obtained from many Irish sections in the older Irish Sea till. The inference is that such weathering and disturbance represents a full interglacial followed by another glacial period (= Weichsel). Away from the vicinity of the clay pits the local head or solifluction 'earth' contains no erratics or striated stones, which are abundant in layers 2 and 3 in Brannam's Claypit.

The gravels capping the Hele–Ellerslie ridge (Fig. 11.5) are highly

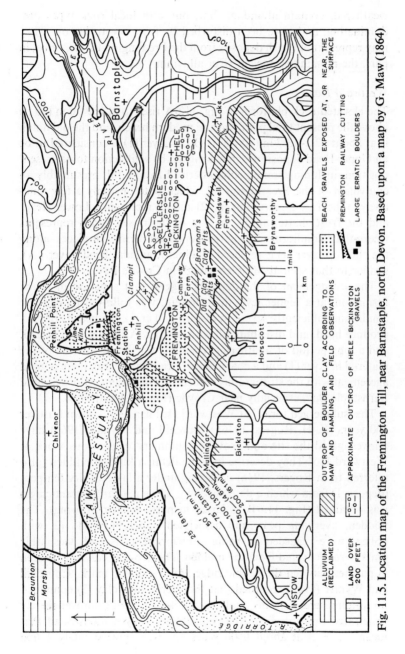

Fig. 11.5. Location map of the Fremington Till, near Barnstaple, north Devon. Based upon a map by G. Maw (1864)

weathered, contain abundant flint, but with local rock types predominating, and have been disturbed by frost cryoturbation. They may represent outwash deposits from the Fremington till ice, or be, in part, the remains of ancient river alluvium.

### Westward Ho!

The elevated rock platform in Fig. 11.6A rises to at least 6 m (20 ft) above the present tidal platform and modern beach, and the old notch is at about 12·2–13·7 m (40–45 ft) O.D. Layer (1) is considered to be an Upper Head deposit which has moved by solifluction to its present position, including some pebbles derived from a till, presumably an equivalent of the Fremington till, which outcrops only a few miles away in the Taw estuary. Layers (2) and (3) represent a Lower Head, and the frost shattering on the top of the raised beach represents freeze-thaw conditions during the same cold period. In places, along the cliff-top layer (1) is replaced by blocky, angular head (layer 1A), similar to that seen at Middleborough in Croyde Bay.

The arrangement of the deposits suggest that a period of beach formation, at a higher level than present, was followed by climatic conditions sufficiently severe to disturb and crack the beach cobbles, produce head, and distribute till with erratics. A further cold period, and a long period of weathering, accounts for the weathering of the material in layer (1) and a coarse upper head. In general, there is good agreement with the stratigraphy recorded from other sections around Barnstaple Bay, especially near Croyde (Fig. 11.3 and 11.4).

The 'submerged forest' at Westward Ho! has been known for over 300 years (E. H. Rogers, 1946). Churchill and Wymer (1965) have reinvestigated the site and provide a summary of previous work at Westward Ho! They describe in some detail a kitchen midden (also referred to in Rogers, 1946) occurring under a layer of dry fen peat (deposit 1) from which a date of 6585 ± 130 B.P. (Early Atlantic age) was obtained (Fig. 11.6B, is based in part on their drawing). The midden, overlying peat, and associated upper blue clay (2, with pollen) were regarded as being contemporaneous and Mesolithic flints were recovered from the midden. The deposits can still be seen in patches between the pebble ridge and low-water mark. Churchill and Wymer (1965) recorded that the 'sterile blue clay (3) contained no foraminifera, diatoms, mollusca, pollen or seeds', but was merely penetrated by roots from the peat above. They demonstrated that mean sea-level at about 6500 years B.P. was approximately 4–6 m

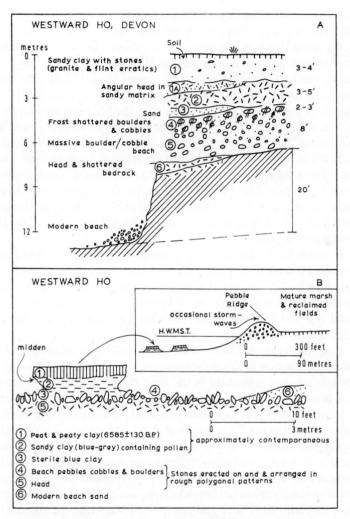

Fig. 11.6. Coastal sections at Westward Ho!, north Devon.

(13·7–20 ft) below that at present, which is consistent with conditions at Margam (South Wales) and in Somerset; their results implied an absence of crustal warping in the Bristol Channel since that time.

In 1966 the scouring of the beach by winter storms permitted fresh observation to be made of deposits (4) and (5). The postglacial peats and clays, and sterile clay overlaid a definite horizon of beach cobbles (4). The majority of the beach cobbles were erected on end, some being embedded in sterile blue clay (3) but most in the head (5). There was a rough polygonal arrangement of the cobbles and stones in the head, making a kind of patterned ground (Fig. 11.6B). The lower intertidal rock platform forms the base for these deposits, and is thus in part at least a relic feature, older than the head resting directly on it. The difference in height of the beach deposit in sections A and B is significant, but the contrast in the stratigraphy makes correlation difficult. It seems likely that these are beaches of a different age. Head layer (2) in the cliff section (Fig. 11.6A) is correlated tentatively with head layer (5) in Fig. 11.6B resting on the foreshore rock platform. Nowhere can it be shown that the beach cobbles (4) on the foreshore are overlain by a head deposit, only by sterile blue clay, and datable postglacial deposits.

It seems likely that the cryoturbation of the head (5) and beach cobbles (4) took place during the Weichsel cold period (cf. frost disturbance of deposits at Saltmills, Co. Wexford; Mitchell, 1962) and thus the beach gravels may be of Eemian age. The highly weathered head (5), and probably head (2) in the cliff section could then be regarded as Saale in age.

Postglacial shingle beaches have never been observed to be disturbed by frost action in this way, for many of the cobbles and boulders are erected, and embedded in the head (5). This 'Eemian' beach (4) has been responsible for the trimming of an older rock platform, eroding and stripping away part of a head (5) deposit resting upon it, and also cutting back a new rock cliff in places. There is no evidence to suggest that the height of beach formation and cutting of a platform across rock and head differed significantly from that of the present day. It seems likely that the post-Weichsel Flandrian transgression simply reworked beach gravels already in existence on a pre-existing platform of rock and head. Temporary lagoons allowed the fossiliferous clay (2) and peat (1) to form, before a particular shingle bar was rolled landward across it, at some time after 6500 years B.P. One of these postglacial shingle bars now forms the famous 'Pebble Ridge' of Westward Ho! [about 3 km (2 miles) long and 6–9 m (20–30 ft) high above its base] its crest rising some 4·6 m (15 ft) above high-water mark (Spring tides). It has been estimated that the southwestern end of the ridge occupied a position 180 m (600 ft) seawards of its present position in 1863. In position and height (crest at about

7·6–9 m: 25–30 ft O.D.) it probably resembles the form of the Eemian beach ridges, for it rests upon head (5) in at least one place, and rock near the southern end of the ridge.

### Hartland Quay and Damehole Point

Of all the possible ice-marginal or subglacial channels which have been described briefly (Stephens, 1966), those near Hartland Quay and Damehole Point are perhaps the most convincing examples (Fig. 11.2). At both sites channels have been incised in such a way as to isolate a small hill on the seaward side of the channel, and throughout their length the channels 'hang' above the sea. The channels are characterised by flat floors, frequently overlain by 1·8–2·4 m (6–8 ft) of coarse, blocky head.

The 'in and out' channels at Hartland Quay and Damehole Point are not marine-cut features, but probably form part of the seaward ends of a system of coastal valleys, which have been linked together and widened appreciably by water flowing along the edge, or below the margin, of an ice mass impinging against the coast. It is conceivable therefore that the channels were formed when ice was depositing till at Fremington, Trebetherick Point and on the Scilly Isles; subsequently they received a layer of head during the last glacial period.

### CORNWALL

Glacial deposits of till or outwash gravels have not been recorded between Barnstaple Bay and Trebetherick Point (Fig. 11.1), nor have any further systems of dry channels (of possible marginal or subglacial origin) been mapped. Thus it is possible only to indicate the position of the southern margin of the Irish Sea ice in a general way. There is no actual proof that ice crossed the present coast except at Trebetherick and in the Scilly Isles, but many of the modern beaches contain copious quantities of erratic pebbles, other than the ubiquitous flint, as for example at Widemouth Bay (abundant granites) and Portgaverne, between Pentire Point and Tintagel (mostly limestone pebbles). On the other hand, great thicknesses of head deposits are well exposed in many coastal sections, and together with raised beaches have been described in the *Memoirs of the Geological Survey*, and by other authors (e.g. Guilcher, 1949). Space permits only brief comments to be made on the deposits at two sites in north Cornwall–Trebetherick and Godrevy (Fig. 11.7) and in the Scilly Isles (Mitchell and Orme, 1967).

*Trebetherick Point*

Trebetherick Point is situated about 2·4 km (1½ miles) north of
Padstow, within the shelter of the mouth of the Camel estuary
(Fig. 11.2). The sections have been described by Reid *et al.* (1910), by
Arkell (1943) and by Clarke (1965a). The present writer has examined
the sections going north from Daymer Bay towards the open sea, and
two are reproduced in Fig. 11.7.

At the base of the deposits there are two well-planed rock platforms,
the lower 2·4–3 m (8–10 ft) above the back of the modern beach cuts
into the higher, 6–7·6 m (20–25 ft) above high-water mark (Spring
tides) at the northern end of the exposed sections. Their age is
unknown.

In section (Fig. 11.7A) blown sand and top soil (1) overlies a dark
earthy clay (2) containing fine slate slivers, quartz pebbles and some
erratic pebbles. A fossil ice-wedge passes into the top of the main
head below (cf. Middleborough, Croyde Bay, north Devon). Coarse
slate head with sandy-clayey matrix (3) (= main head), a lens of sand,
and a few slate fragments form the rest of the section resting upon a
rock platform some 2·4–3 m (8–10 ft) above the back of the modern
beach.

The other section (Fig. 11.7B) shows a sandy soil (1) resting upon
an upper head (2) of fine slate slivers, quartz fragments and pebbles,
and with a sandy-clay matrix. Immediately below is another head
deposit of slates and large quartz fragments, together with pebbles
and boulders (4) (= Arkell's boulder bed). This material is partly
bedded, especially the lower 0·9–1·2 m (3–4 ft); it resembles an outwash deposit which has incorporated some head, but in no way can it
be termed a raised beach. The top of the deposit shows frost convolutions (3) and small fossil ice wedges. Several feet of sand or sandrock
(5) then overlie a coarse head and the shattered bedrock of the top of
the high platform.

Arkell (1943) showed this head to interdigitate with the raised
beach, and this section, therefore, compares closely with that recorded
at Pencil Rock, north Devon (Stephens and Synge, 1966). Whatever
the precise age of this old head, whether marking the beginning or end
of a glacial phase, it seems clear that it accumulated contemporaneously with the raised beach upon a rock platform already in existence,
and therefore the platform was not formed necessarily by the waves
responsible for the deposition of the raised beach.

Arkell (1943) interpreted the Trebetherick boulder gravel (= boulder bed) as a raised beach (Eemian interglacial age), but although there is great variation in the deposit the boulder bed

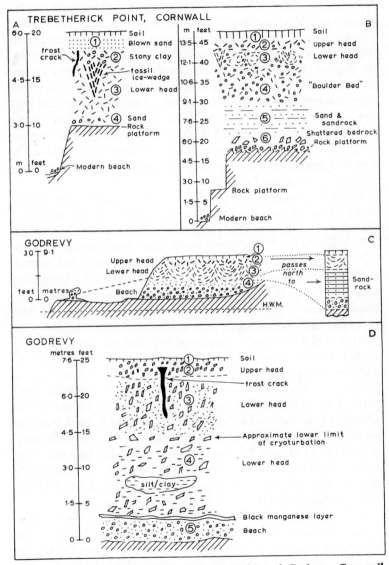

Fig. 11.7. Coastal sections at Trebetherick Point and Godrevy, Cornwall

resembles an outwash gravel, being well bedded in its lower part, containing erratic material (cf. Mitchell and Orme, 1967), and filling chance hollows in the bedrock. It is associated closely with the lower (main) head, deposition of both taking place contemporaneously. According to Arkell (1943, p. 147) it is 'full of slate fragments and in places resembles rearranged head mixed with sand and erratic pebbles...the gravels and head inosculate'. Clarke (1967) comments that 'the large stones are partly water worn and some perfectly smooth; the absence of local rock as head (slate slivers) is noteworthy and the matrix is unsorted, consisting of gravel, sand and the finest rock flour. It seems most likely to be a mixture of outwash gravel and Irish Sea ice till, which has been subjected to later frost action.' Synge (1966) recorded a possible till deposit with erratic pebbles, 'pushed against lower head' in the vicinity of these sections; such a till must be regarded as deposited by Irish Sea ice and equated tentatively with the Ballycroneen and Fremington tills.

Arkell demonstrated two solifluction deposits resting either on the rock platform or on bedded sand and fine shingle of the raised beach. He referred to these as the 'Head of Daymer Bay' and 'the pebbly solifluction deposit'. The lower coarse slate head resting upon the raised beach is equivalent to the Main Head elsewhere in Cornwall and Devon (Stephens, 1961a; Clarke, 1963, 1965a and b). The pebbly solifluction clay (Clarke refers to 'brown solifluction clay') with abundant quartz pebbles, angular slate pieces and slivers is a later (younger) head consisting of the weathered products of the main (lower) head and remanié deposits of Irish Sea till. The clay matrix could have been more easily derived from till than head or bedrock, and nothing like it is seen lower down in the sections, nor at inland localities. It fills hollows and caps the main head along the lower parts of the cliff, where it might be expected to accumulate in the absence of marine/river submergence. It is quite distinct from the lower (main) head (cf. sections at Westward Ho! and Godrevy) (Figs. 11.6, 11.7C and D). During the second cold phase the upper surface of the main head was disturbed by frost action at some points (Fig. 11.7B), but only in places where it was not capped by the pebbly solifluction clay.

A possible third cold phase is shown by the frost cracks, fossil ice wedges, and convolutions which disturb the pebbly solifluction clay and the upper head (Fig. 11.7A). These disturbances are closely comparable to those recorded at Middleborough (Fig. 11.4B).

*Godrevy, near Hayle*

At Godrevy a broad shallow valley has been filled to a depth of about 15 m (50 ft) with head deposits, which outcrop in a series of finely exposed sections (Fig. 11.7C and D) between the Red River and Godrevy Point (Reid and Flett, 1907; Robson, 1946). The upper head (deposit 2) is best displayed towards the northern end of the sections where fresh, angular slate and quartz fragments form a definite layer 0·6–1·2 m (2–4 ft) thick over the weathered and highly convoluted head below; this is regarded as a young solifluction deposit of Weichsel age.

Much of the main (lower) head (deposit 3, and deposit 4 in Fig. 11.7D) consists of pieces of quartz rock, the shales and slates having been reduced to a sandy clay, although it is not possible to say with certainty that all the weathering took place after it had accumulated by solifluction processes. The processes involved in the moving of the head rubble may have set in motion a large quantity of pre-existing weathered debris. The depth of cryoturbation of the main (lower) head varies, but frequently exceeds 1·5 m (5 ft) (deposit 3, Fig. 11.7D). The main head, and the cryoturbation of the upper 1·5–2·4 m (5–8 ft) of the deposit, is considered to be of Saale age.

Erratics are found in the beach deposits (deposit 4, Fig. 11.7C), but not in the main head. The large greenstone boulder referred to in Robson's (1946) account may not be a true erratic, but derived from a nearby source. The existence of some shattered bedrock in contact with the rock platform, and yet below the raised beach, is believed to indicate a cold phase prior to the deposition of the raised beach. This implies the existence of a platform on which possible 'erratics' could accumulate and where any earlier head could collect (cf. Trebetherick).

As one proceeds northward towards Godrevy Point well-cemented sand-rock appears between the lower (main) head and the raised beach (Fig. 11.7C) and the section resembles closely those already described around Croyde Bay. No glacial till or outwash gravels were recorded.

*Scilly Isles*

A sequence of pre-Holocene deposits, including beach conglomerate, erratic gravels with striated pebbles, and head, were described by Barrow (1906), and have been re-examined subsequently by Mitchell and Orme (1965, 1967); Fig. 11.8 is based on their most recent paper.

They have provided an outline of the stratigraphy of the superficial deposits above the deeply weathered basal granites.

Wave-cut rock platforms are present at 3–7·6 m (10–25 ft) above mean sea-level, but generally do not maintain a constant height for any great distance on the well-jointed granite, just as at St Loy near Land's End, and Carnsore Point south of Wexford. Old beach deposits

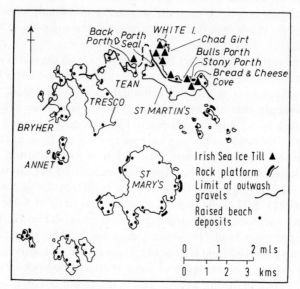

Fig. 11.8. Pleistocene marine platforms and coastal deposits in the Scilly Isles (after Mitchell and Orme, 1967)

resting on the rock platforms may contain erratic debris and sometimes interdigitate with fine head. Occasionally the beach deposits are strongly cryoturbated, and are always overlain by lower or main head, which is usually coarse in texture and consists of angular blocks of granite set in quartz sand, without erratics.

A fine, laminated head forming a distinct deposit with included erratics may rest on the lower head. Cryoturbation convolutions occur in the top of the fine head and large granite blocks may lie on the surface, derived presumably from the granite slopes above as a 'clitter' (cf. St Loy, Carnsore Point and Dartmoor) in the last stages of solifluction activity. At many sites the fine laminated head and granite blocks form the uppermost deposit (e.g. Bread and Cheese Cove on St Martin's), but north of the line delimiting the southern limit of outwash gravels, the latter rests upon the fine head and inter-digitates with it. No definite morainic features were recorded by Mitchell and Orme (1967), but at several sites, including Bread and Cheese Cove, and Back Porth on Tean Island, a compact brown clay is present between the lower and upper head deposits. This clay has been interpreted as a weathered facies of Irish Sea ice till, closely resembling the Ballycroneen till in Ireland, and Fremington till in north Devon. A preferred orientation of pebbles in the clay at Stony Porth gave a predominant direction of between 0–30° true north (Synge, 1967).

Thus there seems to be no doubt that pockets of weathered till, to-gether with erratic-bearing outwash gravels extending up to 45·7 m (150 ft) are present on the north-facing coasts of Bryher, Tresco, St Helen's, Tean, White and St Martin's island, overlying a sequence of beach gravels and lower (main) head, without erratics. The till is over-lain by erratic-bearing head. These glacial deposits constitute good evidence for the former existence of a Pleistocene ice-sheet on the northern Scillies, probably as part of the same advance which carried Scottish erratics (e.g. Ailsa Craig micro-granite) to Cork and north Devon. This ice advance may be assigned, tentatively, a Saale age, for it far exceeds the proven limits of the Weichsel ice sheets, and this allows fairly close correlations to be made for the various head deposits and raised beach gravels, lying above and below the till in the Scillies, with those elsewhere in south-west England and southern Ireland.

It should be noted that the absence of till and outwash deposits on St Mary's Island, at Godrevy, and at St Erth (Mitchell, 1967), pro-vides further support for the contention that ice has not overridden the peninsula, and was not responsible for emplacement of the Porthleven erratic (Stephens, 1961a, 1966). The steep submarine slope off north Cornwall may well have exerted considerable physical control of the position of the outer edge of the southward moving ice, until the granite 'shelf' of the Scillies was reached.

On Annet Island Mitchell and Orme (1967) recorded a beach younger than the glacial deposits and lower head, claiming it to be of Eemian age, and lying at about 7·6 m (25 ft) O.D. At Porth Seal on St Martin's Synge (1967) recorded two beach deposits (Fig. 11.9) inter-bedded with head; both the upper and lower head deposits may be of

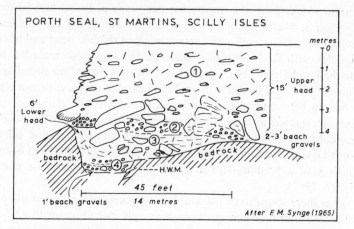

Fig. 11.9. Coastal sections at Porth Seal on St Martin's, Scilly Isles. Reproduced by permission of F. M. Synge from an unpublished field sketch

Saale age, thus suggesting that the beaches (deposits 2 and 4) may overlap the Hoxnian–Saalian periods.

*Discussion*

Examination of sections at St Loy, Prah Sands, Porthleven, and Pendower (Stephens and Synge, 1966; Stephens, 1966a) established that beyond reasonable doubt several periods of head formation were represented, postdating the raised beach and a wave-cut rock platform. The different depths of head may in themselves mean very little except that the Lower or Main Head is always the thickest deposit. After the formation of the main (lower) head there may have been little rock exposed and very little in the way of a slope down which a fresh head could move seawards. But the different types of head do seem to be important: the upper head of coarse material indicating perhaps a shorter, but very sharp period of frost action, whereas the lower 'finer' head may represent the movement in a previous cold period of all the material which had already been substantially reduced by weathering before a mechanism was provided by which extensive mass movement could take place. A further point of importance is the absence of erratic material at levels above the raised beach, in contrast to the situation in the Scilly Islands and along the north coasts of Cornwall and Devon, where erratics are recorded in the head deposits. No sections in glacial tills are known on the English Channel coasts and available erratics are confined to the modern beach, or to the raised beaches sealed below the multiple head deposits.

The amount and scale of the cryergic activity in west Cornwall must mean that we are dealing with a relic periglacial landscape and the coastal sections suggest that several periods of solifluction have taken place. These processes do not appear to have been interrupted by ice movement across the peninsula, a conclusion which has received additional support from Mitchell's careful investigation at St Erth (Mitchell, 1967). Many sections are known in the Penwith peninsula where the raised beach and various deposits of head have been recorded (Fig. 11.2). These are well documented in the *Memoirs of the Geological Survey* (Reid and Flett, 1907), Robson (1946), and Guilcher (1949), in which full bibliographies may be obtained.

Many other sections have been examined between Falmouth and Exmouth, and some of the principal sites are shown in Fig. 11.2. Space does not permit any recording here of the detailed stratigraphy at Mevagissey, Looe, Downderry, Plymouth Hoe and elsewhere, but there are few variations, except in detail, from the sections already described in west and north Cornwall and north Devon (Stephens, 1966a). Orme (1961, 1962, 1964) has described a number of sections in south Devon, and comments upon the findings of earlier workers (e.g. Green, 1943). He reported a similar sequence of head deposits overlying raised beach shingle, and considered that there was evidence for several different levels of raised beach and rock platforms, of varying age. His work in Ireland (1966), especially along parts of the southern coast, nearest to south-west England, confirms that there too raised beach shingle of the old raised beach interdigitates with head. Orme argues that the planation of the platform, on which a mixture of beach and head rests, must have taken place while freeze-thaw cycles were making debris available for beach formation; and he emphasises the problems of attempting to date rock platforms. Bryant's (1966) work in south-west Ireland also indicates the difficulties encountered in attempting to correlate beach gravels and rock platforms under head with those found under glacial drifts.

## The Bristol–Mendip area

What is known of the Pleistocene history of this area, which has lain for the most part within a periglacial environment during the glacial periods, depends to a large extent upon the interpretation of certain coastal sections (e.g. Brean Down, Clevedon and Weston-super-Mare) and the investigations of the cave and allied coombe deposits of the Mendips (Palmer, 1931; Donovan, 1955; ApSimon *et al.*, 1961). No glacial till or erratics have been recorded from inland areas, but on the high ground above 52 m (170 ft) O.D. there are patches of flint and chert gravel, sometimes associated with quartz pebbles. These form a 'plateau drift' of unknown origin, but may prove to be associated with some early stages of the Severn terrace system (Stephens, Chapter 5).

The gravels of the 30 m (100 ft) terrace of the Bristol Avon have yielded Middle Acheulian implements, but in the absence of a fauna of Great Interglacial (Hoxnian) age, the terrace itself cannot be correlated with the Severn terraces or the Thames valley terraces (Donovan, 1955). If altimetric means only are used then the 30 m (100 ft) terrace of the Bristol Avon could be correlated either with the Bushley Green terraces or the Kidderminster terrace of the Severn valley, the former probably of Saale age and the latter of Last Interglacial age. Palmer (1931) still provides the most useful account of the lower terraces of the Bristol Avon, concluding that they are younger than the 30 m (100 ft) terrace, and reflect sea-level changes during the Upper Pleistocene; the mammalian remains (mammoth, woolly rhinoceros, horse, ox and pig) and artefacts suggest that some parts of the lower terraces are of Weichselian age and probably related to the Main and Worcester terraces of the Severn.

Some fissures, or widened joints, in the Carboniferous Limestone plateaux of the area (e.g. Durdham Down, near Bristol; Donovan, 1955) have yielded 'warm' faunas, including *Elephas antiquus*, *Hippopotamus*, cave lion and hyaena, which may be of Last Interglacial age, in contrast to the 'cold' fauna of the terraces *below* the 30 m (100 ft) level. Part of the infilling of the fissures, which were probably open and receiving debris for a long time, may therefore pre-date the lower river terraces. Accurate and meaningful correlation between the terraces, the fissure infillings and the cave deposits is difficult, but it seems likely that the Last Glaciation (Weichsel) is represented by numerous cave faunas in the Mendips. However, it should be remembered that whereas implements and faunal remains sealed in terraces can remain sealed in the same stratigraphical horizon or can descend to lower levels as new (lower) terrace systems replace older (higher) ones, these remains rarely ascend from a lower to a higher terrace. In contrast, cave deposits can contain 'mixtures' of faunal remains and artefacts which require careful excavation (e.g. Gough's Cave, Cheddar, by Donovan, 1955; Fig. 11.10B).

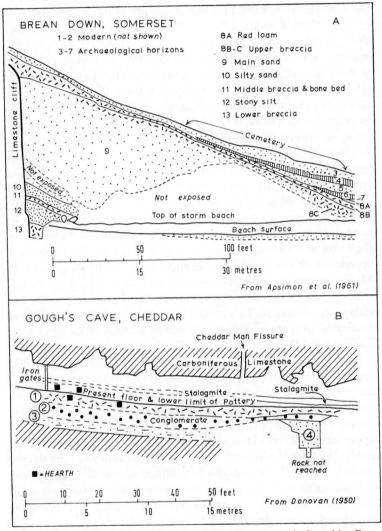

Fig. 11.10. Quaternary sections at Brean Down and Gough's Cave, Somerset. Section A is redrawn from ApSimon *et al.*, 1961. In Section B, at Gough's Cave; Layer 1 is a breccia (with hearth sites); Layer 2 is a gravel conglomerate deposited by water; Layer 3 is composed of sand and clay, also waterlaid. Redrawn from Donovan, 1955

Some pre-Creswellian (proto-Solutrean) flint remains have been recovered from Soldiers' Hole Cave, in association with mammoth and hyaena, but the Last Glaciation is perhaps best represented by the Creswellian culture of which the chief example was found in Gough's Cave, Cheddar (Fig. 11.10B). The cave was probably used by man and animals throughout the Weichsel period, but the Creswellian is generally taken to represent the upper or more recent part of that cold period, and is sometimes referred to as the 'Late Glacial'. The stratigraphy of the deposits in Gough's Cave illustrates the complexity of cave stratigraphy, and shows the Creswellian culture to be associated with a 'cave breccia' (= head = deposit 1), probably indicating severe climatic conditions at least in winter, and Magdalenian (= Upper Weichselian) implements. These are underlain by limestone gravel (= deposit 2) representing cold climate conditions (frosts and spring flooding of the cave), and overlain by a stalagmite layer with intercalated sand and clay, and red cave earth with Romano-British and Early Iron Age remains. The lower deposits represent tundra conditions according to Donovan (1955), and it is noteworthy that the horse was the commonest food animal during the Creswellian occupation of the cave. The transition to the postglacial, that is to Mesolithic and Neolithic, is not well preserved on Mendip, although reindeer is recorded in close association with a Neolithic horizon, and stalagmite is absent. In contrast, the Mesolithic may have been associated with the uppermost stalagmite horizon in Gough's Cave and Soldiers' Hole.

Fans of sub-angular gravel debouched from the mouths of the now dry limestone coombes of the Mendips, and hill-slopes west of Bristol are mantled with breccias, which probably also accumulated during the Weichsel cold period. These breccias may be correlated with those (layers 8–12) at Brean Down (Fig. 11.10A), while near Weston the upper surface of the breccia showed frost convolutions. In some of the valleys the breccia was carried right across the valley bottom to constitute a 'valley-fill', presumably by solifluction, and such gravels may be less angular as a result of such movement. Where sections are available these hillside breccias or head deposits are seen to pass below the postglacial alluvial deposits of the Somerset Levels.

Remnants of wave-cut rock platforms and raised beach deposits occur between the mouth of the Bristol Avon and Brean Down. A '15 m' (50 ft) beach has been recorded, and good sections are available at the Old Pier at Weston and at Kewstoke, although this does not imply that sea-level was 15 m (50 ft) higher than at present on account of the large tidal range. At Middle Hope the beach is lower and has been referred to in the literature as the '3 m' (10 ft) beach. The increased tidal range in the constricted estuary allows the sea to reach 9–12 m (30–40 ft) O.D. during exceptional spring tides, and consequently the shoreline of the '15 m' (50 ft) beach must have swung inland south of Brean Down, along the southern foot of the Mendips between Bleadon and Cheddar, Wookey and Westbury. In other words, the greater part of the Somerset Levels, up to about the 15 m (50 ft) contour may have been submerged at high water of spring tides, aided by only a very modest increase of mean sea-level during the Last Interglacial, of perhaps no more than 3 m (10 ft). The stretch of low hills between Portishead and Clevedon may well have been separated from the mainland.

The fossiliferous Burtle Beds (Bullied and Jackson, 1937), which occur as 'sand bars' and rise above the peat lands of the Somerset Levels, pose many problems. The sands contain a temperate marine fauna (McMillan, 1964), and were well exposed in pits near Middlezoy, reaching a maximum height of between 9–15 m (30–50 ft) O.D. Thus the Burtle Beds may be equivalent to the '15 m' (50 ft) raised beach, and this is given some support by the presence of a discontinuous capping of brickearth and occasional evidence of frost disturbance of the sands. The raised beach at Middle Hope also appeared to be frost disturbed and has been assigned to the Last Interglacial; it was not overlain by great thicknesses of head comparable with those seen in Devon, Cornwall and southern Ireland. In fact, no sections were detected where beach gravels were buried by masses of solifluction debris, but much further mapping is required.

If marine action was carried to nearly 15 m (50 ft) O.D. during the Last Interglacial, under the special tidal conditions operating on this coast, then it is extremely unlikely that any part of Brean Down deposits (ApSimon *et al.*, 1961), except perhaps the massive lowermost breccia (Deposit 13, Fig. 11.10A), could have remained *in situ*. Consequently, one is inclined to accept the argument for a Last Glaciation (Weichsel) age for deposits 8–12, but to consider it possible that the lowermost breccia may be older, and represent a remanié head deposit, from which the smaller debris has been cleaned away by wave action. Thus the layers of breccia and sands (deposits 8–12) would equate with the fan gravels, hillside breccias and cave breccias, and stalagmite layers of the Mendips, and can be correlated with the upper head deposits in Devon, Cornwall and southern Ireland.

The age of the various rock platforms, at Brean Down (0 m O.D.), the Upper (12–15 m: 40–50 ft O.D.) and Lower (9–10 m: 30–33 ft O.D.), Woodspring platforms, and fragments preserved elsewhere, for the most part on Carboniferous Limestone, present the same difficulties of dating as we have seen exist in Devon and Cornwall.

### The Postglacial period

Spectacular changes in the physical landscape have taken place as a result of the eustatic recovery of sea-level from the low level of the last glacial period. This rise in sea-level (Fairbridge, 1961) drowned old land surfaces upon which postglacial vegetation had established itself all round the coasts of the British Isles (Godwin, 1943; Steers, 1946). Submerged peats are intercalated with marine deposits in many harbour and estuarine sections, and in the Bristol Channel one of the lowest levels recorded is at 13 m (42 ft) below datum at Barry Docks (North, 1964). Some of the deposits form part of the infilling of the buried rock channels which are known to extend off-shore in the main estuaries (e.g. Lee Valley, Co. Cork, Farrington, 1959; Barnstaple Bay, Devon, McFarlane, 1955), and many of which extend to over 30 m (100 ft) below datum. Just as in the case of the rock platforms these buried channels cannot be dated accurately, but are presumed to have been cut during the glacial periods which accounted for major withdrawals of the sea to levels several hundred feet below the present.

Major physical changes have also occurred as a result of the silting-up of rias and estuaries, and the closing off of former tidal inlets (e.g. Loe Bar, near Porthleven; at Hallsands and Slapton; and at several sites in south Wexford). The growth of spits and associated dune systems, such as Northam Burrows and Braunton Burrows in Barnstaple Bay, at Dawlish Warren near Exmouth, and along the Wexford–Waterford coast, has allowed considerable natural silting, and artificial reclamation has taken place (Steers, 1946). In the Somerset Levels the large tidal range has permitted great thicknesses of marine silt to accumulate, much of which has been reclaimed. Large areas of the country around Bridgwater and along the valleys of the Parrett, Prue and Axe would be inundated but for the existence of artificial controls. Godwin (1941, 1963) has demonstrated that a fluctuation of sea-level has occurred, superimposed upon the general postglacial (Flandrian) marine transgression.

The peat areas of the Somerset Levels represent the remnants of an extensive complex of ombrogenous raised bogs. The stratigraphy along a line from Burnham-on-Sea to Shapwick Heath is shown schematically in Fig. 11.11. The progression of the Flandrian transgression was recorded by the peat layer (deposit 3) between marine clays (deposits 1 and 2) at Burnham. On the surface of the lower marine clay (at about 0 m O.D.) peat growth began about $5405 \pm 130$ B.P. There can have been little change of land and sea-level relationships for some considerable time because the raised bogs continued to

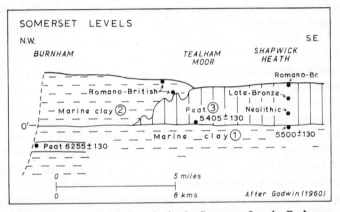

Fig. 11.11. Postglacial deposits in the Somerset Levels. Redrawn from Godwin, 1960

grow throughout the Neolithic, Bronze and Iron Ages, although Godwin (1960) has shown that there was an increase in the wetness of the bog surfaces in Late Bronze and Early Iron Age times (between 0 and 1000 years B.C.).

In Romano-British times there was a limited marine transgression of the seaward (western) margins of the raised bogs, when a second marine clay was deposited, the reasons for which are not known precisely, although it is generally assumed to have resulted from eustatic causes. This involves the assumption that the area south of the Bristol Channel has remained stable throughout postglacial time, some 10,000 years.

The postglacial Flandrian transgression has also permitted the erosion of cliffs to be resumed following the withdrawal of the sea during the last glacial period and the modification of the cliff profiles by cryergic processes. The sea has brought about some considerable recession of the rock cliffs between Hartland Point and Bude, and near Portreath. Elsewhere, the base of the hog's-backed and bevelled cliffs has been re-etched and all trace of glacial and periglacial deposits removed, to reveal a steep rock face below the more gently sloping segments of the subaerially developed slopes (Savigear, 1962). Little in the way of superficial deposits remain below the hog's-backed cliffs between Foreland Point and Ilfracombe, except in a few sheltered bays, such as Lee Bay.

Cliffs in head deposits, the underlying raised beach gravels and dune sand (sandrock = aeolianite), are common throughout the West Country, South Wales and southern Ireland, occasionally exceeding 15 m (50 ft) in height, and postglacial cliff recession has been remarkable in places. Only minimum values can be computed; for example, by projecting the sloping surface of the head 'terrace' seawards until it intersects the wave-cut surface of the intertidal platform or beach a very rough estimate of the amount of recession can be made. Another method is to measure the distance of outliers of head from the cliff-foot; these may be cemented to the rock surface, and at Westward Ho! they occur up to 91 m (300 ft) from the cliff-foot, at Croyde Bay, 61–91 m (200–300 ft), and at Godrevy 30 m (100 ft). Similar figures have been obtained from the Irish coast, although it is likely that total cliff recession in superficial materials has exceeded 300 m (1000 ft) at many points since the sea regained its present level some 5000 years ago. For these reasons it seems likely that some part of the Pleistocene record, especially of the Last Interglacial (Eemian) marine beaches, aud of the Last Glacial (Weichsel) head deposits, may well be missing from critical sections. Undoubtedly much detailed research remains to be done in the field of Quaternary studies.

### CONCLUSION AND CORRELATIONS

Wave-cut rock platforms occur at several different levels in north Devon, Cornwall, southern and eastern Ireland and South Wales. Some of them inter-cliff one another and appear to be older features than the raised beach gravels, which contain erratics and head, resting on the platforms. No boulder clay has been recorded in close association with the erratics in the beaches, nor with the large erratic blocks at Saunton, Middleborough and Porthleven; the latter are believed to have been moved into position by floating ice during the early Pleistocene, and to predate the raised beach, the main (lower) head and the

Fremington till (= Eastern General (Farrington, 1944) or Bally-croneen (Mitchell, 1962)) tills in Ireland. No till has been recorded from these coasts (Fig. 11.1) in contact with the rock platforms yet stratigraphically *below* the raised beach and main (lower) head. Thus all these erratics appear to be part of a remanié deposit, merely re-worked by wave-action.

The precise relationship of the wave-cut platforms, and the raised beaches with erratics, to the buried rock channels of the coastal estuaries is unknown at present (Farrington, 1959; McFarlane, 1955). However, the existence of rock platforms at the back of large bays (Barnstaple Bay, Tor Bay, Courtmacsherry Bay), and in some of the larger inlets and rias (Cork Harbour, Waterford, Camel estuary below Wadebridge, and Fal estuary between Falmouth and St Mawes) indicates that the present coastline was, in large measure, delineated by marine erosion in early Pleistocene times; and that in places the erosion followed a period of valley incision and deepening to levels well below the present base-level of the rivers.

Mitchell (1960, 1962) has provided similar correlations to those shown in Table 11.1, but for a much larger area within the British Isles, and West (1963, 1968) has given some alternative interpreta-tions. No attempt will be made in this conclusion to duplicate or to discuss these wider correlations.

The rock platforms in Devon, Cornwall and Somerset have their counterparts on the coasts of southern and eastern Ireland and South Wales, although in the case of the former area there is one dominant platform (Wright and Muff, 1904; Stephens, 1957; Farrington and Stephens, 1964; Farrington, 1966). Glacial erratics and a consider-able amount of angular rock debris (= oldest head deposit) are mixed with the shingle in many localities (e.g. various sites in Co. Cork; Fethard, Co. Wexford; and Croyde Bay, north Devon—Fig. 11.1). Above the beach deposits there is always one or more of the follow-ing deposits—blown sand with head or main (lower) head, which may exceed 9 m (30 ft) in thickness. In southern Ireland, at Garryvoe, Ballycroneen till, or an equivalent local boulder clay at other sites, rests upon the main head, and the Kilbeg plant beds (Watts, 1959, assigned them to the Elster–Saale (Hoxnian) Interglacial; see also Watts, 1964) underlie the 'local' Waterford boulder clay. This till was deposited during the same cold period as the Ballycroneen and Greater Cork–Kerry boulder clays, and thus the main head is regarded as Saale in age. At Fethard the 'local' boulder clay which rests upon the main head is deeply weathered and disturbed by frost cryoturbations. Because the last general ice advance (Weichsel), known in Ireland as the Midland General glaciation, did not reach the south coast of Ireland it is believed that the last phase of cryoturbation of the 'local' boulder clay can be correlated with the Weichsel glacial period. At the same time ice of Welsh origin reached the coast of South Wales at only a few localities, and according to Synge (1963) Irish Sea ice failed to reach as far south as north Devon and Cornwall by at least 160 km (100 miles), impinging on the Pembrokeshire coast near Fishguard, and on the Irish coast, in the Screen Hills at Curraghcloe, north of Wexford town. The upper head deposits in County Cork are found outside the limits of the Lesser Cork–Kerry Glaciation (= Midland General glaciation = Weichsel glacial period; Farrington, 1954, 1966), and constitute a complex deposit (multilayered, and in places strongly cryoturbated) very similar to sections in the upper head recorded at Pendower, Cornwall, and Croyde, north Devon. Thus the upper head in Devon and Cornwall, and in Somerset, is regarded as the equivalent periglacial deposit to the upper head and cryoturbated boulder clay in the south of Ireland outside the limit of Midland General glaciation (Mitchell, 1960, 1962; Farrington and Stephens, 1964). Outside the Weichsel limit in southern and eastern Ireland, and in south-west England the upper head deposits seldom exceed 1·8 m (6 ft) in thickness and are generally little weathered (Synge, 1964b), in complete contrast to the main head and the 'older' (Eastern General–Ballycroneen–Fremington) boulder clay, which has sometimes been weathered to a depth of 1·8–4 m (6–14 ft). Marginal drainage channels have been recorded where the Irish Sea ice is believed to have pressed against the north Devon coast during the Saale glacial period, and gravel deposits at Hele and Ellerslie, near Fremington (Fig. 11.5), at Trebetherick (Fig. 11.7), on Lundy Island (Mitchell, 1968), and in the Scilly Isles (Fig. 11.8), are now interpreted as outwash deposits from the receding Irish Sea ice of that time. 'Lake Maw' may have discharged southwards through the Chard gap, giving rise to extensive terrace gravels in the lower Axe valley, (see Chapter 5).

Maw's map and sections (Maw, 1864) of the areal extent of the Fremington boulder clay suggests that the whole of the area east of Barnstaple Bay below the 61 m (200 ft) contour may have been 'plugged' with the clay, especially the east–west trending valleys which are today notable relief features of the area. The absence of the same clay from adjacent parts of the coast can be accounted for by: (1) non-deposition or the weathering and erosion of the clay, (2) the barriers of high steep cliffs which opposed the ice advance, and (3) the deposits of upper head which lie on the coastal slopes and make up part of the cliff profiles, implying a period or periods of intense solifluction which would have aided the removal of any boulder clay. The presence of erratics and striated pebbles in the soil of Braunton Great Field implies distribution of material from the boulder clay which had been eroded by the Taw, and probably a former greater extent of the boulder clay itself. Similarly, completely weathered till is recorded in Croyde Bay at Middleborough (Fig. 11.3), at Westward Ho!, from Trebetherick and the Scilly Isles.

Analysis of the Fremington boulder clay and samples of Irish Sea till from Garryvoe, Killiney (a Dublin suburb), and Gwbert, Cardiganshire (Stephens, 1966b) show them to be closely comparable and correlation is made therefore between the Ballycroneen till, the 'Older' Irish Sea Drift in south-west Wales and the Fremington till. The remnants of weathered boulder clay at Westward Ho! and Croyde Bay are also correlated with the Fremington till.

The Fremington till rests upon raised beach shingle in north Devon, as does the glacial outwash (and till) at Trebetherick and the Scillies. At Garryvoe, County Cork, the Ballycroneen till overlies main (lower) head and raised beach shingle. In the cliff sections west of Fremington Quay, and in Croyde Bay, weathered Fremington till rests upon contorted beach gravels. Thus, if it is accepted that the same Irish Sea drift is involved in all these sections a correlation can be made between these raised beaches in Devon, Cornwall and southern Ireland and the beach deposits can be regarded as pre-dating the Older or Saale (Gipping) glacial period.

However, it can be argued (Mitchell, 1967) that where a raised beach deposit is present and not overlain by lower (main) head, by Irish Sea ice till or outwash gravels, the beach could be of Last Interglacial (Eemian) age. If these conditions are satisfied, and the beach notch nowhere exceeds 6–7·6 m (20–25 ft) O.D. then it is possible that the cutting of some of the lower rock platforms may be also assigned to the Eemian interglacial. The best example is perhaps at Westward Ho! (Fig. 11.6), where the beach gravel is cryoturbated and in places overlies head resting upon a well-planed low rock platform, which is now an inter-tidal feature of the coast; the beach is covered, not by head, but by postglacial peats.

11

Mitchell and Orme (1967) have described a similar sequence of events in the Scilly Isles, and it seems likely that the comparatively low raised beaches seen at several sites (e.g. Old Pier at Weston, Middle Hope) between Weston-super-Mare and Clevedon on the Bristol Channel coast are also Eemian in age. Likewise, the Burtle Beds of the Somerset Levels can be regarded as Eemian. The height range of the raised beaches at all these localities east of Barnstaple Bay is in part conditioned by the increasing tidal range as the Bristol Channel constricts to become the Severn estuary.

The attempted analysis of the terraces of the lower Severn valley (Chapter 5) also suggests that some of the low raised beaches—and there are no known high beaches exceeding 12 m (40 ft) O.D., in the Severn estuary—are Eemian in age. If it is accepted that the Kidderminster terrace (6 m: 20 ft O.D. mean sea level suggested by Wills, 1938) is Eemian, its gravel containing a warm fauna, including *Elephas antiquus* and *Hippopotamus*, a correlation with the Burtle Beds (+ 9 m: 30 ft O.D. and also with *Hippopotamus*), and the Middle Hope raised beach (9 m: 30 ft O.D.) seems possible, and is supported by altimetric considerations. Similarly, there is reasonable correlation with other fragments of raised beach along the south coast of Devon (Orme, 1962), and in Cornwall (Mitchell and Orme, 1967). West and Sparks (1960) have correlated the Selsey raised beach deposits in Sussex with the Eemian (Ipswichian) interglacial marine transgression (called Normannian II in Brittany). The Selsey deposits, like those of the Kidderminster terrace, also contain *Elephas antiquus* and *Hippopotamus*, while the associated shingle deposits rise to 7·6–9 m (25–30 ft) O.D. (tidal range, about 8 m: 26 ft).

It appears that the Eemian (Ipswichian) sea-level need have been little different from that of today, if allowance is made for variation in exposure and exceptional storm waves (Stephens and Synge, 1966). There remains the problem of determining the actual mean sea-level of the last interglacial where only beaches are available and of distinguishing Eemian beaches where there is no frost disturbance of the gravels or overlying head deposit.

It can be argued of course that the Irish Sea till (Fremington = Eastern General = Ballycroneen = Older Drift of South Wales = Older Welsh and Gipping line—Fig. 11.1), may not be of the same age everywhere, and thus does not constitute a means to indicate a pre-Saale age for portions of raised beaches in widely separated localities. Moreover, other authors (e.g. Baden-Powell, 1955; Green, 1943;

Orme, 1960; West, 1963; West and Sparks, 1960) have presented evidence to justify the correlation of some of the raised beaches described above with the Eemian Interglacial period, thereby indicat- ing a Weichsel age for the till and head deposits resting upon the beach shingle. However, it has been argued elsewhere (Stephens and Synge, 1966) that correlation by height above datum, by comparisons of molluscan fauna, or by other means, may not be sufficient in the absence of reliable absolute dates, which are not yet available for the pre-Weichsel deposits of the British Isles, except where the beaches can be correlated closely with identifiable pollen horizons of one or other of the interglacial periods; and even then much care is needed (e.g. Mitchell, 1962; Stevens, 1958; West and Sparks, 1960). The age of the raised beaches on the coasts of southern Ireland, South Wales and the south-west peninsula of Devon and Cornwall must for the moment remain in some doubt and question marks be placed against each correlation table.

The Fremington boulder clay has also been regarded as a Weichsel deposit, thus apparently supporting the view expressed by Zeuner (1959), who also alleged that the raised beach in north Devon is Monasterian or Last Interglacial in age. The complexity of events during the Weichsel glacial period has been stressed, and suggestions made that perhaps some of the 'older' drifts of South Wales and southern Ireland (and by implication the Fremington–Trebetherick and Scilly Isles tills) may be much younger than is at present admitted by the Irish workers (notably, Farrington, Mitchell, Synge and Watts). More recently John (1965) has published a short account giving the results of carbon 14 dating of marine shells contained in outwash deposits in Pembrokeshire. He maintains that the dates demonstrate a Weichsel age of the so-called 'Older Drift' deposits of the Irish Sea ice in this area, which runs directly counter to the arguments published by Synge (1963, 1964a; see also Shotton, 1967b) who would place the last glacial limit of the Irish Sea ice along the north-facing slopes of the Lleyn peninsula in North Wales and thence looping southwards across Cardigan Bay to reach and cover only a very narrow coastal zone near Fishguard (Fig. 11.1).

Argument and counter-argument, backed by intensive field work, will continue, but for the moment much remains unproven, river terraces and raised beaches still defy positive identification and dat- ing, and it is admitted that all maps of ice limits for the Last as well as for the preceding glaciations still involve some speculation.

## REFERENCES

APSIMON, A. A., DONOVAN, D. J. and TAYLOR, H. (1961) 'The stratigraphy and archaeology of the late-glacial and postglacial deposits at Brean Down, Somerset', *Proc. Univ. Bristol spelaeol. Soc.* **9**, 67–136.

ANDERSON, J. G. C. and BLUNDELL, C. R. K. (1965) 'The sub-drift rock surface and buried valleys of the Cardiff district', *Proc. Geol. Ass.* **76**, 367–78.

ARBER, E. A. N. (1911) *The Coastal Scenery of North Devon*. London.

ARBER, M. A. (1960) 'Pleistocene sea-levels in north Devon', *Proc. Geol. Ass.* **71**, 169–76.

ARKELL, W. J. (1943) 'The Pleistocene rocks at Trebetherick Point, north Cornwall: their interpretation and correlation', *Proc. Geol. Ass.* **54**, 141–70.

BADEN-POWELL, D. F. W. (1955) 'The correlation of the Pliocene and Pleistocene marine beds of Britain and the Mediterranean', *Proc. Geol. Ass.* **66**, 271–92.

BALCHIN, W. G. V. (1952) 'The erosion surfaces of Exmoor and adjacent areas', *Geogr. Jl.* **188**, 453–76.

BARROW, G. (1906) *The geology of the Isles of Scilly.* Mem. geol. Surv. U.K.

BOWEN, D. Q. (1966) 'Dating Pleistocene events in south-west Wales', *Nature, Lond.* **211**, 475–6.

BROECKER, W. S. (1965) 'Isotope geochemistry and the Pleistocene climatic record', *The Quaternary of the United States* (published for the VII INQUA Congress), ed. H. E. Wright, Jr., and D. G. Frey. Pp. 737–54.

BRYANT, R. H. (1966) 'The "Pre-glacial" raised beach in south-west Ireland', *Ir. Geogr.* **5**, 188–203.

BRUNSDEN, D. (1964) 'The origin of decomposed granite on Dartmoor', in *Dartmoor Essays*, ed. I. G. Simmons. Devonshire Assoc., Exeter, pp. 97–116.

BULLIED, A. and JACKSON, J. W. (1937) 'The Burtle sand beds of Somerset', *Proc. Somerset Arch. Nat. Hist. Soc.* **83**, 171–95.

CHARLESWORTH, J. K. (1957) *The Quaternary Era.* E. Arnold.

CHURCHILL, D. M. and WYMER, J. J. (1965) 'The kitchen midden site at Westward Ho!, Devon, England: ecology, age and relation to changes in land and sea level', *Proc. prehist. Soc.* **31**, 74–84.

CLARKE, B. B. (1963) 'Erosional and depositional features of the Camel estuary as evidence of former Pleistocene and Holocene strandlines', *Proc. Ussher Soc.* **1**, 57–9.

CLARKE, B. B. (1965a) 'The superficial deposits of the Camel estuary and suggested stages in its Pleistocene history', *Trans. Roy. geol. Soc. Cornwall*, **19**, 257–79.

CLARKE, B. B. (1965b) 'The upper and lower surfaces, and some structural features of the frost soils of the Camel estuary', *Proc. Ussher Soc.* **1**, 192–3.

CLARKE, B. B. (1967) Private letter.

CLAYDEN, B. and FINDLAY, D. C. (1960) 'Mendip derived gravels and their relationship to combes', *Proc. Roy. geol. Soc. Cornwall*, pp. 24–5.

COOPE, G. R., SHOTTON, F. W. and STRACHAN, I. (1961) 'A late Pleistocene fauna and flora from Upton Warren, Worcestershire', *Phil. Trans. Roy. Soc.* B, **244**, 379–417.

COOPE, G. R. and SANDS, C. H. S. (1966) 'Insect faunas of the last glaciation from the Tame valley, Warwickshire', *Proc. Roy. Soc.* B, **165**, 389–412.

COTTON, C. A. (1951) 'Atlantic gulfs, estuaries, and cliffs', *Geol. Mag.* **87**, 113–28.

COTTON, C. A. (1958) 'Eustatic river terracing complicated by seaward downflexure', *Trans. Edinburgh geol. Soc.* **17**, 165–78.

DAVIES, J. L. (1964) 'A morphogenic approach to world shorelines', *Zeit. für Geomorph.* **8**, 127–42.

DE LA BECHE, H. T. (1839) *Report on the Geology of Cornwall, Devon and West Somerset*. Geol. Surv. London.

DEWAR, H. S. L. and GODWIN, H. (1963) 'Archaeological discoveries in the raised bogs of the Somerset levels, England', *Proc. prehist. Soc.* **29**, 17–49.

DEWEY, H. (1910) 'Notes on some igneous rocks from north Devon', *Proc. Geol. Ass.* **21**, 429–34.

DEWEY, H. (1913) 'The raised beach of north Devon: its relation to others and to Palaeolithic man', *Geol. Mag.* **10**, 154–63.

DEWEY, H. (1935) 'South west England', *British Regional Geology*. Geol. Surv. Pub. H.M.S.O. London.

DONOVAN, D. T. (1954) 'An occurrence of mammoth and other bones at Whatley, Somerset', *Proc. Univ. Bristol spelaeol. Soc.* **7**, 51–3.

DONOVAN, D. T. (1955) 'The Pleistocene deposits at Gough's Cave, Cheddar, including an account of recent excavations', *Proc. Univ. Bristol spelaeol. Soc.* **7**, 76–104.

DONOVAN, D. T. et al. (1961) 'Geology of the floor of the Bristol Channel', *Nature, Lond.* **189**, 51–2.

DURY, G. H. (1958) 'Tests of a general theory of misfit streams', *Trans. Inst. Br. Geog.* **25**, 105–18.

EVANS, J. (1897) *The Ancient Stone Implements, Weapons and Ornaments of Great Britain*, 2nd edn. London.

EVERARD, C. E., LAWRENCE, R. H., WITHERICK, M. E. and WRIGHT, L. W. (1964) 'Raised beaches and marine geomorphology' (from 'Present views on some aspects of the geology of Cornwall and Devon'), *Proc. Roy. geol. Soc. Cornwall*, pp. 283–310.

FAIRBRIDGE, R. W. (1961) 'Eustatic changes in sea-level', *Physics Chem. Earth*, **4**, 99–185.

FARRINGTON, A. (1944) 'The glacial drifts of the district around Enniskerry, Co. Wicklow', *Proc. Roy. Irish Acad.* **50**B, 133–57.

FARRINGTON, A. (1954) 'A note on the correlation of the Kerry–Cork glaciations with those of the rest of Ireland', *Ir. Geogr.* **3**, 47–53.

FARRINGTON, A. (1959) 'The Lee basin. Part 1: Glaciation', *Proc. Roy. Irish Acad.* **60**B, 153–66.

FARRINGTON, A. (1965) 'A note on the correlation of some of the glacial drifts of the south of Ireland', *Ir. Nat. Jr.* **15**, 29–33.

FARRINGTON, A. (1966) 'The early-glacial raised beach in County Cork', *Sci. Proc. Roy. Dub. Soc.* **2**, 197–219.

FARRINGTON, A. and STEPHENS, N. (1964) 'The Pleistocene geomorphology of Ireland', in *Field Studies in the British Isles*, ed. J. A. Steers. Nelson. Pp. 445–61.

FLETT, J. S. and HILL, J. B. (1912) *Geology of the Lizard and Meneage.* Mem. Geol. Surv. U.K.

GEORGE, T. N. (1932) 'The Quaternary beaches of Gower', *Proc. Geol. Ass.* **43**, 291–324.

GODWIN, H. (1943) 'Coastal peat beds of the British Isles and North Sea', *Jl. Ecol.* **31**, 199–247.

GODWIN, H. (1960) 'Prehistoric wooden trackways of the Somerset levels: their construction, age and relation to climatic change', *Proc. prehist. Soc.* **26**, 1–36.

GODWIN, H. (1961) 'The Croonian lecture. Radiocarbon dating and quaternary history in Britain', *Proc. Roy. Soc.* **153**B, 287–320.

GREEN, J. F. N. (1943) 'The age of the raised beaches of south Britain', *Proc. Geol. Ass.* **54**, 129–46.

GUILCHER, A. (1949) 'Aspects et problèmes morphologiques du massif de Devon–Cornwall comparés à ceux d'Amorique', *Rev. Geogr. Alp.* **37**, 689–717.

GUILCHER, A. (1950) 'Nivation, cryoplanation et solifluction quaternaires dans les collines de Bretagne occidentale et du nord du Devonshire', *Rev. Geomorph. Dynamique*, **1**, 55–77.

HUGHES, T. MCKENNY (1887) 'On the ancient beach and boulders near Braunton and Croyde in north Devon', *Q. Jl geol. Soc.* **43**, 657–70.

JOHN, B. S. (1965) 'A possible Main Würm Glaciation in west Pembrokeshire', *Nature, Lond.* **207**, 602–23.

KIDSON, C. (1964) 'The physiographic evolution of the Severn estuary', *Proc. 4th Int. Harbour Conference*, Antwerp, pp. 1–11.

KIRBY, R. P. (1967) 'The fabric of head deposits in south Devon', *Proc. Ussher Soc.* **1**, 288–90.

MCFARLANE, P. B. (1955) 'Survey of two drowned river valleys in Devon', *Geol. Mag.* **92**, 419–29.

MAW, G. (1864) 'On a supposed deposit of boulder clay in north Devon', *Q. Jl geol. Soc.* **20**, 445–51.

MCMILLAN, N. FISHER (1964) Personal communication on the shell content of the Burtle Beds.

MITCHELL, G. F. (1960) 'The Pleistocene history of the Irish Sea', *Advmt Sci.* **17**, 313–25.

MITCHELL, G. F. (1962) 'Summer field meeting in Wales and Ireland', *Proc. Geol. Ass.* **73**, 197–213.

MITCHELL, G. F. (1965) 'The St Erth beds—an alternative explanation', *Proc. Geol. Ass.* **76**, 345–66.

MITCHELL, G. F. (1966) Private letter.

MITCHELL, G. F. (1967) Lecture to Section C, Br. Ass. Advmt Sci. (Leeds).

MITCHELL, G. F. (1968) 'Glacial gravel on Lundy Island', *Trans. Roy. geol. Soc. Cornwall*, pp. 65–8.

MITCHELL, G. F. and ORME, A. R. (1965) 'The Pleistocene deposits of the Scilly Isles', *Proc. Ussher Soc.* 1, 190–2.

MITCHELL, G. F. and ORME, A. R. (1967) 'The Pleistocene deposits of the Isles of Scilly', *Q. Jl geol. Soc.* 123, 59–92.

NORTH, F. J. (1964) *The Evolution of the Bristol Channel*, 3rd edn. Nat. Museum of Wales, Cardiff.

ORME, A. R. (1960) 'The raised beaches and strandlines of south Devon', *Field Studies*, 1, 109–30.

ORME, A. R. (1961) 'The geomorphology of the South Hams', unpublished Ph.D. thesis, Univ. of Birmingham.

ORME, A. R. (1962) 'Abandoned and composite sea cliffs in Britain and Ireland', *Ir. Geogr.* 4, 279–91.

ORME, A. R. (1964) 'Planation surfaces in the Drum Hills, Co. Waterford', *Ir. Geogr.* 48–72.

ORME, A. R. (1966) 'Quaternary changes of sea-level in Ireland', *Trans. Inst. Br. Geogr.* 39, 127–40.

PALMER, L. S. (1931) 'On the Pleistocene succession of the Bristol district', *Proc. Geol. Ass.* 42, 345–61.

PALMER, L. S. (1934) 'Some Pleistocene breccias near the Severn estuary', *Proc. Geol. Ass.* 45, 145–61.

PALMER, J. and NIELSON, R. A. (1960) 'The origin of granite tors on Dartmoor, Devonshire', *Proc. Yorks. geol. Soc.* 33, 315–40.

PÉWÉ, T. L. (1964) 'Palaeoclimatic significance of fossil ice wedges', contribution to discussion at the Exeter Symposium, I.G.U. Symposium on Pleistocene Geomorphology. *Biul. Peryglac.* 15, 63–149.

PRESTWICH, J. (1892) 'The raised beaches and head of the south of England', *Q. Jl geol. Soc.* 48, 263–343.

REID, C. and REID, E. M. (1904) 'On a probable Palaeolithic floor at Prah Sands (Cornwall)', *Q. Jl geol. Soc.* 60, 106–12.

REID, C. and SCRIVENOR, J. B. (1906) *Geology of Newquay*. Mem. Geol. Surv. U.K.

REID, C. and FLETT, J. S. (1907) *Geology of the Land's End District*. Mem. Geol. Surv. U.K.

REID, C., BARROW, G. and DEWEY, H. (1910) *Geology of the Country Around Padstow and Camelford*. Mem. Geol. Surv. U.K.

ROBSON, J. (1946) 'The recent geology of Cornwall: a review', *Trans. Roy. geol. Soc. Cornwall*, 17, 132–63.

ROGERS, E. H. (1946) 'The raised beach, submerged forest and kitchen midden of Westward Ho! and the submerged stone row of Yelland', *Proc. Devon. archaeol. Soc.* 3, 109–35.

ROGERS, I. (1908) 'On the submerged forest at Westward Ho! A history of Northam Burrows', *Trans. Devon Ass.* 40, 249–59.

ROGERS, I. and SIMPSON, B. (1937) 'The flint gravel deposit of Orleigh Court, Buckland Brewer, north Devon', *Geol. Mag.* 74, 309–16.

SALTER, A. E. (1899) 'Pebbly and other gravels in southern England', *Proc. Geol. Ass.* 15, 264–86.

312     *The Glaciations of Wales*

SAVIGEAR, R. (1960) 'The seaward and valley slopes and cliffs at Porth Naven, West Penwith', *Trans. Roy. Geol. Soc. Cornwall*, pp. 22–3.

SAVIGEAR, R. (1962) 'Some observations on slope development in north Devon and north Cornwall', *Trans. Inst. Br. Geogr.* **31**, 23–42.

SEDGWICK, A. and MURCHISON, R. I. (1840) 'Description of a raised beach in Barnstaple or Bideford Bay, on the north-west coast of Devonshire', *Trans. Geol. Soc. Lond.* **5**, 279–88.

SHOTTON, F. W. (1953) 'The Pleistocene deposits of the area between Coventry, Rugby and Leamington, and their bearing upon the topographic development of the Midlands', *Phil. Trans. Roy. Soc.* **237**, 209–60.

SHOTTON, F. W. (1962) 'The physical background of Britain in the Pleistocene', *Advmt Sci.* **19**, 1–14.

SHOTTON, F. W. (1967a) 'The problems and contributions of methods of absolute dating within the Pleistocene period', *Q. Jl geol. Soc.* **122**, 357–83.

SHOTTON, F. W. (1967b) 'Age of the Irish Sea glaciation in the Midlands', *Nature, Lond.* **215**, 1366.

SIMPSON, S. (1953) 'The development of the Lyn drainage system and its relation to the origin of the coast between Combe Martin and Porlock', *Proc. Geol. Ass.* **64**, 14–23.

SISSONS, J. B. (1964) 'The Glacial Period', in *The British Isles*, ed. J. W. Watson with J. B. Sissons. Nelson. Pp. 131–51.

SMITH, A. J., STRIDE, A. H. and WHITTARD, W. F. (1965) 'The geology of the western approaches of the English Channel', IV, 287–302, in *Submarine Geology and Geophysics (Colston Papers* 17), ed. W. F. Whittard and R. Bradshaw. Butterworth.

SPARKS, B. W. and WEST, R. G. (1963) 'The interglacial deposits at Stutton, Suffolk', *Proc. Geol. Ass.* **74**, 419–32.

STEERS, J. A. (1946) *The Coastline of England and Wales*. Cambridge University Press, 2nd edn 1964.

STEPHENS, N. (1957) 'Some observations on the "interglacial" platform and the early post-glacial raised beach on the east coast of Ireland', *Proc. Roy. Ir. Acad.* **58**B, 129–49.

STEPHENS, N. (1961a) 'Re-examination of some Pleistocene sections in Cornwall and Devon', Abstracts Proc. 4th Conf. Geol. Geomorph. S.W. England, *Trans. Roy. geol. Soc. Cornwall*, pp. 21–3.

STEPHENS, N. (1961b) 'Pleistocene events in north Devon', *Proc. Geol. Ass.* **72**, 469–72.

STEPHENS, N. (1966a) 'Geomorphological studies in Ireland and western Britain with special reference to the Pleistocene Period', unpublished Ph.D. thesis, The Queen's University, Belfast.

STEPHENS, N. (1966b) 'Some Pleistocene deposits in north Devon', *Biul. Peryglac.* **15**, 103–14.

STEPHENS, N. and SYNGE, F. M. (1966) 'Pleistocene shorelines' in *Essays in Geomorphology*, ed. G. H. Dury. Heinemann Educational, pp. 1–51.

STEVENS, L. A. (1958) 'The interglacial of the Nar valley, Norfolk', *Q. Jl geol. Soc. Lond.* **115**, 291–315.

STRIDE, A. H. (1961) 'Quaternary sedimentation around S.W. England', *Trans. Roy. geol. Soc. Cornwall*, pp. 16–17.

STRIDE, A. H. (1962) 'Low Quaternary sea-levels', *Proc. Ussher Soc.* **1**, 6–7.

STRIDE, A. H. (1963) 'North-east trending ridges of the Celtic Sea', *Proc. Ussher Soc.* **1**, 62–3.

SYNGE, F. M. (1963) 'A correlation between the drifts of S.E. Ireland with those of W. Wales', *Ir. Geogr.* **4**, 360–6.

SYNGE, F. M. (1964a) 'The glacial sequence in west Caernarvonshire', *Proc. Geol. Ass.* **75**, 431–44.

SYNGE, F. M. (1964b) 'Some problems concerned with the glacial succession in south-east Ireland', *Ir. Geogr.* **5**, 73–82.

SYNGE, F. M. (1966) Private letter.

TAYLOR, C. W. (1956) 'Erratics of the Saunton and Fremington areas', *Rep. Devon Ass. Adv. Sci.* **88**, 52–64.

TE PUNGA, M. T. (1957) 'Periglaciation in southern England', *Tijdschr. Kon. Ned. Aardr. Gen.* **64**, 401–12.

TOMLINSON, M. E. (1925) 'River terraces of the lower valley of the Warwickshire Avon', *Q. Jl geol. Soc.* **81**, 137–63.

TOMLINSON, M. E. (1935) 'The superficial deposits of the country north of Stratford-on-Avon', *Q. Jl geol. Soc.* **91**, 423–60.

TRICART, J. (1956) 'Cartes des Phenomenes Periglaciaires Quaternaires en France', *Memoires Carte Geologique Detaillée Ministère de l'Industrie et du Commerce*, pp. 1–40.

D'URBAN, W. S. M. (1878) 'Palaeolithic implements from the valley of the Axe', *Geol. Mag.* **5**, 37–8.

WATERS, R. S. (1960a) 'The bearing of superficial deposits on the age and origin of the upland plain of east Devon', *Trans. Roy. Geol. Soc. Cornwall*, pp. 26–8.

WATERS, R. S. (1960b) 'Pre-Würm periglacial phenomena in Britain', *Les congrès et colloques de l'Université de Liège*, **17**, 163–76.

WATERS, R. S. (1961) 'Involutions and ice-wedges in Devon', *Nature, Lond.* **189**, 389–90.

WATERS, R. S. (1962) 'Altiplanation terraces and slope development in Vest-Spitsbergen and south-west England', *Biul. Peryglac.* **11**, 89–101.

WATERS, R. S. (1964) 'The Pleistocene legacy to the geomorphology of Dartmoor', in *Dartmoor Essays*, ed. I. G. Simmons, Devonshire Assoc. Exeter, pp. 73–96.

WATERS, R. S. (1965) 'The geomorphological significance of Pleistocene frost action in south-west England', in *Essays in Geography for Austin Miller*, ed. J. B. Whittow and P. D. Wood, University of Reading, pp. 39–57.

WATSON, E. (1965) 'Periglacial structures in the Aberystwyth region of Central Wales', *Proc. Geol. Ass.* **76**, 443–62.

WATSON, E. and WATSON, S. (1967) 'The periglacial origin of the drifts at Morfa-Bychan, near Aberystwyth', *Jl. Geol.* **5**, 419–40.

WATTS, W. A. (1959) 'The interglacial deposits at Kilbeg and Newtown, Co. Waterford', *Proc. Roy. Ir. Acad.* **60**B, 79–134.

WATTS, W. A. (1964) 'Interglacial deposits at Baggotstown, near Bruff, County Limerick', *Proc. Roy. Ir. Acad.* **63**B, 167–89.

WEST, R. G. (1963) 'Problems of the British Quaternary', *Proc. Geol. Ass.* **74**, 147–73.

WEST, R. G. (1968) *Pleistocene Geology and Biology*. Longmans.

WEST, R. G. and SPARKS, B. W. (1960) 'Coastal interglacial deposits of the English Channel', *Phil. Trans. Roy. Soc.* **243**, 95–133.

WHITTARD, W. F. and BRADSHAW, R. (1965) *Submarine Geology* and *Geophysics*. Butterworth.

WILLIAMS, R. B. G. (1965) 'Permafrost in England during the last Glacial period', *Nature, Lond.* **205**, 1304–5.

WILLS, L. J. (1938) 'The Pleistocene development of the Severn', *Q. Jl geol. Soc.* **94**, 161–242.

WRIGHT, L. W. (1967) 'Some characteristics of the shore platforms of the English Channel coast and the northern part of the North Island, New Zealand', *Zeit. für Geomorph.* **11**, 36–46.

WRIGHT, W. B. and MUFF, H. B. (1904) 'The Pre-glacial raised beach of the south coast of Ireland', *Sci. Proc. Roy. Dub. Soc.* **10**, 250–324.

ZEUNER, F. E. (1959) *The Pleistocene Period*, 2nd edn. Hutchinson.

# CHAPTER TWELVE

# The Pleistocene Period in Wales

F.M. Synge, B.A., M.Sc

Throughout the Pleistocene period, glaciation on both sides of the Irish Sea basin was dominated by ice from various sources—from an ice-shed that extended across the north-western part of the British Isles; from a snowfield built up on the highlands of Wales; and from smaller snowfields that grew on the Cumbrian Mountains and on the Wicklow Mountains (Fig. 12.1). In Wales the 'struggle' between local and extraneous ice is clear from a study of the glacial deposits. The invasion by extraneous ice was most intense in the northern part of Wales, as shown by the widespread deposition of shelly drift derived from the Irish Sea. Ice from this source inundated the west side of the Lleyn peninsula, the Vale of Clwyd, and the foothills of Flintshire. Only the Vale of Conway apparently escaped this invasion of Irish Sea ice, except at its mouth, as the northward thrust of local Welsh ice was sufficiently powerful to prevent invasion by coastal ice. As the power of the different ice bodies varied from time to time, the limits of both Welsh and extraneous ice likewise altered. For instance, Irish Sea ice was once sufficiently powerful to reach some 430 m (1400 ft) on the western foothills of Snowdonia, where a mass of shelly drift has survived a subsequent expansion of Welsh ice westwards to the Menai Strait.

In west Wales the extraneous ice was debarred from the coastal tract between Criccieth and Aberaeron owing to the presence of local ice emanating from the valleys of Glaslyn, Ffestiniog, Wnion, Mawddach, and Dovey. Further south the pressure of Welsh ice was weaker, and Irish Sea ice was able to extend across the present coastline. At some stage ice streamed across Pembrokeshire in a south-easterly direction, covering the Gower peninsula, and depositing shelly drift in the Glamorgan lowland. In the eastern marchlands of Wales the extraneous ice from the north did not extend much south of Church Stretton in Shropshire. The westward penetration of this ice

315

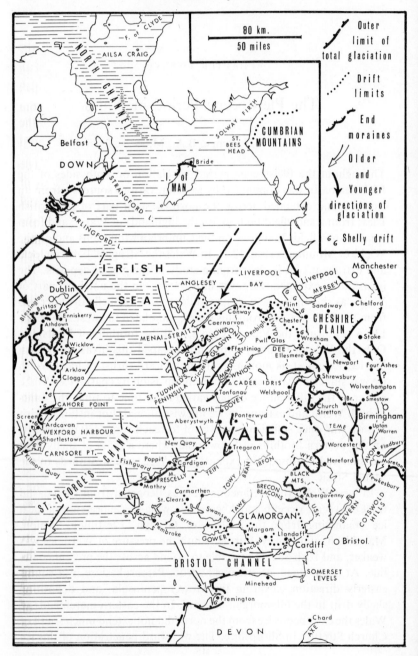

was limited to Shrewsbury on account of the presence of Welsh ice in the valley of the upper Severn.

This basic pattern of glaciation was first recognised just a century ago by officers of the Geological Survey. Following the decline of the 'Monoglacial Theory' research was aimed at finding a satisfactory means of subdividing the glacial sequence into an 'older' and 'younger' drift series. One of the first bases used for such a division was morphological (Wright, 1937). In theory, the 'older' drifts, having been subjected to longer phases of weathering and erosion, can be distinguished from 'younger' drift which was deposited during the Last Glaciation. The 'younger' or Newer Drift is characterised by fresh glacial landforms such as eskers, kames, kettle-holes, and moraine ridges, hardly modified since they were formed. Although the more extensive Older Drift was associated with similar landforms, these have generally been modified by weathering and erosion.

The southern limit of fresh glacial landforms in the British Isles was first recognised by Carvill Lewis, an American geologist. Throughout parts of its length his line was later recognised as the limit of the Last Glaciation by a number of researchers. In Wales, Lewis placed his limit across Caernarvonshire, from Clynnog to Bryncir, and eastwards close to the north coast of Wales to the Cheshire Plain. This limit was equated with one that passed off the coast at the mouth of the Humber (Lewis, 1894). Much later, but quite independently, Charlesworth mapped a Newer Drift limit that follows, in general, a similar course across England. In west Wales this limit was placed much further south than that of Carvill Lewis, along the north coast of Pembrokeshire. This line was carried east across South Wales by Tregaron, Swansea, and Abergavenny. In east Wales the line was continued north by Hereford and Welshpool, and east into Shropshire by Church Stretton and the vicinity of Bridgnorth. Further east this limit crosses the Vale of York at Escrick, and extends south,

---

Fig. 12.1. The glaciation of the Irish Sea Basin. During the Gipping glaciation Wales was probably covered almost entirely by ice. At the time of the Last Glaciation the ice cover was less extensive. The position of the ice limit at that time is now established in the midlands of England, as shown; but in Wales the ice limit has not been definitely established. The Irish Sea ice may have covered Pembrokeshire, or terminated at Mathry, or even further north, at Bryncir in Caernarvonshire, while the contemporaneous Welsh ice may have terminated as far south as Swansea and Margam, or further north, at Tregaron

close to the coast, from the Humber to Hunstanton in Norfolk. To the west Charlesworth extended his limit across the Irish Sea from Mathry in Pembrokeshire to Wexford Harbour in Ireland, outlining the snout of a great glacier that flowed from the Firth of Clyde and deposited erratics of microgranite from Ailsa Craig on both sides of St George's Channel (Charlesworth, 1928, 1929).

According to the researches of Farrington in Ireland, some modifications to this limit were found necessary in the vicinity of Dublin. As these researches are particularly important with regard to the drift succession throughout the entire basin of the Irish Sea, they will be outlined here. They are based both on the analysis of the drift content, as well as on a study of the morphology of both the deposits and associated erosion features. At Brittas, County Dublin, in the area first selected for these investigations, the Newer Drift limit of the 'British Isles Ice-Sheet' is represented by terminal moraine deposits composed largely of limestone, derived from the lowland area west of Dublin, and carried southwards on to the foothills of the Wicklow Mountains, here composed partly of shale, and partly of granite. At an earlier date a moraine composed of granitic material was deposited by local ice emanating from the Wicklow Mountains. This relationship is seen by the fact that outwash rich in limestone material, associated with the Newer Drift limit of the ice-sheet, infills a valley eroded through the moraine of the local ice advance. The presence of chert, all that remains from the former presence of limestone erratics, and some igneous boulders from source rocks further north, within the granitic drift, shows that this local ice advance followed an earlier and more extensive invasion by a general ice-sheet (Farrington, 1942).

An extension of these investigations to the east side of the Wicklow Mountains in the vicinity of Enniskerry indicated that this earlier invasion by general ice is represented by an extensive sheet of Irish Sea drift that covers the coastal lowland between Wicklow Head and Carnsore Point. As this drift surface appeared to be more weathered than that of the later ice invasion from the north-west, and also conformed with a major erosional break in the succession, an interglacial period was believed to have intervened between these two ice advances (Farrington, 1944). The earlier, or Eastern General glaciation from the north-east was associated with an advance of local ice (the Brittas Mountain glaciation) which attained a maximum somewhat later. Likewise, the Newer Drift limit of the general ice, marking the greatest extent of Farrington's Midland General glaciation, was also associated with a later advance of local ice (the Athdown advance). This was shown by the fact that outwash associated with the terminal moraine of the local glaciation descends to a level below that of the deltaic outwash deposited in an extensive proglacial lake impounded during the maximum advance of the Midland General ice in the vicinity of Blessington, on the western side of the Wicklow Mountains (Farrington, 1934, 1966). This drift succession is shown in the adjoining diagram (Fig. 12.2).

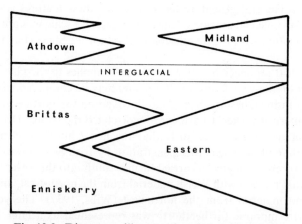

LOCAL WICKLOW ICE          GENERAL ICE

Fig. 12.2. Diagram to illustrate the relationship between the local mountain glaciations and the general ice-sheet glaciations on the flanks of the Wicklow Mountains. Note that local and general ice were at no time in contact.
(After Farrington, 1957)

The very extensive distribution of the blue-grey or purple shelly 'stoneless' Eastern General till along the south-east coast, and on the south coast of Ireland, as far west as Cork Harbour just west of Bally-croneen Bay, shows that a massive ice stream must have pushed down the Irish Sea during the older glaciation. At that time also, ice stream-ing from the midlands of Ireland (the 'Munster General glaciation') prevented the Irish Sea glacier from invading considerable stretches of the south coast. This ice thrust seems to have attained its maximum slightly later than that of the Irish Sea glacier, as the ground moraine of the former overlaps the latter at Kilmore Quay and other localities on the Wexford coast. The presence of Leinster granite has caused the upper drift of south Wexford to be correlated with the local Brittas advance (Mitchell, 1960) rather than with an expansion of ice from the Irish midlands. According to the latter interpretation, this advance, termed 'the Bannow advance', was penecontemporaneous with the deposition of shelly till at Ballycroneen (Synge, 1964a).

During deglaciation, marginal, submarginal, and subglacial drainage, associated with the decay of the Irish Sea glacier, was particularly pronounced. Impressive rock-cut gorges, up to 70 m (220 ft) deep and 2 km (1 mile) long, are particularly prevalent below the limit of the Eastern General drift in east Wicklow. The fact that the Midland General glaciation has overridden these features is significant. Fluvioglacial erosion on such an impressive scale was not found in association with the dissolution of the Last Ice Sheet (Farrington, 1944).

The drift sequence described above also applies to the glacial succession on the west flanks of Wales, and may be compared with that present in the Cheshire–Shropshire lowland on the north-east flank. As mentioned earlier, Charlesworth correlated the Newer Drift limit with an upper Irish Sea drift that reached Church Stretton and Wolverhampton. The earlier penetration by Irish Sea ice was represented by erratic material carried further south into the lower Severn basin that merged with Welsh material from the north-west, and East Midland material from the north-east (Wills, 1937). The younger Ellesmere moraine further north was represented as a later retreat stage of the Newer Drift ice, co-extensive with an advance of Welsh ice in the upper Severn as far east as Shrewsbury.

By 1960, and earlier, the basic division of the drifts into an older and newer series was recognised. The correlation of these drifts in Wales, Ireland, and England, is represented in Table 12.1.

In view of the inadequacies in determining the limits of the Last Glaciation by morphological means, greater emphasis was subsequently placed on the detailed stratigraphy of drift successions containing interbedded organic beds deposited during interglacial or interstadial phases. The nature and quantity of the pollen present in such deposits has enabled botanists to determine the type of vegetation cover present during such phases of the Pleistocene. In some cases, particular interglacials could be recognised by this method. This technique of dating was used both in England and in Ireland before being applied to Wales. The first site to be examined in this

TABLE 12.1. *Divisions of the drifts of Wales, Ireland, and England, suggested before 1960*

| *Wales* (Charlesworth, 1929) | *Ireland* (Farrington, 1949) | *England* (Wills, 1937; Suggate and West, 1959) |
|---|---|---|
| Newer Drift limit; Mathry–Swansea–Abergavenny–Hereford–Welshpool | Athdown local glaciation Midland general glaciation—limit, Blessington–Brittas–Enniskerry | Little Welsh advance, and Ellesmere moraine Newer Drift limit; Church Stretton–Wolverhampton–Escrick–Hunstanton |
| Older Drift: total glaciation by Welsh ice, contemporaneous with penetration of Irish Sea ice to Glamorgan | Brittas local glaciation Eastern general glaciation | Older Drift: glaciation as far south as Tewkesbury and Moreton and north London |

way, with a bearing on the succession of the Irish Sea basin, was discovered in a drift cliff bordering Wexford Harbour, at Ardcavan. Here, a thin organic layer was found in an old kettle-hole filled with solifluction deposits. The presence of pollen of alder, birch, hazel, and willow, with very small amounts of pine, oak, and elm, indicated that the deposit was laid down under temperate conditions believed to be those of the Last or Ipswichian Interglacial (Mitchell, 1948). The suggestion that this deposit is late- or postglacial in age now renders this site unimportant (Wilson, 1968). Fresh morainic relief at Screen Hills, only 5 km (3 miles) north-east of Ardcavan, has been correlated with the furthermost advance of the ice during the Last Glaciation (Mitchell, 1948, 1960).

According to Watts the older drift of the south of Ireland is dated to the preceding glaciation of the British Isles, the Gipping. This is partly because the older Irish midland drift of the Munster General glaciation overlies organic deposits at Kilbeg and Newtown, in County Waterford, that have affinities with deposits of the Hoxne Interglacial of England, and partly because those sites are located south of the accepted limit of the Midland General ice advance (Watts, 1959). Similar interglacial deposits near Gort in County Galway (Jessen *et al.*, 1955), and at Baggotstown in County Limerick (Watts, 1964), lie well within the limits of the Newer Drift.

Recently attempts have been made to correlate the drift succession on both sides of the Irish Sea. Wirtz noted that fresh drift was absent in west Wales along much of the coast adjoining Cardigan Bay (Wirtz, 1953). This discovery, taken in association with the recognition of fresh Irish Sea drift in the vicinity of the town of Cardigan, caused Wirtz to suggest a Newer Drift limit extending south from the western part of the Lleyn peninsula, across Cardigan Bay, to the mouth of the Teifi. Later, comparison of drift sections on the coast of west Wales with those in south-east Ireland, appeared to confirm the absence of fresh drift between the Dovey estuary and the vicinity of Cardigan (Synge, 1963). The view was also put forward that the drift surfaces between Cardigan and Mathry were developed on older, rather than newer drift, as Wirtz suggested (Mitchell, 1960). In these investigations the most significant marker horizons used were the Irish Sea till (termed Eastern General by Farrington, and Ballycroneen by Mitchell), and a 'Preglacial' beach associated with a marine wave-cut platform (Wright and Muff, 1904). In order that the different tills of the succession might be more precisely defined, Mitchell (1960) proposed a new nomenclature based on type sites, shown in Table 12.2.

At that time Mitchell recognised a single Irish Sea till in west Wales and on the Irish coast, which could be correlated with the Gipping glaciation of England. The Cardigan moraine, considered the southern limit of the Newer Drift by both Charlesworth and Wirtz, was regarded as an older deposit by Mitchell because of the apparent absence of a later boulder clay capping the Fremington or Ballycroneen type of boulder clay in coastal sections. The presence of the organic mud found in a kettle-hole of 'Brittas' boulder clay at Ardcavan, and the development of a soil on the surface of the equivalent Welsh, or Pencoed boulder clay, was used as evidence by Mitchell for placing an interglacial between the deposition of the Pencoed boulder clay and the formation of the Llandaff moraine. This correlation is now questioned as Mitchell now considers that the Ardcavan deposit is late- or postglacial in age.

Below the Ballycroneen boulder clay, on the south coast of Ireland, thick deposits of head bury an ancient beach gravel reaching some 7 m (22 ft) above high-water mark. This deposit, the 'Preglacial' beach of the earlier geologists, generally rests upon a smooth, sloping, wave-cut platform. These features are particularly well developed at Howe's Strand (W 560 427), on Courtmacsherry Bay in County Cork (Wright and Muff, 1904). The presence of some erratic pebbles in these ancient beach gravels may indicate the destruction of

TABLE 12.2. *Correlation table of the Pleistocene stratigraphy in Wales, Ireland, and England, as suggested by Mitchell* (1960). (*Farrington's nomenclature for the Irish succession is added for comparison*)

| Wales (Mitchell, 1960) | Ireland (Mitchell, 1960) | (Farrington, 1954) | England (Mitchell, 1960) |
|---|---|---|---|
| Solifluction-earth (Younger Dryas Period, equivalent to Pollen Zone III) | Solifluction-earth | | Solifluction-earth |
| Welsh advance to Shrewsbury moraine | | | |
| Llandaff moraine | Tipperary and Athdown moraines | Midland General glacial maximum | Smestow–Escrick–Flamborough moraine |
| | | | Chelford mud |
| | | | Adswood and Hunstanton (?) boulder clays |
| Llansantffraid soil | Ardcavan mud | Ardcavan Interglacial (now regarded as Late- or Postglacial) | Ipswich mud, Selsey clay |
| Pencoed boulder clay | Brittas and Garryvoe boulder clays | Brittas advance | Solifluction-earth |
| Fremington boulder clay | Ballycroneen boulder clay | Eastern General glacial maximum | Gipping (and Fremington) boulder clay |
| New Quay boulder clay Solifluction-earth | Enniskerry boulder clay | Clogga advance | |
| Gower beach | Kilbeg and Newtown muds Courtmacsherry beach | | Hoxne mud Fremington beach and Kirmington marine clay |
| Gower erratics | Courtmacsherry erratics | | Lowestoft boulder clay Porthleven erratics |
| Gower shore-platform | Courtmacsherry shore-platform | | Cromer Forest Bed Series |
| | Killincarrig Crag (redeposited) | | Red Crag and higher crags and St Erth beds, Hele and Scilly gravel |

a pre-existing glacial drift, even though no such deposit has been found *in situ*. Longshore drift, carriage by seaweed, and ice-rafting in bergs or floes have been also suggested as a means whereby such material could have been transported. The presence of large erratics, buried by this beach and resting on the rock platform at certain localities in north Devon and Cornwall, suggests that a phase of ice-rafting took place between the cutting of the platform, and the deposition of the beach.

Although not found on the south coast of Ireland, boulder clays that predate the deposition of the Ballycroneen and Fremington tills have been observed further north on both sides of the Irish Sea. Do these early drifts of Welsh and Irish origin represent an early phase of the Gipping glaciation; or do they represent an earlier glaciation that predates the formation of the Courtmacsherry beach? There is some evidence to indicate that drift deposition by local ice took place both during an earlier glaciation, as well as during the time that immediately preceded the advance of the Irish Sea glacier during the Gipping glaciation. On the east coast of Ireland the oldest drift recorded is the Clogga boulder clay, associated with a glacial movement due east across the Courtmacsherry shore-platform. This drift has generally been included within the Gipping glaciation because no weathering horizon intervenes between it and the overlying Ballycroneen or Eastern General till (Farrington, 1954). But if the removal of such a weathering horizon by the erosive action of the later ice advance is considered possible, as may be observed in East Anglia, where fresh Lowestoft till is truncated by the Gipping advance (West, 1968), then a Lowestoft age for the Clogga till is possible. The presence of beach gravel containing derived Clogga material at Cahore (T 222 478) in County Wexford, beneath Irish Sea till believed to be Ballycroneen in age supports the view that a major unconformity separates the Clogga drift from the later glacial deposits (Synge, 1964a).

Along much of the coast of west Wales the succession is somewhat different, as the local ice was sufficiently powerful to prevent the access of Irish Sea ice. However, when the Irish Sea ice attained its maximum, local Welsh ice was pushed back temporarily some distance further east. Thus in the marginal zone of these two ice streams there is a threefold division of the drift succession into Lower Welsh (New Quay) till, Irish Sea (Fremington) till, and Upper Welsh (Pencoed) till. The position of the Irish Sea till can be represented stratigraphically as a wedge of extraneous drift forced into the Welsh succession from the west (Fig. 12.3). At Aberarth the New Quay till overlies a solifluction deposit derived from an earlier boulder clay, which may be the correlative of a Clogga till of Lowestoft age.

The absolute chronology of the drift sequence round the southern part of the Irish Sea basin depends largely on the stratigraphical position of the organic muds at Kilbeg and Newtown in County

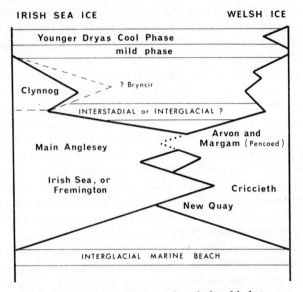

**IRISH SEA ICE**                                    **WELSH ICE**

Younger Dryas Cool Phase
mild phase

Clynnog                      ? Bryncir

INTERSTADIAL or INTERGLACIAL ?

                                              Arvon and
Main Anglesey                                 Margam (Pencoed)

Irish Sea, or
Fremington                                    Criccieth

                            New Quay

INTERGLACIAL  MARINE  BEACH

Fig. 12.3.  Diagram to illustrate the relationship between Irish Sea ice and Welsh ice. In west Wales both ice bodies were in contact, as shown in the diagram

Waterford, pollen dated to the Hoxne Interglacial. At Kilbeg till of the Munster General glaciation deeply buries interglacial muds at an inland site, whereas at Newtown the organic sequence underlies a coastal cliff composed of a similar till. At this latter locality part of the organic sequence was picked up and incorporated in the base of the overriding till; there is little evidence to suggest that the position of the till that overlies the organic horizon can be ascribed to later landslip action to reverse the natural succession of younger beds over older ones (Watts, 1959). The above evidence places all the tills south of the Newer Drift limit as indicated by the moraines at Tipperary and

Screen Hills in Ireland, and at Tregaron and Llandaff in Wales, within the area of the Gipping and older glaciations.

Recently the suggestion has been put forward that the Newer Drift in County Wexford extended some 18 km (11 miles) further south of Screen, to overlie a shallow water deposit of marine clay at Shortlestown, near Killinick. Although pollen analysis of organic material found in these clays indicates interglacial conditions, the evidence is insufficient to show which interglacial is represented (Mitchell, G. F., and Colhoun, E. A., personal communication). If the Last Interglacial is represented at this locality the limit of the Newer Drift will have to be extended to the south of Wexford Harbour, to embrace the shelly Irish Sea till, the uppermost drift deposit in that area. This till, part of Farrington's Eastern General drift sheet, could then be regarded as an earlier phase of the Last Glaciation, while the similar till at Ballycroneen, much further west, would still be regarded as Gipping in age (Mitchell, G. F., personal communication). Yet according to the series of absolute dates obtained by means of the radiocarbon assay of marine shells and organic material collected from the Irish Sea drifts of west Wales and the Cheshire–Shropshire basin, a very extensive spread of the Irish Sea glacier during the Last Glaciation is envisaged (John, 1967; Shotton, 1967b). The implications of a glaciation of this age, extending, according to some, even south of St George's Channel to the Scilly Isles, will be discussed later.

The broad correlation, mentioned above, was first presented by Mitchell (1960) in his 'Pleistocene history of the Irish Sea'; his correlation table forms the background to the detailed regional studies presented in the preceding chapters. Such a table must be regarded as a convenient working hypothesis; changes will have to be made, and new questions posed, as more information becomes available from time to time. Probably future work will tend to confirm the general succession, but the age of each major event may have to be altered as more absolute dates become available.

In the following sections of this chapter, the results of the regional studies of the Welsh Quaternary described in the preceding chapters will be discussed. In the final section, the whole question of absolute chronology will be examined.

### REGIONAL DESCRIPTION

Before the full history of the Quaternary of Wales can be appreciated the limits of each of the main glacial episodes must be established, particularly those of the last two glaciations. Almost all the regional descriptions presented in this volume deal with this problem. These and related questions concerning the Quaternary of Wales and the adjoining parts of England will be discussed under the following regional headings: The North-west Midlands of England, North-east Wales, North-west Wales, West Wales, South-west Wales, South Wales, South central Wales, the Severn Valley, and South-west England. In a number of cases it will be shown that alternative interpretations to those already presented in the previous chapters are possible, and that, in a few cases, differences of opinion exist regarding even the interpretation of individual sections and landforms.

### NORTH-WEST MIDLANDS OF ENGLAND

The evidence discussed by Worsley (Chapter 4) in the Cheshire–Shropshire lowland would appear to dispose of the view that the Ellesmere moraine can any longer be regarded as marking the maximum advance of the Last Glaciation. The presence of till associated with the advance of the ice southwards, overlying bedded sands and gravels containing organic material radiocarbon dated to approximately 30,000–40,000 years B.P. shows that an ice-sheet reached the vicinity of Wolverhampton during the Last Glaciation. This ice limit is associated with outwash gravels at Smestow which merge with those of the Main Terrace of the Severn south of Bridgnorth (Shotton, 1968). West of Wolverhampton this limit has been traced as the southern edge of the Irish Sea or Northern drift of Shropshire between Bridgnorth and Church Stretton (Wills, 1937). In the vicinity of the latter place there is some evidence that the Northern drift was contemporaneous with Welsh drift (Mackney and Burnham, 1966).

The importance of the Quaternary sequence in this part of the English Midlands lies in the fact that here is the most complete succession for the earlier part of the Last Glacial period yet found in the British Isles. In this area periglacial conditions endured from the end of the Last Interglacial until some time after 30,000 years ago when the ice-sheet extended over the region from the north. These climatic conditions are represented by a series of silts, sands, and gravels of lacustrine, fluvial and aeolian origin, interbedded by horizons of frost wedge casts, and patterned ground. The presence of organic beds within these deposits, at Upton Warren, Four Ashes, and Chelford has shown, by radiocarbon dating, that these conditions persisted for, at least, 15,000 years if not longer.

The subsequent advance of the ice to the Wolverhampton 'line' laid down a cover of fresh drift on top of the deposits described above. These deposits are associated with an extremely fresh drift landscape in many places—lake-filled kettle holes in the vicinity of Four Ashes; a sharp esker ridge south of Newport; and kettled drift at Berrington. The greater amounts of decalcification of the drift surface recorded south of the Ellesmere moraine, a subsequent retreat stage, in contrast to that found further north, may not be very significant. The great variations encountered in depths of decalcification may reflect local differences of drainage, and location, rather than differential age of the deposit, as was originally believed.

Less is known about the earlier or Gipping glaciation. Wills mapped the southern limit of till with Irish Sea and Northern erratics as a zone that diverged south-east from the Wolverhampton 'moraine' in the vicinity of Bridgnorth. The westward continuation of this 'limit' across the Severn valley has never been defined.

### NORTH-EAST WALES

Investigations of the drift succession in north-east Wales by Embleton extend westwards those glacial limits observed in Shropshire and Cheshire. In this area there is evidence of a widespread, but weak, glaciation by Welsh ice that flowed from west to east across Denbighshire. As the Last Glaciation was believed to have covered the area completely, all moraine stages mapped have been represented as retreat stages. The most significant of these marks the southern limit of an Irish Sea or Northern drift, deposited at a time when no local ice was present in the immediate vicinity (Embleton, 1964). Between Halkyn Mountain and Wrexham this limit has been traced at 120–240 m (400–700 ft) above sea-level (Peake, 1961). At higher altitudes to the west of this limit an earlier invasion of Irish Sea ice left shelly till up to about 300 m (1000 ft); the manner in which this occurs and its association with Welsh till, suggests that both ice-masses were in contact at the time. The earlier drifts are considerably dissected, and are associated with solifluction deposits.

There is some support for the suggestion that the younger of the two Northern drifts represents the maximum of the Last Glaciation. The younger drift appears to block caves containing Middle Aurignacian implements near Tremeirchion (Boswell, 1932), at the north end of the Clwydian Hills; a significant fact, as it is known that the

Aurignacian culture came to a close with the culmination of the Last Glaciation.

The extent of Welsh ice seems to have been rather limited during this late advance of Northern ice. The fresh drift mounds and ridges of Pwll Glas and Bryneglwys may mark the northern edge of Welsh ice that extended down the Dee valley at this time. The massive morainic deposits at Pwll Glas have the appearance of a major ice limit, but the relationship between these deposits and those in the Vale of Clwyd is not clear. Further west, Welsh and Northern ice appear to have been in contact, as ice flowing down the Conway valley would seem to have kept the extraneous ice north of the present coast.

Three distinct glacial episodes have been recognised in North Wales—an early spread of Welsh ice eastwards from the Arenigs and Snowdonia; the expansion of contiguous Irish Sea or Northern ice southwards to its maximum position; and finally a later readvance of Irish Sea and Welsh ice. During this last episode, the Irish Sea ice reached the Vale of Clwyd, while Welsh ice extended down the valleys of the Dee and the Conway. All three glacial episodes are generally placed within the Last Glaciation, although some authorities prefer to place only the latest of these episodes within that glaciation.

## NORTH-WEST WALES

With the extension of the glacial sequence further west, the critical area embracing Caernarvonshire and Anglesey will now be considered. The simple tripartite division of the drifts of Caernarvonshire into Lower boulder clay, Middle sands and gravels, and Upper boulder clay, was established by Jehu in 1909, from an examination of the principal drift sections exposed round the coast of the Lleyn peninsula. In this succession the Lower boulder clay, Welsh in origin to the east, and from the Irish Sea in the west, is overlain by a sand and gravel series that represents outwash produced as the ice-sheet decayed. The Upper boulder clay, commonly termed a rubble drift, represented a later advance of Welsh ice. According to subsequent work the view was put forward that the Upper boulder clay terminated to the south against the hills between Clynnog and Bryncir at 150–200 m (500–650 ft), but was represented in the southern part of Lleyn by the cryoturbation of older drifts and by the formation of head or solifluction deposits (Synge, 1964b). Such a limited distribution of the Upper boulder clay has not been generally accepted; both Whittow and Ball (Chapter 2), Saunders (1968a, b, c) and Simpkins (1968) all produced evidence to support Jehu's contention that an Upper boulder clay of the same age throughout, covers the greater part of the Lleyn peninsula. However, the writer still maintains his original view, that true Upper boulder clay was limited to the northern part of the county, and is of the opinion that the bulk of the contortions seen in the cliff on the coast at Clynnog, and further north, are due to moving ice rather than cryoturbation. These structures differ in type from those of periglacial origin observed at Criccieth and other places on the south coast of Lleyn. Also such a distribution of the Upper boulder clay accords with the pronounced kame surface that terminates so abruptly, towards the south, at Bryncir.

During the formation of the Lower boulder clay series, the Irish Sea glacier affected the western part of the Lleyn peninsula. According to analysis made on the till fabric (Saunders, 1968a) this ice moved across Lleyn from the north-west. Glacial striae are few, and some of these, such as those found running NW–SE at Porth Dinllaen, appear to be misleading as they may reflect a local orographic deflection of the basal ice. The distribution of shelly till as far east as Pwllheli harbour shows that Irish Sea ice once covered practically the whole of Lleyn, while on the basis of till fabric analyses its influence extended even further east, to Portmadoc (Saunders, 1968a). At that time Welsh ice, emanating westwards from the valleys of Snowdonia, was deflected sharply southwards. With the later waning of the Irish Sea ice the zone of contact between the two streams was displaced westwards to St Tudwal's peninsula. This westward expansion of Welsh ice, represented by the deposition of a blue-grey slaty argillaceous till, has been termed the Criccieth Advance by Whittow and Ball. Like the Irish Sea till with which it is associated, this till series overlies head deposits that bury a raised marine platform.

According to Whittow and Ball, Saunders, and Simpkins, a period of erosion and weathering took place before the deposition of an upper boulder clay, as evidenced by the presence of a weathering horizon underlain by frost wedge casts between the lower and upper boulder clays. The writer, on the other hand, maintains that in the Criccieth area both the lower and upper Welsh boulder clays belong to the same glacial advance; the lower till representing ground moraine laid down by the advancing ice, while the upper till was laid down during the waning of the ice-sheet, probably by ice moving in a slightly different direction. In this case the weathering horizon is interpreted as a layer produced by the downward penetration of percolating water to the surface of the impervious ground moraine below, and the wedge-like structures are regarded as cracks induced in the compact basal till by overloading (Synge, 1964b).

During the deposition of the Upper boulder clay of Caernarvonshire, Northern ice crossed Anglesey, according to Whittow and Ball, and terminated in a lobe that crossed the present coastline between Porth Dinllaen and Nefyn. This Main Anglesey Advance is regarded as being contiguous with the western margin of Welsh ice of the Arvon Advance that extended from Clynnog to St Tudwal's peninsula. Saunders suggested a similar limit for this glacial advance, but advocated a more extensive spread of Northern ice that embraced the entire west coast of Lleyn, and covered Bardsey island. This later advance of Northern or Irish Sea ice he equated with an upper grey stony till, differentiated from a lower less stony till beneath, which he correlated with the previous advance (Saunders, 1968a).

The Welsh till of this Arvon Advance is distinguished from that of the earlier Criccieth Advance by both colour, texture, and contained erratics. Generally more gravelly in texture, the Arvon drift contains material from South Snowdon and the Vale of Ffestiniog, and is buff or yellow in colour. Correlation of the Arvon and Main Anglesey Advance with the Last Glaciation is based on the freshness of the landforms developed on the drift, and the presence of kettle holes containing a continuous pollen sequence going back to Pollen Zone I. These kettle holes are located in the vicinity of Glanllynau, some 4 km (2½ miles) west of Criccieth (Simpkins, 1968). The apparent absence of such kettle holes elsewhere in south Caernarvonshire, and the possible association of these enclosed hollows with outwash that emanated from the Bryncir area, would, if substantiated, make the tills in this area older.

The succession outlined above can be compared with that described from north-east Wales. During the Arvon stage, or last main advance of Welsh ice in Caernarvonshire, in the Conway valley and in the Dee valley, the Irish Sea ice, which entered the Vale of Clwyd further east, was deflected south-west across Anglesey by ice flowing north from Snowdonia. A later stage of glaciation might be represented by the red Triassic till deposited along the coast from the Clwydian Hills, intermittently, to north-east Anglesey. The manner in which this drift limit descends in height westwards, and the absence of any obvious morainic feature suggests that it represents the southern limit of Triassic material dredged up from Liverpool Bay by basal ice.

After the coastal ice had melted, the local glaciers withdrew into the valleys radiating from Snowdonia. During the final climatic phase of Younger Dryas (Pollen Zone III) times, glaciation was very limited. According to pollen analysis the youngest and innermost moraines of the larger corries of Snowdonia were formed at that time (Seddon, 1962).

### WEST WALES

In the chapter on west Wales the difficulties of establishing a satisfactory chronology of the drifts between the Mawddach estuary at the mouth of Afon Wnion and the Teifi valley have been pointed out. No distinctive rock types break the monotony of the mudstones and grits that cover such large tracts of central Wales. This makes directions of ice movement difficult to ascertain, particularly as the preservation of glacial striae is rare, both because of the nature of the rock, and its susceptibility to frost action which was particularly prevalent in that area. This widespread development of solifluction deposits, readily identified from true till by means of stone orientation analysis, and the absence of fresh glacial depositional features, suggests that most of the area escaped the Last Glaciation.

During an older glaciation Welsh ice flowed down the Wnion and Dovey valleys to be deflected southwards in the coastal tract by the barrier of Irish Sea ice further to the west. Further south this ice stream and that of the adjoining Irish Sea glacier pushed south-east into the Teifi valley. The tills of this glaciation were deposited after a phase of widespread head formation. In some areas, as in the cliffs of Morfa Bychan near Aberystwyth, this till was rearranged by solifluction before the overlying head was formed. With the waning of this glaciation, large submarginal channels were cut in the bedrock, particularly between Borth and Machynlleth.

According to the evidence presented, the coastlands were only covered by the Last Glaciation in the northern part of this area. The underwater ridge of Sarnbwch, opposite Tonfanau, was suggested as the terminal moraine of that glaciation, contemporaneous with Whittow and Ball's Arvon limit on St Tudwal's peninsula. The more southerly sarn, Sarncynfelyn, was placed within the older drifts.

The remarkable absence of fresh drift forms in the valleys that adjoin such highlands as that of Cader Idris is surprising. Outside the confines of the freshly glaciated corries the valley sides are characterised by scree deposits and landslips. This suggests that the western margin of the Welsh ice cap during the Last Glaciation lay further east. Further south the moraines of the Rheidol valley at Ponterwyd, and at Tregaron in the upper Teifi valley, appear to mark the western margin of this ice-cap. At that time corrie glaciers were present on the north-east facing slopes of Cader Idris, and nivation hollows were formed in certain favoured sites.

There is some suggestion that the remarkable kames at Banc y Warren, just east of Cardigan town, represented an ice lobe that extended across the present coastline from the north-west (Wirtz, 1953). The suggestion has been made that these kames represent the eroded remnants of thick fluvioglacial deposits (Mitchell, 1960; Synge, 1963). However, the view of these features as seen from the east, emphasises their depositional nature, as previously noted by Williams and Charlesworth (Fig. 12.4). If these gravels belong to a distinct and later glacial advance than that which deposited Irish Sea erratics in the Teifi valley it should eventually be possible to relate them to a particular upper drift horizon which is not represented further to the north. The presence of organic matter found inter-bedded with the Banc y Warren sands and gravels promised to give a valid maximal date to the deposit. This hope was not realised; even though the date obtained suggested deposition during the middle part of the last major cold period, the presence of Tertiary pollen would seem to render the date invalid.

## SOUTH-WEST WALES

West of Cardigan, and along the north and west coasts of Pembroke-shire, John has recognised two types of till, both associated with Irish Sea ice—a clay till with few stones, but containing shell fragments; and a stony till containing local rock material. The latter is regarded as the land facies and the former the marine facies, of the same glacia-tion. On the basis of radiocarbon dates obtained from marine shell, organic material, and wood fragments collected from this drift, it appears that glaciation last affected this coastline shortly after the period between 38,000 and 32,000 years B.P. If these dates are accept-ed, then the Irish Sea glaciation of Pembrokeshire, termed by John the Dewisland Glaciation, represents an early Last Glaciation maxi-mum to the south of St George's Channel.

According to John there is no evidence for a separate glacial stage or even for an end-moraine along the north coast of Pembrokeshire as advocated by Charlesworth. No evidence was found to support the suggestion that the large rock-cut channels that parallel the coast between Jordanstown and Newport could be the outlets of a large lake impounded in the Teifi valley by Irish Sea ice (Charlesworth, 1929; Bowen, 1966a). But the suggestion put forward by John that these channels predate his Dewisland glaciation seems difficult to

Fig. 12.4. Sketch of the Cardigan 'moraine' (the gravel ridge at Banc y Warren), as seen from the east at Penparc (SN 222483). The ridge is seen standing on the plateau surface at about 300 m (500 ft), perched above the deep trench of the Teifi valley. Note that large gravel pits breach the west face of the ridge

accept. The presence of lower head and till in some of the channels is used as evidence for their antiquity, while the absence of such infill elsewhere is explained by the later re-excavation of channels by meltwater during the Dewisland glaciation. Perhaps the presence of both head and till could be explained by later slumping; or by contemporaneous oscillations of the ice margin.

The fact that the alignment of the majority of channels is transverse to the direction of glaciation is also used as an argument for their greater antiquity. But as subglacial drainage tends to seek a lower base level beneath the ice, it is controlled by the relief of the subglacial surface. Even with the original ice movement across the coast to the south-east, an alignment of the subglacial drainage parallel to the coast during deglaciation might be expected. The influence of relief on subglacial drainage can be considerable (Sissons, 1967).

Evidence for a still earlier glaciation is based on the presence of erratics found in beach gravels up to 7 m (22 ft) above sea-level, beneath the Dewisland drift. Head deposits intervene between this drift and the beach gravels. No till has been found in Pembrokeshire which might be correlated with these erratics.

As the Dewisland glaciation is believed to have covered Pembrokeshire completely, the question of an upper altitudinal limit to the drift did not arise. Nevertheless, on the north side of Mynydd Prescelly, the lower limit of tors does seem to coincide with an upper limit of glacial erosion. Above Dinas the tors are well developed above 250 m (800 ft), while below that level, in the vicinity of Castell Corwynt (979378) the tors have been ice moulded, and sometimes removed completely. This limit seems to accord roughly with the height of the highest fluvioglacial channels, mentioned earlier.

On this rockbound coast drift sections are limited. Within the Newer Drift limit suggested by Wirtz between Cardigan and Mathry only two have been described; at Poppit and Aber Mawr. At the latter place contorted sand and gravel derived from the underlying shelly till is overlain by loose brown sandy drift containing shell fragments (Fig. 12.5). John suggests that these contortions were produced by the decay of dead ice during deglaciation of the Dewisland ice. The contortions are not considered to result from a later invasion of coastal ice as was suggested by Charlesworth and Wirtz.

**ABER  MAWR**    **Pembrokeshire**

7   Blown Sand
6a  Brown Sandy Till with shells    6b Blocky Till
5   Contorted Sands and Gravels
4   Purple Clay–till with shells
3   Blocky Head
2   Soliflucted Drift
1   Coarse Head

Fig. 12.5. The drift succession at Aber Mawr in Pembrokeshire as seen in the north cliff. The section varies in height from about 15 m to 24 m (50–80 ft). Bed 6*a* may represent a separate later advance of the ice

## SOUTH WALES

In the chapter dealing with the glaciation of South Wales, between Carmarthen Bay in the west, and the Usk in the east, Bowen presents evidence for a Last Glaciation maximum more extensive than that suggested by Charlesworth. The drift of this glacial advance, deposited by local Welsh ice flowing south, is shown to terminate south along a limit extending east from St Clears to Mumbles Head, Margam, and Llandaff. The relationship of this Margam Advance to the rest of the Quaternary deposits of South Wales is shown from the succession observed in the drift sections exposed in the cliffs of the Gower peninsula and at Marros.

The drift of the Margam Advance is regarded as Newer Drift because it is associated with depositional forms such as moraines, kames, drumlins, eskers, and kettle holes. Late Glacial deposits dating back to Pollen Zone I, but not older, are found in kettle holes, and red soil formed during temperate conditions in limestone areas were found only outside, but not within, the Margam limit. Finally, beach deposits containing marine shells and rhinoceros remains found up to 11 m (37 ft) above sea-level in the Minchin Hole cave date to the Last Interglacial on faunal evidence.

Bowen believed that the red soil, produced from weathered limestone, formed during the Last Interglacial; yet similar red soil in County Wexford has been suggested to predate the Gipping glaciation (Gardiner, 1963). In addition, Bowen places considerable emphasis on the Gower beaches. But it is difficult to show that, for example, the *Patella* beach at Minchin Hole correlates with Mitchell's Courtmacsherry beach or with John's beach at Poppit. Nevertheless Bowen has shown that two glaciations occurred in South Wales. During the Pencoed glaciation ice swept east along the coast of Glamorgan from the Irish Sea, in contact with Welsh ice pressing southwards. During the Margam glaciation, local Welsh ice flowed out of the mountains to a limit that roughly corresponds with the limit for the local ice in the preceding glaciation.

## SOUTH CENTRAL WALES

The study of the drifts in Brecon by Lewis, and in Hereford by Luckman, demonstrate a Newer Drift limit that is less extensive than that of an earlier glaciation. The later glaciation was fed from a source area to the north of the Bran-Irfon depression, and caused large glaciers to flow down the Usk and Wye valleys. At that time the ice margin reached 375 m (1250 ft) on the northern flank of the Black Mountains and glaciers developed on the Brecon Beacons, merging with the Usk glacier.

The fresh drift surfaces derived from this glaciation have long been known to extend east as far as Hereford and Abergavenny (Charlesworth, 1929). The more detailed mapping undertaken by the above authors confirms, in general, this Newer Drift limit. Well-defined re-advance moraines, associated with the retreat of the Usk and Wye glaciers, depict the manner in which the ice withdrew to the source area on the plateau north of the Bran-Irfon depression. But the last area to be freed of ice was the northern face of the Brecon Beacons. A suite of block moraines was dated to the Younger Dryas cold phase (Pollen Zone III) by pollen analysis, and indicates a reoccupation of certain corries by ice in Late Glacial times.

## THE SEVERN VALLEY

The Severn valley is important in the interpretation of the Welsh Quaternary by virtue of its position. Openly exposed to invasion by extraneous ice from the northern part of the Irish Sea basin, as well as acting as the natural discharge route of ice or glacial meltwaters from Wales in the west, and from the English Midlands in the east, this region has become the keystone of the upper part of the Quaternary succession of England.

In Chapter 5 Stephens has presented the evidence for a Gipping ice limit as far south as Moreton on the northern edge of the Cotswold Hills, associated with ice invasion from the north-east. This great invasion of 'eastern' ice forced the ice-stream that emanated from North Wales, and from the Irish Sea, to turn south-east and south into the tract of country between Birmingham and the Severn valley. The latter valley was probably initiated at that time, because it forms the boundary between the drifts of this ice advance, and the unglaciated hills to the west. The southern limit of this ice advance reached Tewkesbury. This drift is characterised by a lack of fresh constructional forms, is patchy in distribution, and often occurs as hilltop patches in association with widespread periglacial deposits. The position of this drift sheet, extending south of the Newer Drift limit near Wolverhampton, and overlying organic lacustrine deposits of Hoxnian age at Nechells, near Birmingham (Kelly, 1964), shows that

12

it is Gipping in age. No break in the succession could be found in the whole area between the drift limit at Moreton, and the Wolverhampton 'moraine'. The story of deglaciation, closely bound up with the formation and evolution of glacial Lake Harrison, a proglacial lake impounded in the area of the present Avon valley against the Cotswold Hills in an interlobate re-entrant between Welsh and Eastern ice, has already been told (Shotton, 1953).

This Gipping drift succession, termed Wolstonian by Shotton, is cut off to the north by the much later advance of northern ice to the vicinity of Wolverhampton. This break does not coincide with any clear change in the relief, but can be correlated with the most southerly exposures of shelly drift; with a concentration of northern erratic boulders; and with the glacial outwash terraces at Smestow. Evidence has already been presented to show that this line probably represents the limit of the Last Glaciation. This advance postdates the deposition of organic matter some 30,000 years old at Four Ashes; it is associated with the major phase of deposition of outwash that forms the Main Severn terrace; and it is associated with fresh depositional landforms.

Although this ice limit does not coincide with a dramatic physical boundary, it does separate terrain to the south that has been severely subject to periglacial activity from that to the north, which has been much less affected. But even in the 'older' drift terrain some constructional features have managed to survive. At Kingswinford an esker occurs (Wills, 1937); and at Wood End and Kingswood drift hummocks survive on the Tame–Avon watershed (Tomlinson, 1925).

Stephens has already pointed out the complexities and difficulties of terrace correlation. For instance, the Main Severn terrace appears to be made up of at least three different elements—a high terrace, M1, containing a warm fauna; the Main Terrace proper, associated with the glacial outwash at Smestow and the valley of the Worf; and the terrace at Upton Warren on a tributary of the Severn, containing organic material dated 39,000 to 40,000 years B.P. Clearly the Main Severn terrace, if it has been positively dated everywhere, must embrace a lengthy period of time. At first deposition could take place during the mild conditions at the close of the Last Interglacial, which is represented by the warm fauna of the Kidderminster terrace, or during a later mild interstadial phase. After this, the bulk of the terrace gravels were laid down by outwash while the ice stood at the position of maximum advance near Wolverhampton.

### SOUTH-WEST ENGLAND

Extension of the drift chronology southwards from the Severn valley into Somerset, Devon, and Cornwall, suggests that the greatest glaciation occurred at the time of the Gipping glaciation of East Anglia. At that time Irish Sea ice extended to the northern edge of the Scilly islands, and overlapped the present coastline of north-west Devon. There is no evidence of any ice invasion of the country between Tewkesbury and Minehead. In this connection, evidence of a vast lake impounded by an ice barrier in the Bristol Channel is important (Maw, 1864). The dry gorge at Chard, the lowest possible outlet that could have functioned, does show evidence of powerful fluviatile erosion. The dating of this event by means of waterworn implements of mid-Acheulian age found in the terrace gravels issuing from this gorge, and extending down the valley of the Axe, would seem valid. Implements of this type cannot be younger than the Hoxne Interglacial. As these implements must have been picked up from their position on the floor of the Axe valley by the flood of water that issued from Lake Maw through the Chard gap, they effectively date the ice dam in the Bristol Channel to the Gipping glaciation.

The stratigraphical position of the till in north-west Devon can be related to the suite of coastal deposits that are characteristic of the coastline from Bristol to Land's End. Sections in the vicinity of Fremington clearly show that the till postdates thick head deposits that seal an ancient beach gravel some 11 m (35 ft) above Ordnance Datum. In this beach the presence of erratics, some of them very large, suggests an early glacial phase. Such erratics could hardly be derived from a still older drift as they are only found near the coast at low altitudes, and are distributed widely, even as far as Brittany. The suggestion that they are ice-rafted seems reasonable. Some of them appear to have originated in the West Highlands of Scotland.

The coastal sections thus furnish evidence for two glaciations, both predating the Last Glaciation. Evidence for several marine phases also seems clear. The oldest of these is represented by marine clays containing shells at St Erth at 30 m (100 ft), evidently *in situ* (Mitchell, 1960). The species of shell present are largely extinct; the deposit may be Tertiary, or Early Pleistocene. Evidence for a high sea-level before the Gipping glaciation is recorded by marine sand and gravel up to about 15 m (50 ft) beneath shelly till at Fremington in north-west Devon (Mitchell, 1960). Later evidence for a marine incursion is shown by the Burtle Beds. The absence of an overburden of thick

12*

head places them later than the pre-Gipping beaches, but the considerable cryoturbation present in the upper layers suggests that they predate the Last Glaciation. They are therefore regarded as belonging to the Last or Ipswichian Interglacial.

Evidence of possible fluctuations of the postglacial sea-level has been observed from a study of the marine clays in the Somerset Levels. The presence of two peat horizons interleaved with marine clays shows that two regressions of the sea may have taken place; the first at about 6200 years ago, and the second some 5400 years ago. The latest extension of the peat took place after the culmination of a transgression in Roman times. There is, however, not enough evidence to show whether this was due to actual changes of world sea-level, or to the local flooding and draining of lagoons associated with the construction and destruction of beach bars. Some of the main storm beaches in existence today may be relict forms, originally constructed long ago, at the time of the Ipswichian transgression.

## CONCLUSION

The description of the Quaternary history of the various regions of Wales, and the neighbouring parts of England and Ireland, summarised in Table 12.3, shows that two, or possibly three, major glaciations have occurred. But in our present state of knowledge no complete agreement can be reached with regard to the extent and limit of the Last Glaciation. The older, or Gipping glaciation, appears to have covered Wales almost entirely, with the exception perhaps, of some nunataks in the south. The youngest, or Last Glaciation, has been represented by a southern limit of Irish Sea ice to the south of Pembrokeshire (John, 1965, 1967), by a limit along the north coast of Pembrokeshire (Charlesworth, 1929, 1963; Wirtz, 1953), and by one extending from the north Pembrokeshire coast to the Lleyn peninsula (Wirtz, 1953). An even more limited glaciation of Wales at that time, has also been advocated. According to this view, the general ice sheet failed to extend into south Caernarvonshire, but terminated in the vicinity of Clynnog and Bryncir (Carvill Lewis, 1894; Synge, 1964b), while further south the ice did not encroach on the Welsh coast, but filled the main trough of the Irish Sea (Mitchell, 1960; Synge, 1963).

Although north-west Wales is generally regarded as having been completely ice covered during the Last Glaciation, during a later stage the Northern ice terminated along the west coast of Lleyn (Saunders, 1968c) at the same time that a lobe of Welsh ice filled Tremadoc Bay

TABLE 12.3. *Correlation table of the Pleistocene stratigraphy in Wales, England, and Ireland, according to various contributors to this volume (1969)*

| North-west Midlands of England (Worsley) | North-west Wales (Whittow and Ball) | South-west Wales (John) | South Wales (Bowen) | East Ireland (Farrington, 1954 with Mitchell, 1960 and 1968) | South-west England (Stephens) |
|---|---|---|---|---|---|
| Postglacial deposits | Postglacial deposits | | Flandrian (sea-level, 4-6 m or 15 ft) | | Flandrian |
| ← LAST GLACIATION → | Inner Corrie moraines, Snowdonia | Head | ← LAST GLACIATION → | Inner Corrie moraines, Wicklow Mts | ← LAST GLACIATION → |
| | Outer Corrie moraines, Snowdonia | | | Outer Corrie moraines, Wicklow Mts | ← GIPPING → |
| Welsh Advance Ellesmere-Whitchurch m. | Liverpool Bay/Welsh Intermediate Moraines phase | | Valley end-moraines | Midland General Glaciation | |
| Wolverhampton limit | Main Anglesey/Arvon Advance deposits | Irish Sea (Dewisland) Glaciation | Margam Glaciation | ← LAST GLACIATION → MITCHELL 1960 / LAST GLACIATION — MITCHELL 1968 | |
| Upton Warren Interstadial complex | Weathering horizon (interstadial) | | Paviland interstadial | ? Ardcavan | Cryoturbation |
| | Cryoturbation | | | ← GIPPING → | Burtle beds (beach gravel) |
| | | | | Brittas mt Advance | Solifluction-earth |
| Chelford Interstadial | Irish Sea/Criccieth Advance deposits Lowest Head | Lower Head | Head | Eastern General Glaciation Clogga Advance | Fremington till Main Head |
| | Raised beach Porth Oer | Raised beach | *Neritoides* beach *Patella* beach (Minchin Hole intergl.) | Newtown Shortlestown | Raised beach |
| ? Early Glaciation | | Early Glaciation | Pencoed (Irish Sea) Glaciation | Ballycroneen till | |

and covered Lleyn as far west as St Tudwal's peninsula (Saunders, 1968a; Whittow and Ball, in Chapter 2). Watson has considered the possibility that this glacial limit represents the maximum of the Last Glaciation. In South Wales the local ice, at that time, was considered to have reached Swansea, Margam, and Llandaff (Charlesworth, 1929; Bowen, in Chapter 9), and in the eastern marchlands this ice limit was established at Abergavenny, Hereford, and in the vicinity of Welshpool (Wills, 1937; Lewis, in Chapter 7; Luckman, in Chapter 8).

These different correlations depend on the different methods of analysis used, whether this be the freshness of the drift surfaces, or *morphological approach*; the recognition of the different drifts and periglacial deposits by means of geological composition, texture, and relationship to one another in section, or *litho-stratigraphical approach*; by the relationship of the different drifts to organic deposits which are dated by pollen analysis, or *bio-stratigraphical approach*; and finally, by the radiocarbon dating of organic materials associated with the drift succession, or *'absolute-dating' approach*. Any interpretation of the drift sequence based on one of these methods alone, to the exclusion of the others, must be open to question.

The morphological approach is somewhat subjective, as the 'degree of freshness' of the drift features can be differently defined. The view is widely held that morainic ridges, ice-contact slopes, lake shorelines, drumlins, kames, eskers, and kettle-holes cannot survive from a glaciation that is as old as the Gipping. Such features have, therefore, been thought diagnostic of the Last Glaciation. Yet in localities that have escaped the widespread erosional activity of subsequent times, such features may survive from an even earlier glaciation, even though they may be modified by periglacial erosion and slumping. For example, in England, a morainic ridge survives at Moreton; ice-contact outwash gravels survive at Kelling in association with kames near Blakeney in Norfolk (West, 1961); a shoreline of glacial Lake Harrison survives in the Avon basin (Shotton, 1953); eskers survive at Kingswinford, south-west of Birmingham (Wills, 1937); and at Saleen, near Cork Harbour in Ireland an esker also survives (Lamplugh *et al.*, 1905).

Caution must be exercised in differentiating fresh constructional forms from those that owe their sharpness to subsequent erosion. The 'Cromer moraine', although a prominent feature of the landscape on the north Norfolk coast, is strongly dissected by steep-sided periglacial erosion gullies, to produce a 'hummocky' landscape (West,

1961). Gravel areas within the 'older' drifts tend to stand up as higher features because the surrounding clays have been more easily eroded. Alternatively, the absence of fresh drift forms may not be the result of subsequent erosion. In some cases a level till plain was left by the ice initially. Some areas that undoubtedly lay within the Last Glaciation limit are characterised by that type of landscape, such as parts of Anglesey, and much of Counties Meath and Dublin, in Ireland.

As mentioned earlier, the intensity of periglacial activity was greater outside the limits of the Last Glaciation, than inside these limits. In these areas the upper 1·5–3·0 m (5–10 ft) of the till surface has been disturbed in such a way that the contained stones and boulders have been rearranged so as to lie with their long axes vertical. In some instances these structures, together with frost wedges, are continuous along considerable lengths of coastal cliff sections. In Ireland, apart from the occasional frost wedge, such structures have not been found within the limit of the Midland General glaciation. Here, the last phase of severe climatic conditions associated with the Younger Dryas period of 10,000 years ago, are only represented as minor solifluction deposits noted in numerous lake sites.

The suggestion put forward by Wirtz (1953), and Watson (Chapter 6), for a rather limited expansion of the ice in north-west Wales during the Last Glaciation has not been generally accepted because of the belief that a great glacier flowed south from the Firth of Clyde to south Wexford at that time. Such a glacier could hardly fail to inundate the whole of Caernarvonshire. Furthermore the proximity of Snowdonia, an area today of high precipitation, would be expected to nourish a large local ice cap that merged with ice in the Irish Sea. Evidence in Ireland, however, does not support the assumption that any large glacier flowed south from Scotland at the time of the Midland General glaciation. The upper drift of the east coast of Ireland, north of Dublin, was associated with ice that streamed east and south-east from the Irish midlands, without any evidence of severe counter pressure from ice in the Irish Sea basin, except in County Down (Hill and Prior, 1968). The coastal hills such as the Carlingford and Mourne Mountains acted as 'barriers' around which Irish ice was forced, causing the level of the ice to drop rapidly towards the south-east. Ice thus accumulated in the basin of the Irish Sea to form a large sluggish piedmont glacier. The nunatak form of the Isle of Man further supports this concept, and explains why the interior of that island is covered in thick periglacial deposits and structures suggestive of 'older drift'. The very prominent Bride's Hill moraine, that loops round the north end of the island is generally correlated with the Carlingford Readvance, a retreat stage of the Last Glaciation (Charlesworth, 1955). As the north-east continuation of this moraine passes off the Irish coast at the mouth of Strangford Lough, it would appear to be younger than the Bride's Hill moraine. Thus there is a possibility that the latter may represent the maximum of the Last Glaciation (Fig. 12.1).

If the Irish Sea acted as a great accumulation basin for glacier ice derived from an ice cap that straddled the north-western part of the British Isles, from a centre in the Cumbrian Mountains, and from the Welsh highlands, the formation of a sluggish piedmont glacier might be expected. Such a glacier would spread laterally to impinge across the present coastline in the same way that the Baltic glacier behaved with relation to the deposition of the arcs of end-moraine in Denmark.

Comparison between sections observed on the south coast of Ireland and those in west Wales support Watson's thesis that the drifts bordering the central part of Cardigan Bay lay outside the limit of the Last Glaciation. On both coasts thick deposits of head underlie the tills. These tills are dated to the Gipping glaciation on the south coast of Ireland because they lie well outside the limit of the Last Glaciation, and because they are associated with tills that overlie organic beds of Hoxnian type at Kilbeg and Newtown in County Waterford. Unlike the lower head, the overlying head is much thinner than that which underlies the till; also it never spreads far from the old pre-glacial cliff. Near Aberystwyth, even the till itself shows signs of having been moved by solifluction before the deposition of the overlying head (Watson, Chapter 6).

Biostratigraphical correlations cannot, at present, be made in Wales because of the dearth of interglacial deposits that have been discovered to date. No good sequence of the Hoxne or Ipswichian Interglacial is known. The dating of the Quaternary deposits has been attempted by means of radiocarbon assay of marine shells, and also from fragments of organic matter such as wood and peat, which have been found in the drifts. Marine shells found in the Irish Sea tills and outwash give dates ranging from 54,000 to 28,000 years B.P. Yet the latest analysis of dates obtained from the Midlands of England does not indicate such an early maximal advance for the Last Glaciation. This analysis would seem to accord with the north European concept of a maximal advance about 22,000 years B.P. If the shelly till in west Wales was deposited at that time, the majority of contained shells would have to be considered as having lived at a much earlier time. Furthermore, if, as John (Chapter 10) believes, only one shelly till is present throughout, in Pembrokeshire, south Wexford, and on the coasts of County Cork and north-west Devon, then this date would represent the onset of a glacial advance that reached the Scilly islands. In that case the Courtmacsherry beach would represent the Ipswichian Interglacial along with the organic beds near Gort and Kilbeg (John, 1967). But according to pollen analysis these beds are Hoxnian.

The possibility that the shells are even older than the age ascribed to them by radiocarbon analysis has been considered by Shotton. Contamination, often very difficult to detect and eradicate, tends to give dates younger than they should be (Shotton, 1967a). Yet, even allowing for this, the dates do seem to show a pattern, becoming generally younger from south to north (Fig. 12.6). Clearly more work on absolute dates will have to be carried out before any more definite conclusions can be reached.

According to the evidence presented above, and correlated in Table 12.3, there are three interpretations as to the limit of the Last Glaciation. First, that it reached the Scilly islands, and is associated with the Ballycroneen till in Ireland; with the Fremington till in north-west Devon; the Pencoed till in South Wales; the Dewisland till in Pembrokeshire; and Irish Sea drift in north-west Wales, all associated with the same glacial advance of about 30,000 years ago by John. Secondly, that the Last Glaciation reached south as far as a line from Fishguard to Wexford, and was associated with the Irish Sea till of west Wales, and the Eastern General till of the east of Ireland; the similar Ballycroneen till of the south coast of Ireland would then be regarded as belonging to the preceding or Gipping glaciation. Thirdly, that the Last Glaciation was even more limited in extent, and was represented only by the Main Anglesey Advance and the Arvon Advance deposits in north-west Wales, and by the Midland General Advance in Ireland; in this case all Irish Sea tills would be regarded as Gipping in age. The suggestion has also been made that the Last Glaciation even failed to reach the south side of the Lleyn peninsula. On the other hand, the possibility that the Arvon drift in Caernarvonshire is younger than 16,500 years B.P. has been suggested (Saunders, 1968b). Curiously enough, the Newer Drift that extended down the east coast of England to Hunstanton has been similarly dated, as it overlies moss beds, radiocarbon dated to about 18,000 years B.P. at

Dimblington on the Yorkshire coast (Penny, L. F., personal communication). This date is considered too young by several authorities.

The use of the Last or Ipswichian Interglacial beach as a key stratigraphical horizon to elucidate the glacial succession of South

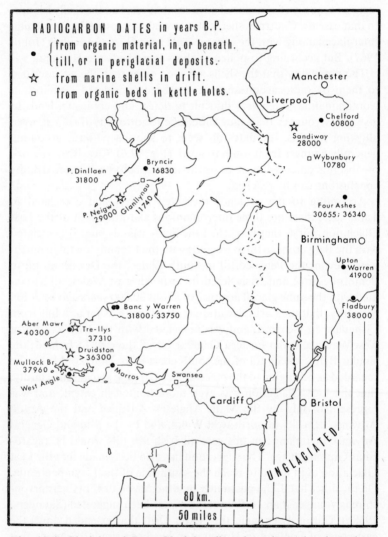

Fig. 12.6. Glacial and Late Glacial radiocarbon dates that have been determined in Wales and the western part of the midlands of England

Wales has proved difficult. On this coast Bowen has interpreted all the low level beaches that occur below glacial drift or head deposits as Ipswichian. Even though beach deposits can be dated by fossil evidence at Minchin Hole, there is little evidence to indicate how this deposit fits into the general stratigraphy of the Irish Sea basin as a whole. Sections on this coast are small and isolated, so that stratigraphical relationships are not always clear. Furthermore, as Stephens has pointed out, the presence of different beaches of various ages at the same height increase the difficulties of making accurate correlations. Generally the presence of only one fossil beach at any particular locality is readily explained; the later beach being associated with the phase of marine erosion that has removed any earlier beach at that level. In the Bristol Channel area beaches at the same height may be pre-Gipping, Ipswichian, Flandrian or recent in age.

Although knowledge of the Quaternary has increased considerably in the past four decades, since Charlesworth published the first detailed description of the possible limits of the Last Glacial maximum, many questions remain to be solved. For instance, one may ask what is the significance of the enormous rock-cut channels associated with the margins of the Irish Sea drift near Fishguard, south of Machynlleth, in west Lleyn, along the edge of the Flintshire hills, at the north end of the Stiperstones, and the Long Mynd, in Shropshire? In the vicinity of Dublin such channels have been assigned to the Gipping glaciation, or its equivalent (Farrington, 1944), while on the opposite side of the Irish Sea, in Lleyn, similar features have been dated to the Last Glaciation (Saunders, 1968c). In Ireland there is no indication that such feats of fluvioglacial erosion were ever accomplished during the final deglaciation.

The origin and survival of tors as a significant feature of the landscape is another problem to be solved. Tors occur in many parts of Wales. On Mynydd Prescelly and neighbouring hills to the northwest they are common, while nearer the coast they appear to descend below some limit of glacial erosion, below which they have been modified or removed. No satisfactory explanation of their survival in some places, and destruction in others, has been given.

These are but some of the questions still to be answered before we can have a complete understanding of the glaciation of Wales. Yet, in these chapters the significant advances of knowledge that have been made in the mapping and dating of the drifts in the Cheshire–Shropshire lowland may be appreciated, as there is now a basis for the

linking of the Welsh drift succession to that of the English Midlands.
Also important, is the delimitation and mapping of the various drifts
of the north-west corner of Wales; the distribution and significance
of the extensive periglacial deposits that margin Cardigan Bay; the
attempts to place the drift chronology of west Wales on an 'absolute-
date' basis; and finally by indicating in much greater detail than
before the manner of glaciation and deglaciation in South Wales.

## REFERENCES

BOSWELL, P. G. H. (1932) 'The contacts of geology: the Ice Age and early man in Britain', *Rep. Br. Ass. Advmt Sci.* C, 57–88.

BOULTON, G. S. and WORSLEY, P. (1965) 'Late Weichselian glaciation in the Cheshire–Shropshire basin', *Nature, Lond.* **207**, 704–6.

BOWEN, D. Q. (1966a) 'On the supposed ice dammed lakes of South Wales', *Trans. Cardiff Nat. Soc.* **93**, 4–17.

BOWEN, D. Q. (1966b) 'Dating Pleistocene events in south-west Wales', *Nature, Lond.*, **211**, 475–6.

BOWEN, D. Q. and GREGORY, K. J. (1966) 'A glacial drainage system near Fishguard, Pembrokeshire', *Proc. Geol. Ass. Lond.* **76**, 275.

CARVILL LEWIS, H. (1894) *The Glacial Geology of Great Britain and Ireland.* Longmans.

CHARLESWORTH, J. K. (1928) 'The glacial retreat from central and southern Ireland', *Quart. Jl. geol. Soc. Lond.* **84**, 295–300.

CHARLESWORTH, J. K. (1929) 'The South Wales end-moraine', *Quart. Jl. geol. Soc. Lond.* **85**, 335–58.

CHARLESWORTH, J. K. (1955) 'The Carlingford readvance between Dundalk, Co. Louth, and Kingscourt and Lough Ramor, Co. Cavan', *Ir. Nat. Jl.* **11**, 1–4.

CHARLESWORTH, J. K. (1963) 'Some observations on the Irish Pleistocene', *Proc. Roy. Irish Acad.* **62**B, 295–322.

EMBLETON, C. (1964) 'Subglacial drainage and supposed ice-dammed lakes in north-east Wales', *Proc. Geol. Ass. Lond.* **75**, 31–8.

FARRINGTON, A. (1934) 'The glaciation of the Wicklow mountains', *Proc. Roy. Irish Acad.* **42**B, 173–209.

FARRINGTON, A. (1942) 'The granite drift near Brittas, on the border between Co. Dublin and Co. Wicklow', *Proc. Roy. Irish Acad.* **47**B, 279–91.

FARRINGTON, A. (1944) 'The glacial drifts of the district of Enniskerry, Co. Wicklow', *Proc. Roy. Irish Acad.* **50**B, 133–57.

FARRINGTON, A. (1949) 'The glacial drifts of the Leinster mountains', *Jl. Glaciol.* **1**, 220–5.

FARRINGTON, A. (1954) 'A note on the correlation of the Kerry–Cork glaciations with those of the rest of Ireland', *Ir. Geogr.* **3**, 47–53.

FARRINGTON, A. (1966) 'The last glacial episode in the Wicklow mountains', *Ir. Nat. J.* **15**, 226–9.

GARDINER, M. J. (1963) *The Soils of County Wexford.* Nat. Soil Surv. Bull., No. 1. Dublin, An Foras Taluntais.

HILL, A. R. and PRIOR, D. B. (1968) 'Directions of ice movement in north-east Ireland', *Proc. Roy. Irish Acad.* **66**B.

JEHU, J. J. (1909) 'The glacial deposits of western Caernarvonshire', *Trans. Roy. Soc. Edin.* **47**, 17–56.

JESSEN, K. *et al.* (1955) 'The interglacial deposit near Gort, Co. Galway, Ireland', *Proc. Roy. Irish Acad.* **60**B, 1–77.

JOHN, B. S. (1965) 'A possible Main Würm glaciation in west Pembrokeshire', *Nature, Lond.,* **207**, 622–3.

JOHN, B. S. (1967) 'Further evidence for a Middle Würm interstadial and a Main Würm glaciation of south-west Wales', *Geol. Mag.* **104**, 630–3.

JOHN, B. S. (1968a) 'Age of raised beach deposits of south-western Britain', *Nature, Lond.,* **218**, 665–7.

JOHN, B. S. (1968b) 'Directions of ice movement in the southern Irish Sea basin during the last major glaciation: an hypothesis', *Jl. Glaciol.* **7**, 507–10.

KELLY, M. R. (1964) 'The Middle Pleistocene of north Birmingham', *Phil. Trans. Roy. Soc.* B, **247**, 533–92.

LAMPLUGH, G. W. *et al.* (1905) *The Geology of the Country around Cork and Cork Harbour.* Mem. Geol. Surv. Ireland. 135 pp.

MACKNEY, D. E. and BURNHAM, C. P. (1966) *The soils of the Church Stretton district of Shropshire.* Mem. Soil. Surv. Great Britain, Harpenden.

MAW, G. (1864) 'On a supposed deposit of boulder clay in north Devon', *Quart. Jl. geol. Soc. Lond.* **20**, 445–51.

MITCHELL, G. F. (1948) 'Two interglacial deposits in S.E. Ireland', *Proc. Roy. Irish Acad.* **52**B, 1–14.

MITCHELL, G. F. (1960) 'The Pleistocene history of the Irish Sea', *Rep. Br. Ass. Advmt. Sci.* **68**, 313–25.

MITCHELL, G. F. (1968) see Wilson (1968).

PEAKE, T. M. (1961) 'Glacial changes in the Alyn river system, and their significance in the glaciology of the North Welsh border', *Quart. Jl. geol. Soc. Lond.* **117**, 335–66.

SAUNDERS, G. E. (1968a) 'A fabric analysis of the ground moraine deposits of the Lleyn peninsula of southwest Caernarvonshire', *Geol. Jl.* **6**, 105–18.

SAUNDERS, G. E. (1968b) 'Glaciation of possible Scottish readvance age in north-west Wales', *Nature, Lond.* **218**, 76–8.

SAUNDERS, G. E. (1968c) 'A reappraisal of glacial drainage phenomena in the Lleyn peninsula', *Proc. Geol. Ass. Lond.* **79**, 305–24.

SEDDON, B. (1962) 'Late glacial deposits at Llyn Dwythwch and Nant Ffrancon, Caernarvonshire', *Phil. Trans. Roy. Soc. Lond.* B, **244**, 459–81.

SHOTTON, F. W. (1953) 'The Pleistocene deposits of the area between Coventry, Rugby and Leamington and their bearing upon the topographic development of the Midlands', *Phil. Trans. Roy. Soc. Lond.* B, **237**, 209–60.

SHOTTON, F. W. (1967a) 'The problems and contributions of methods of absolute dating within the Pleistocene period', *Quart. Jl. geol. Soc. Lond.* **122**, 356–83.

SHOTTON, F. W. (1967b) 'Age of the Irish Sea glaciation of the Midlands', *Nature, Lond.* **215**, 1366–7.

13

SIMPKINS, K. (1968) 'Aspects of the Quaternary history of central Caernarvonshire', unpublished Ph.D., University of Reading.

SISSONS, B. (1967) *The Evolution of Scotland's Scenery*. Archon Books, Hamden, Connecticut.

SUGGATE, R. P. and WEST, R. G. (1959) 'On the extent of the Last Glaciation in eastern England', *Proc. Roy. Soc.* B, **150**, 263–83.

SYNGE, F. M. (1963) 'A correlation between the drifts of south-east Ireland and those of west Wales', *Ir. Geogr.* **4**, 360–6.

SYNGE, F. M. (1964a) 'Some problems concerned with the glacial succession in south-east Ireland', *Ir. Geogr.* **5**, 73–82.

SYNGE, F. M. (1964b) 'The glacial succession in Caernarvonshire', *Proc. Geol. Ass. Lond.* **75**, 431–44.

SYNGE, F. M. and STEPHENS, N. (1960) 'The Quaternary Period in Ireland: an assessment', *Ir. Geogr.* **4**, 121–30.

TOMLINSON, M. E. (1925) 'The superficial deposits of the country north of Stratford on Avon', *Quart. Jl. geol. Soc. Lond.* **81**, 423–60.

WATTS, W. A. (1959) 'Interglacial deposits at Kilbeg and Newtown, Co. Waterford', *Proc. Roy. Irish Acad.* **60** B, 79–134.

WATTS, W. A. (1964) 'Interglacial deposits at Baggotstown, near Bruff, Co. Limerick', *Proc. Roy. Irish Acad.* **63** B, 167–89.

WEST, R. G. (1961) 'The glacial and interglacial deposits of Norfolk', *Trans. Norfolk and Norwich Nat. Soc.* **19**, 365–75.

WEST, R. G. (1968) *Pleistocene Geology and Biology*. Longmans.

WHITTOW, J. B. (1957) 'The Lleyn Peninsula, North Wales: a geomorphological study.' Unpublished Ph.D., University of Reading.

WILLIAMS, K. E. (1927) 'The glacial drifts of western Cardiganshire', *Geol. Mag.* **64**, 205–27.

WILLS, L. J. (1937) 'The Pleistocene history of the West Midlands', *Rep. Br. Ass. Advmt Sci.* C, 71–94.

WILSON, H. E. (1968) 'Geology of the Irish Sea area', *Ir. Nat. Jl.* **16**, 102–5.

WIRTZ, D. (1953) 'Zur Stratigraphie des Pleistocäns im Westen der Britischen Inseln', *Neues Jb. Geol. u Palaeont.* **96**, 267–303.

WRIGHT, W. B. (1937) *The Quaternary Ice Age*. Macmillan, London (2nd ed.).

WRIGHT, W. B. and MUFF, H. B. (1904) 'The Pre-Glacial raised beach of the south coast of Ireland', *Sci. Proc. Roy. Dublin Soc.* **10**, 250–324.

# INDEX

For ease of reference the index has been divided into three sections: A, authors; B, place-names; C, general. Features and deposits that are mentioned with great regularity, such as 'boulder clay', have not been included. To save space it has also proved necessary to group features and deposits under common headings, but we hope that this will not inconvenience the reader too much.

## A. *Authors*

## B. *Place names*

## DATE DUE